SMART, RESILIENT AND TRANSITION CITIES

SMART, RESILIENT AND TRANSITION CITIES

EMERGING APPROACHES AND TOOLS FOR A CLIMATE-SENSITIVE URBAN DEVELOPMENT

Adriana Galderisi
University of Campania Luigi Vanvitelli, Aversa (CE), Italy

Angela Colucci
Co.O.Pe.Ra.Te. ldt, Pavia, Italy

Elsevier
Radarweg 29, PO Box 211, 1000 AE Amsterdam, Netherlands
The Boulevard, Langford Lane, Kidlington, Oxford OX5 1GB, United Kingdom
50 Hampshire Street, 5th Floor, Cambridge, MA 02139, United States

Copyright © 2018 Elsevier Inc. All rights reserved.

No part of this publication may be reproduced or transmitted in any form or by any means, electronic or mechanical, including photocopying, recording, or any information storage and retrieval system, without permission in writing from the publisher. Details on how to seek permission, further information about the Publisher's permissions policies and our arrangements with organizations such as the Copyright Clearance Center and the Copyright Licensing Agency, can be found at our website: www.elsevier.com/permissions.

This book and the individual contributions contained in it are protected under copyright by the Publisher (other than as may be noted herein).

Notices
Knowledge and best practice in this field are constantly changing. As new research and experience broaden our understanding, changes in research methods, professional practices, or medical treatment may become necessary.

Practitioners and researchers must always rely on their own experience and knowledge in evaluating and using any information, methods, compounds, or experiments described herein. In using such information or methods they should be mindful of their own safety and the safety of others, including parties for whom they have a professional responsibility.

To the fullest extent of the law, neither the Publisher nor the authors, contributors, or editors, assume any liability for any injury and/or damage to persons or property as a matter of products liability, negligence or otherwise, or from any use or operation of any methods, products, instructions, or ideas contained in the material herein.

Library of Congress Cataloging-in-Publication Data
A catalog record for this book is available from the Library of Congress

British Library Cataloguing-in-Publication Data
A catalogue record for this book is available from the British Library

ISBN: 978-0-12-811477-3

For Information on all Elsevier Publications visit our website at
https://www.elsevier.com/books-and-journals

Publisher: Candice Janco
Acquisition Editor: Laura S Kelleher
Editorial Project Manager: Emily Thomson
Production Project Manager: Prem Kumar Kaliamoorthi
Designer: Greg Harris

Typeset by TNQ Technologies

Contents

List of Contributors — xiii
Foreword — xv
Introduction — xvii

I METAPHORS TO ENHANCE CITIES' CAPACITY TO COPE WITH CLIMATE CHANGE

1. The Smart City Metaphor to Foster Collaborative and Adaptive Decision-Making Processes in the Face of Climate Issues
ADRIANA GALDERISI

1.1	Introducing the Smart City Metaphor	3
1.2	The Smart City Between Conflicting Approaches and Goals: Looking for New Directions	4
1.3	Smart City Initiatives: From Sectoral Toward Integrated Perspectives	5
1.4	The Governance of the Smart City: From Top-Down Initiatives Toward Collaborative Decision-Making Processes	6
1.5	Information, Knowledge, and Learning for Adaptive Cities	7
1.6	Concluding Remarks	8
References		8

2. The Resilient City Metaphor to Enhance Cities' Capabilities to Tackle Complexities and Uncertainties Arising From Current and Future Climate Scenarios
ADRIANA GALDERISI

2.1	Introducing the Resilient City Metaphor	11
2.2	Tracing The Evolution Of Resilience Across Disciplinary Boundaries: Emerging Perspectives	12
2.3	The Resilient City Initiatives: Strengths and Weaknesses	13
2.4	Concluding Remarks	16
References		17

3. The Transition Approach in Urban Innovations: Local Responses to Climate Change
ANGELA COLUCCI

3.1	Societal Transition Models	19
3.2	Transition Initiatives	21
3.3	The Transition (Towns) Network	22
3.4	Transition Initiatives: Innovations for Facing Urban Climate Change Issues	24
References		26

4. Smart, Resilient, and Transition Cities: Commonalities, Peculiarities and Hints for Future Approaches
ADRIANA GALDERISI, ANGELA COLUCCI

4.1	Three Urban Metaphors	29
4.2	Commonalities and Peculiarities of the Three Metaphors in Scientific Literature	30

4.3	Commonalities and Peculiarities of the Three Metaphors in Current Practices	31
4.4	Concluding Remarks	32
References		34

II LARGE-SCALE STRATEGIES TO COUNTERBALANCE CLIMATE CHANGE

5. European Strategies and Initiatives to Tackle Climate Change: Toward an Integrated Approach
ADRIANA GALDERISI, ERICA TRECCOZZI

5.1	Europe and Climate Change Issues	39
5.2	European Strategies to Counterbalance Climate Change	40
5.3	Concluding Remarks: Which Future for Climate Policies in Europe?	44
References		45

6. The American Approach to Climate Change: A General Overview and a Focus on Northern and Arctic Regions
CATHERINE DEZIO

6.1	Strategies and Initiatives in the Face of Climate Change: The American Approach	47
6.2	Focusing on Northern and Arctic Regions	48
References		51

7. Climate Change Mitigation and Adaptation Initiatives in Africa: The Case of the Climate and Development Knowledge Network "Working With Informality to Build Resilience in African Cities" Project
KWAME NTIRI OWUSU-DAAKU, STEPHEN KOFI DIKO

7.1	Introduction	53
7.2	Background: Resilience as a Convening Concept for Integrating Climate Change Mitigation and Adaptation	53
7.3	Case Study: The Climate and Development Knowledge Network "Working with Informality to Build Resilience in African Cities" Project	55
7.4	Lessons Learned	57
7.5	Conclusion	58
References		58

8. Addressing Climate Change in China: Policies and Governance
GØRILD HEGGELUND

8.1	Introduction: Climate Change and Energy in China	61
8.2	Climate Governance	61
8.3	Strategies and Initiatives in Climate Change	62
8.4	Mitigation and Adaptation in Cities	63
8.5	Concluding Remarks	63
References		65

9. Mitigation and Adaptation Strategies in the Face of Climate Change: The Australian Approach
GIUSEPPE FORINO, JASON VON MEDING, GRAHAM BREWER

9.1	Climate Change Issues in Australia	67
9.2	Significant Mitigation and Adaptation Strategies by the Australian Federal Government	68
9.3	Beyond Mitigation and Adaptation: Climate Change as a Contested Political Issue	69
9.4	Conclusions: Which Approach for Mitigation and Adaptation in Australia?	70
Acknowledgments		70
References		70

III CITIES DEALING WITH CLIMATE CHANGE: INSTITUTIONAL PRACTICES

10. European Cities Addressing Climate Change
GIADA LIMONGI

10.1	The Pivotal Role of Cities in Counterbalancing Climate Change	75
10.2	European Cities Addressing Mitigation Issues: The Covenant of Mayors Initiative	77
10.3	European Cities Addressing Adaptation Issues: The Mayors Adapt Initiative	79
10.4	European Cities Embracing an Integrated Climate Strategy: The New Covenant of Mayors for Climate and Energy	80
10.5	Concluding Remarks	83
	References	84

11. Adaptation and Spatial Planning Responses to Climate Change Impacts in the United Kingdom: The Case Study of Portsmouth
DONATELLA CILLO

11.1	Climate Change Impacts in the United Kingdom	85
11.2	Portsmouth's Climate Change Adaptation and Planning Responses	88
11.3	Toward a Resilient Portsmouth	89
11.4	Conclusions	91
	Disclaimer	92
	References	92

12. Land-Use Planning and Climate Change Impacts on Coastal Urban Regions: The Cases of Rostock and Riga
SONJA DEPPISCH

12.1	Climate Change Impacts and Land-Use Planning in Urban Regions	95
12.2	Challenges, Opportunities, and Barriers for Resilient Land-Use Planning	97
12.3	Conclusions and Outlook	99
	Acknowledgments	100
	References	100

13. Importance of Multisector Collaboration in Dealing With Climate Change Adaptation: The Case of Belgrade
RATKA ČOLIĆ, MARIJA MARUNA

13.1	The Belgrade Local Context	101
13.2	Current Initiatives/Practices	102
13.3	Critical Analysis of Current Initiatives/Practices	106
	References	108

14. Genoa and Climate Change: Mitigation and Adaptation Policies
DANIELE F. BIGNAMI, EMANUELE BIAGI

14.1	The Genoa Case Study: Climatic and Geographical Features	109
14.2	Description of Current Climate Initiatives in Genoa	111
14.3	Conclusions and Critical Analysis	113
	References	113

15. The Evolution of Flooding Resilience: The Case of Barcelona
ANDREA FAVARO, LORENZO CHELLERI

15.1	Introduction: Flooding Resilience and Barcelona	115
15.2	First Patterns of Flooding Resilience: Links Between Drainage Systems Design and Urban Plans	116

15.3	Barcelona Olympic Model Framing the Contemporary Challenges in Flooding Resilience	118
15.4	Sustainable Drainage: Enhancing Decentralization for the Future of Urban Flooding Resilience	120
15.5	Conclusions	121
References		122

16. Athens Facing Climate Change: How Low Perceptions and the Economic Crisis Cancel Institutional Efforts
KALLIOPI SAPOUNTZAKI

16.1	The Profile of Metropolitan Athens: Current Vulnerabilities, Exposure to Climate Change and Resilience Assets	125
16.2	Current Responses to Climate Change: Policies, Initiatives, Practices	129
16.3	Critical Analysis of Current Policies, Initiatives, and Practices	132
References		133

17. Sustainability of Climate Policy at Local Level: The Case of Gaziantep City
OSMAN BALABAN, BAHAR GEDIKLI

17.1	Introduction	135
17.2	The Climate Policy in Turkey	135
17.3	The Case Study of Gaziantep	136
17.4	Concluding Discussion	139
References		139

18. Toward Integration: Managing the Divergence Between National Climate Change Interventions and Urban Planning in Ghana
STEPHEN KOFI DIKO

18.1	Background	141
18.2	Climate Change Issues	143
18.3	Climate Change and Urban Planning in the Kumasi Metropolis	147
18.4	Managing the National and Urban Planning Divergence	149
References		150

19. Spatial Planning for Climate Adaptation and Flood Risk: Development of the Sponge City Program in Guangzhou
MENG MENG, MARCIN DĄBROWSKI, FAITH KA SHUN CHAN, DOMINIC STEAD

19.1	Introduction	153
19.2	The Profile of Guangzhou: Rapid Urbanization and Exposure to Climate Change	154
19.3	Sponge City: Shifting the Spatial Planning Frame	155
19.4	Critical Analysis of the Municipal Interpretation of the Sponge City Program in Guangzhou	159
19.5	Conclusions	161
References		162

20. Climate Change and Australian Local Governments: Adaptation Between Strategic Planning and Challenges in Newcastle, New South Wales
GIUSEPPE FORINO, JASON VON MEDING, GRAHAM BREWER

20.1	Introduction	163
20.2	Climate Change Adaptation Between Land Use and Development	164
20.3	Climate Change Adaptation and the Target on Local Communities	164
20.4	Challenges for Climate Change Adaptation	165
20.5	Conclusion	166
Acknowledgments		166
References		166

IV CITIES DEALING WITH CLIMATE CHANGE: TRANSITION INITIATIVES

21. Transition Initiatives: Three Exploration Paths
ANGELA COLUCCI

21.1	Transition Initiatives: Three Exploration Paths	171
21.2	Transition Network	172
21.3	Transition Initiatives Acting on Urban Commons and Public Life	174
21.4	Resilience Practices Observatory	176
21.5	Concluding Remarks	179
References		180

22. Transition Towns Network in the United Kingdom: The Case of Totnes
ALESSIA CANZIAN

22.1	Totnes Transition Town	183
References		190

23. Transition Towns Network in Italy: The Case of Monteveglio
ALESSIA CANZIAN

23.1	Monteveglio Transition Town	191
23.2	Critical Analysis of the Initiative	193
References		193

24. Model for Integrated Urban Disaster Risk Management at the Local Level: Bottom-Up Initiatives of Academics
MARIJA MARUNA, RATKA ČOLIĆ

24.1	Introduction	195
24.2	Initiative by the Integrated Urbanism Master's Program to Build UDRM Capacity at the Local Authority	196
24.3	Critical Analysis of the Initiative	201
References		202

25. Enhancing Community Resilience in Barcelona: Addressing Climate Change and Social Justice Through Spaces of Comanagement
LUCA SÁRA BRÓDY, LORENZO CHELLERI, FRANCESC BARÓ, ISABEL RUIZ-MALLEN

25.1	Adapting and Mitigating Climate Change From Below	203
25.2	From Community-Led to Comanagement: The Pla BUITS Experience	205
25.3	Discussion: Promises From a New Governance Model Dealing With Comanagement?	207
Acknowledgment		207
References		208

26. Barriers to Societal Response and a Strategic Action Plan Toward Climate Change Adaptation and Urban Resilience in Turkey
FUNDA ATUN

26.1	Introduction	209
26.2	Physical Dimension: Increasing Exposure and Systemic Vulnerability	210
26.3	Societal Concerns: Stakeholder Coordination and Collaboration	210
26.4	Discussion and Conclusion	212
References		212

27. Victims or Survivors: Resilience From the Slum Dwellers' Perspective
DEEPIKA ANDAVARAPU, DAVID J. EDELMAN, NAGENDRA MONANGI

27.1	Pedda Jalaripeta (Large Fishing Village)	215
27.2	Data Sources and Collection	217
27.3	Surviving Disasters Through Collaboration: Fire and Gentrification/Eviction	218
27.4	Collaborating, Empowering, and Educating	219
References		219

28. Bottom-Up Initiatives for Climate Change Mitigation: Transition Town in Newcastle
GIUSEPPE FORINO, JASON VON MEDING, GRAHAM BREWER

28.1	Introduction	221
28.2	Transition and Transition Town	221
28.3	Transition Newcastle	222
28.4	Transition and Climate Change Mitigation: Challenges and Opportunities for Transition Newcastle	223
Acknowledgments		224
References		224

V CROSS-CUTTING ISSUES: HINTS FOR INTEGRATED PERSPECTIVES

29. Integrated Knowledge in Climate Change Adaptation and Risk Mitigation to Support Planning for Reconstruction
SCIRA MENONI

29.1	Cities as Complex Entities	227
29.2	Dealing With Hazards and Risk in Contemporary Cities: A Multidisciplinary Challenge	228
29.3	Resilience: Will This Concept Provide the Necessary Link Between Disaster Scientists, Climate Change Experts, and Planners?	229
29.4	Knowledge Supporting Better and More Adaptive Recovery and Reconstruction	230
29.5	An Example of the Possible Application of a Multirisk, Attentive-to-Climate Change Recovery After a Devastating Earthquake: the Case of the 2016 Central Italy Event	232
References		234

30. Boundaries, Overlaps and Conflicts Between Disaster Risk Reduction and Adaptation to Climate Change. Are There Prospects of Integration?
KALLIOPI SAPOUNTZAKI

30.1	Climate Change and Disaster Risk: Are They Related?	237
30.2	DRR and Adaptation to CC: Convergences and Mismatches in Terminology, Processes, and Tools	240
30.3	The Need for and Prospects of Integration of DRR and Adaptation to CC	243
References		247

31. The Contribution of the Economic Thinking to Innovate Disaster Risk Reduction Policies and Action
GIULIA PESARO

31.1	Urban Systems and Disasters From an Economic Perspective	249
31.2	Coping With Disaster Risk: How Economic Understanding Might Enhance the DRR Action	251
31.3	CBA as an Economic-Based Tool to Enhance Decision-Making Processes in DRR Action and Territorial Resilience	252
31.4	Concluding Remarks	254
References		255

32. Flood Resilient Districts: Integrating Expert and Community Knowledge in Genoa
DANIELE F. BIGNAMI, EMANUELE BIAGI

32.1	Urban Context, Environmental Characteristics and Citizens' Engagement	257
32.2	Main Meteorological Phenomena and Their Impacts	257
32.3	Main Fragilities	258
32.4	The Suggested Local Strategy of Disaster Risk Reduction and Planning of Actions	260
32.5	Conclusions and Critical Analysis	265
References		265

VI TOWARDS A CLIMATE-SENSITIVE URBAN DEVELOPMENT

33. Drawing Lessons From Experience
ADRIANA GALDERISI, ANGELA COLUCCI

33.1	Large-Scale Strategies in the Face of Climate Change: Trajectories and Barriers	269
33.2	Climate Policies at Local Level: Drawing Upon Cities' Experiences	272
33.3	Transition Initiatives in the Face of Climate Change: Hints From Current Practices	276
33.4	Final Remarks	282
References		283

34. Future Perspectives: Key Principles for a Climate Sensitive Urban Development
ADRIANA GALDERISI, ANGELA COLUCCI

34.1	Introduction	285
34.2	The Key Principles for Climate-Sensitive Urban Development	285
References		291

Index 293

List of Contributors

Deepika Andavarapu   University of Cincinnati, Cincinnati, OH, United States
Funda Atun   Politecnico di Milano, Milan, Italy
Osman Balaban   Middle East Technical University, Ankara, Turkey
Francesc Baró   Universitat Autònoma de Barcelona (UAB), Cerdanyola del Vallès, Spain
Emanuele Biagi   Fondazione Politecnico di Milano, Milan, Italy
Daniele F. Bignami   Fondazione Politecnico di Milano, Milan, Italy
Graham Brewer   University of Newcastle, Callaghan, NSW, Australia
Luca Sára Bródy   Gran Sasso Science Institute (GSSI), L'Aquila, Italy
Alessia Canzian   Independent Researcher, Italy
Faith Ka Shun Chan   University of Nottingham Ningbo China, Ningbo, China; University of Leeds, Leeds, United Kingdom
Lorenzo Chelleri   Gran Sasso Science Institute (GSSI), L'Aquila, Italy; Universitat Internacional de Catalunya (UCI), Barcelona, Spain
Donatella Cillo   Environment Agency, England, United Kingdom
Ratka Čolić   University of Belgrade, Belgrade, Serbia
Angela Colucci   Co.O.Pe.Ra.Te. ldt, Pavia, Italy
Marcin Dąbrowski   Delft University of Technology, Delft, The Netherlands
Sonja Deppisch   HafenCity University Hamburg, Hamburg, Germany
Catherine Dezio   Politecnico of Milan, Milan, Italy
Stephen Kofi Diko   University of Cincinnati, Cincinnati, OH, United States
David J. Edelman   University of Cincinnati, Cincinnati, OH, United States
Andrea Favaro   Universidad Politécnica de Madrid, Madrid, Spain
Giuseppe Forino   University of Newcastle, Callaghan, NSW, Australia
Adriana Galderisi   University of Campania Luigi Vanvitelli, Aversa, Italy
Bahar Gedikli   Middle East Technical University, Ankara, Turkey
Gørild Heggelund   Fridtjof Nansen Institute, Lysaker, Norway
Giada Limongi   Engineer, Naples, Italy
Marija Maruna   University of Belgrade, Belgrade, Serbia
Meng Meng   Delft University of Technology, Delft, The Netherlands
Scira Menoni   Politecnico di Milano, Milano, Italy
Nagendra Monangi   Cincinnati Children's Hospital, Cincinnati, OH, United States
Kwame Ntiri Owusu-Daaku   University of West Florida, Pensacola, FL, United States
Giulia Pesaro   Poltecnico di Milano, Milan, Italy
Isabel Ruiz-Mallen   Universitat Oberta de Catalunya (UOC), Barcelona, Spain
Kalliopi Sapountzaki   Harokopio University of Athens, Athens, Greece
Dominic Stead   Delft University of Technology, Delft, The Netherlands
Erica Treccozzi   Building Engineer, Naples, Italy
Jason von Meding   University of Newcastle, Callaghan, NSW, Australia

Foreword

Metaphors, as well as analogies and allegories, are cognitive tools for understanding abstract phenomena. Metaphors are used in the course of everyday communications, as the human conceptual system is heavily metaphorical in nature: "Conceptual metaphors consist of sets of systematic correspondences, or mappings, between two domains of experience and […] the meaning of a particular metaphorical expression […] is based on such correspondences" (Kövecses, 2015, p. ix).

Urban planning has historically used existing concepts available in other realms—such as evolution (Mehmood, 2010)—as metaphorical mechanisms to describe the development of urban forms and communities. Urban studies, and urban studies authors, have used metaphors as a tool in urban theory and urban planning as described by Nientied (2016). Nientied added that "without new metaphors—and the shift in style of thinking that new metaphors bring with them—urban studies will be unable to do proper justice to the heterogeneity and complexity of cities" (Nientied, 2016, p. 1). Nientied echoed Sharon Meager (2015), who advocated the need for new metaphors or figurations that help people to think creatively about urban conditions and the possibilities for political interventions.

Utopian (and dystopian) narrations of potential or imaginative urban futures have used metaphors and allegories to map and depict hypothetical societies and urban forms. Those imaginative futures started from the utopian authors' dissatisfaction with the society and the urban life they lived in. Many thinkers—whether urbanists, architects, social reformers, philosophers, or religious persons—envisioned potential future societies to escape the burdens and the constraints of the time they lived in. When Thomas Moore—the Henry VIII knight later on incarcerated by the same monarch because of his refusal to give his oath to support the King as head of the Church of England—described in 1516 the island "Utopia," located somewhere in the New World. Moore imagined that the society that lived on the island was "based on far-reaching equality but under the authority of wise, elderly men" (Sargent p. 2). The society Thomas Moore depicted "provides a much better life for its citizens than was available to the citizens of England at the time" (Sargent pp. 2, 3).

Ecological and environmental planning has adopted metaphors as well (Haar 2007). Ian McHarg titled one of the chapters of *Design with Nature* (1969) "A world as a Capsule" to emphasize the limitation of the resources available on earth and the attitude humans should have toward their use. McHarg reverberated the economist Kenneth Boulding, who in turn was influenced by Henry George's 1879 *Progress and Poverty*. Boulding wrote in 1966 "The Economics of the Coming Spaceship Earth," in which he compared the *cowboy economy*, "the cowboy being symbolic of the illimitable plains and also associated with reckless, exploitative, romantic, and violent behavior" (Boulding 1966, p. 2), with the *spaceman economy*, "in which the earth has become a single spaceship, without unlimited reservoirs of anything, either for extraction or for pollution, and in which, therefore, man must find his place in a cyclical ecological system which is capable of continuous reproduction of material form even though it cannot escape having inputs of energy" (Boulding 1966, p. 4).

"Cities of resilience" is, in itself, a metaphor (Musacchio and Wu 2002; Pickett et al., 2004), as are "smart cities" and "transition towns," all of them well described by the authors in the first section of this book. The use of metaphors, they explain, "is also related to the need for 'positive' visions, capable to support innovation and define new urban development models, focusing on strategic and synergic actions rather than on regulative (planning and land use) tools" (Chapter 4). The two authors advocated for the role of the metaphors in planning discourse and urban studies as a "positive" device to support innovation in the way we manage our resources, our cities, our *oikos*. The use of metaphors, I would add, does not push us away from reality but, on the contrary, provides the right distance from it, a distance that allows farsightedness and opportunities for speculations.

Italo Calvino, the author of *Invisible Cities* (1974), gave Marco Polo the opportunity to describe to Kublai Khan the cities he visited in his expeditions using several urban metaphors. Polo portrayed cities like Zora that have "languished, disintegrated, disappeared. The earth has forgotten her" (p. 16) or like Fedora, where you can see in crystal globes "the model of a different Fedora. These are the forms the city could have taken if, for one reason or another, it had not become what we see today" (p. 32), or like Ottavia, the spiderweb city "suspended over the abyss" in which "the life of Octavia's inhabitants is less uncertain than in other cities. They know the net will last only so long" (p. 75).

Polo looks like he is using metaphors, allegories, and bold images to describe all the cities he has visited, but at one point the Khan asks Polo why he has never spoken about Venice, his hometown. Marco, smiling to the Khan, answered, "What else do you believe I have been talking to you about? Every time I describe a city I am saying something about Venice" (p. 86). When we use metaphors, we are like Marco Polo; we talk about the cities we know, the risks they are running, and the solutions that can be implemented to improve the life of the communities within.

Danilo Palazzo
College of DAAP
Cincinnati, OH, United States

References

Boulding, K.E., 1966. The economics of the coming spaceship earth. In: Jarrett, H. (Ed.), Environmental quality in a growing economy. Resources for the Future/Johns Hopkins University Press, Baltimore, MD.

Calvino, I., 1974. Invisible Cities. Houghton Mifflin Hartcourt Publishing Company, New York, NY.

Haar, S., 2007. The Ecological City: Metaphor Versus Metabolism. University of Illinois at Chicago, School of Architecture, Great Cities Institute. Publication Number GCP-07–05.

Kövecses, Z., 2015. Where Metaphors Come From: Reconsidering Context in Metaphor. Oxford University Press, New York, NY.

McHarg, I., 1969. Design with Nature. Natural History Press, Garden City, NY.

Meager, S., 2015. The politics of urban knowledge. City 19 (6), 801–819.

Mehmood, A., 2010. On the history and potentials of evolutionary metaphors in urban planning. Planning Theory 9 (1), 63–87. http://dx.doi.org/10.1177/1473095209346495. http://plt.sagepub.com.

Musacchio, L., Wu, J., 2002. Cities of Resilience: Four Themes of the Symposium. Understanding and Restoring Ecosystems: A Convocation. Ecological Society of America, Washington, DC.

Nientied, P., 2016. Metaphor and urban studies-a crossover, theory and a case study of SS Rotterdam. City, Territory and Architecture 3, 21. http://dx.doi.org/10.1186/s40410-016-0051-z.

Pickett, S.T.A., Cadenasso, M.L., Grove, J.M., 2004. Resilient cities: meaning, models, and metaphor for integrating the ecological, socio-economic, and planning realms. Landscape and Urban Planning 69, 369–384. http://dx.doi.org/10.1016/j.landurbplan.2003.10.035.

Sargent, L.T., 2010. Utopianism: A Very Short Introduction. Oxford University Press, New York.

Introduction

In the past decades, the growth of urban population and the "pervasive imprint of urbanization processes" (Brenner and Schmid 2014) led planning theory to largely focus on the need for more in-depth investigation of the emerging urbanization patterns to overcome the traditional dichotomy between urban and rural and to better conceptualize contemporary urbanization geographies. However, current urbanization processes and expected trends also claim for a better conceptualization of the numerous environmental and social challenges they are responsible for. As pointed out few years ago by the Italian planner Secchi (2014), social inequalities and the consequences of climate change are indeed among the most important aspects of the new urban issue.

The negative implications of current urbanization patterns were also clearly remarked on by the sociologist Ulrich Beck, who defined the *current society* as a risk society, characterized by a "pluralization" of risks, which contend each other for the primacy (Beck, 1992, 2017).

In this vein, this book specifically focuses on climate change, considered to be a global problem—with "grave implications: environmental, social, economic, and political and for the distribution of goods" (Encyclical Laudato Si, 2015)—whose negative impacts will affect humans for a long time, although some measures have already been undertaken.

Climate change is, indeed, widely considered not only to be a by-product of urbanization processes, above all, of current urban lifestyles but also to be an engine for significant planetary geographical and social changes. As remarked by Beck (2017), climate change and its related phenomena (such as sea level rise, drought, floods, and heat waves) are creating new landscapes of inequality all over the world. For example, the sea level rise is designing new geographies where key lines are no longer identifiable in administrative boundaries but rather in the sea level altitudes and the changes in mean temperatures are inducing drought and consequent water and food scarcity in some regions, opening new agricultural and economic opportunities in others.

The global dimension of climate change requires a thorough review of current development models and effective strategies to be developed and implemented at different geographical levels, from the global and national levels to the local scale. To better cope with climate issues, both large-scale strategies to counterbalance greenhouse gas emissions and local policies aimed at facing the unavoidable impacts of climate change at local scale are required.

However, at present, even in the absence of national strategies or in case of their inconsistency, cities are increasingly emerging as key actors in counterbalancing climate change: They are giving life to, and participating in, new global carbon ethic geographies, creating new common norms about responsible urban development and regaining a central position similar to the one they had long ago in the prenational world (Beck, 2017). In other words, all over the world, climate issues are entering urban policy agenda, assigning cities a pioneer role in carrying out strategies and initiatives aimed at reducing greenhouse gas emissions, while adapting to the impacts of the increasingly frequent climate-related hazards.

The emerging role of cities in addressing global challenges, including climate change, has two major and interconnected consequences that will be explored in-depth in this book:

- The proliferation of metaphors, which introduce new ways of describing cities and guiding their future development in the face of changing climatic (and more in general environmental, social, and economic) conditions.
- The development of heterogeneous approaches, tools, and governance models for guiding cities toward a climate-sensitive urban development.

In respect to the first point, it is worth reminding that metaphors have been always widespread in planning discipline, but they have been mostly used "when the urban condition is transformed and shifting" (Secchi, 2014), as is occurring nowadays under the pressure of the global challenges discussed here. The role of urban metaphors for strengthening cities' capacities to cope with emerging challenges is crucial: They aim at structuring, indeed, both the way we think and the way we act, playing a central role in building awareness and consensus around a given issue as well as in driving urban planning practices toward a given goal.

In respect to the second point, it is crucial nowadays to explore how cities, different in size and geographical, cultural, and economic contexts, are dealing with the common challenge of climate change, to better grasp synergies and commonalities arising from current urban practices and to select the most promising strategies and tools so far developed to deal with climate issues.

Finally, although Europe is considered one of the world leaders in global mitigation policies and has devoted great efforts in enhancing and coordinating local initiatives toward adaptation, this book aims at overcoming the most foregone North European—centered perspective, to embrace a wider geographical perspective including the less-explored practices in eastern and southern Europe on the one hand and the numerous initiatives that have been undertaken out of Europe (Turkey, Africa, China, India, and Australia) on the other: These initiatives are crucial, indeed, to effectively contribute to the global climate issues and may provide significant clues to reframe the consolidated theoretical and operational approaches in the European context.

According to these premises, the book is structured into six sections that will be briefly described in the following.

The first section presents three of the most widespread urban metaphors: smart city, resilient city, and transition towns. These metaphors are drawing increasing attention from urban planners and decision-makers because they emphasize the need for enhancing cities' capacities to cope with the heterogeneous challenges threatening contemporary cities and their future development and, first, with climate change and climate-related hazards. In detail, the first three chapters analyze and discuss each metaphor, by focusing on their evolution paths across different approaches and disciplinary domains and exploring the most promising conceptualizations in the scientific debate as well as the main initiatives and practices they are inspiring and guiding. Moreover, the potentials of these metaphors to support climate mitigation and adaption strategies as well as the barriers currently hindering their effectiveness in counterbalancing climate change are discussed. Then, the last chapter focuses on commonalities and peculiarities of the three urban metaphors, emphasizing that, despite the significant commonalities linking them, they have been so far developed separately both in theory and in practice, moving as independent labels in addressing climate issues.

The second section focuses on the heterogeneous large-scale strategies so far undertaken in the face of climate change in different geographical contexts, highlighting the heterogeneous approaches, the different weights assigned to mitigation and adaptation policies as well as the main barriers hindering their effectiveness and translation into measurable outcomes. In detail, an overview is provided of both mitigation and adaptation strategies promoted in Europe, the United States, China, Africa, and Australia.

The third section, based on a case study approach, provides an overview of current practices guided by institutional actors (local authorities) and addresses mitigation and/or adaptation issues at the urban scale, with reference to cities selected according to different geographical locations, cultural and economic contexts, and urban sizes. In detail, an overview of current practices in Europe, Turkey, Africa, China, and Australia addressed both to reduce energy consumption and greenhouse gas emissions and to build disaster resilient cities in the face of climate-related hazards will be provided, while pointing out gaps and barriers that limit their development.

The fourth section, which similarly adopts a case study approach, focuses on "bottom-up" initiatives and practices directly enacted by citizens or by larger groups of stakeholders, including local authorities, and addressed to deal with climate issues. The section includes, in addition to an introductory chapter on transition initiatives all over the world, case studies from Europe, Turkey, India, and Australia. These case studies illustrate the renovate and widespread interest of local communities in environmental issues, the opportunities and criticalities arising from the rich, "sprawled," and "blurred" landscape of current processes, and the success factors and weaknesses in building up inclusive and participative governance processes.

The fifth section focuses on the need to overcome currently prevailing sectoral approaches to climate issues in favor of more integrated perspectives. In detail, this section will explore the relationships among disaster risk reduction and climate policies, namely climate adaptation strategies, by focusing both on the potential of more integrated approaches and knowledge bases and on the synergies and obstacles to the development of integrated strategies and policies in these domains. Moreover, the need for innovating economical approaches and for better integrating experts and community knowledge to enhance urban resilience in the face of climate change will be discussed.

The last section will, on the one hand, draw lessons from current strategies and practices in the face of climate change and, on the other hand, provide hints for improving them. In detail, the numerous and heterogeneous case studies will be compared and discussed based on theoretical principles arising from the three considered urban metaphors (e.g., adaptive or transformational approach, role of innovation and creativity, relevance attributed to learning capacity). Furthermore, mobility of climate knowledge and policies among different urban contexts within Europe and between European and extra-European contexts will be deepened.

Finally, also based on the outcomes of the fifth section, current approaches will be reframed to outline an integrated approach to climate issues. In detail, based on mutual links and/or conflicting issues identified in current large-scale and local policies and practices, an overarching framework capable of guiding planners and decision-makers in building up climate-sensitive urban development processes will be outlined. The framework will include strategic and operative guidelines to overcome barriers and critical issues previously highlighted, favoring the identification of cross-sectoral strategies and measures; the capacity to take into account synergies and tradeoffs between mitigation and adaptation strategies and measures in climate policies at an urban scale; and the mutual capacitation and contamination among institutional and bottom-up climate strategies/initiatives.

Adriana Galderisi
University of Campania Luigi Vanvitelli Aversa, Italy

Angela Colucci
Co.O.Pe.Ra.Te. ldt Pavia, Italy

References

Brenner, N., Schmid, C., 2014. The 'urban age' in question. International Journal of Urban and Regional Research 38 (3), 731–755. http://dx.doi.org/10.1111/1468-2427.12115.

Beck, U., 1992. The Risk Society. Towards a New Modernity. Sage Publications.

Beck, U., 2017. The Metamorphosis of the World. Polity Press, UK, USA.

Secchi, B., 2014. A new urban question: when, why and how some fundamental metaphors were used. In: Gerber, A., Patterson, B. (Eds.), Metaphors in Architecture and Urbanism. An Introduction. Transcript Verlag, ISBN 978-3-8376-2372-7.

SECTION I

METAPHORS TO ENHANCE CITIES' CAPACITY TO COPE WITH CLIMATE CHANGE

CHAPTER

1

The Smart City Metaphor to Foster Collaborative and Adaptive Decision-Making Processes in the Face of Climate Issues

Adriana Galderisi
University of Campania Luigi Vanvitelli, Aversa, Italy

1.1 INTRODUCING THE SMART CITY METAPHOR

The concept of a *smart city* was first introduced in the early 1990s, according to an economic perspective and pointing out a urban development more and more dependent on technology, innovation, and globalization phenomena (Gibson et al., 1992). However, using heterogeneous terms (wired, intelligent, digital, etc.) and referring to technologies that largely differ from those currently available, the smart city concept has actually been discussed since the late 1970s (Batty et al., 2012). Nowadays, more and more sophisticated hardware and software technologies (sensors' networks, smart devices, etc.) allow for the collection, systematization, analysis, and integration of large amounts of structured and unstructured information, creating opportunities for more informed decision-making processes in different thematic and geographical areas.

Despite that in the last decade the term *smart city* has gained prominent attention from scholars, practitioners, and decision makers, definitions and approaches are still very heterogeneous, and a common definition is still lacking. The term has been used so far with so many different meanings that it is in danger of becoming another vague urban label (Holland, 2008), a fuzzy concept often improperly used (Nam and Pardo, 2011).

Referring to the rich literature on smart cities published since the year 2000, it is possible to find more than 50 definitions, mostly from 2010 to 2012, from both scholars and representatives of the leading technology companies (e.g., Siemens, Cisco, IBM, etc.).

Although a shared definition is still lacking, starting from the middle of the 2000s, the operational domains of smart cities have been clearly defined. According to Giffinger et al. (2007), "a Smart City is a city well performing in a forward-looking way in six characteristics, built on the 'smart' combination of endowments and activities of self-decisive, independent and aware citizens." These characteristics refer to smart economy, smart people, smart governance, smart mobility, smart environment, and smart living. For each characteristic specific factors and indicators have been provided. Giffinger's model represents a reference point for the following research studies and initiatives in the field of smart cities, although it paved the way for numerous criticisms too. The representation of the six characteristics as separate from each other, for example, leads to promoting a sectoral approach to urban development; moreover, some characteristics, such as quality of life or governance, should not be intended as separate dimensions, since all the actions undertaken in the different operational domains (mobility, environment, etc.) should have the final goal of raising the quality of life and require effective governance mechanisms for their implementation.

Nevertheless, an in-depth comparative analysis of the numerous available definitions of smart city as well as a detailed analysis of its main characteristics are beyond the scope of this chapter. Numerous literature reviews on smart cities, aimed at comparing and discussing existing definitions of the concept, based on different perspectives and addressing heterogeneous goals, have been carried out in recent years (Mosannenzadeh and Vettorato, 2014;

Albino et al., 2015; Anthopoulos, 2015); moreover, the Giffinger model has been largely used both for ranking medium-sized European cities and for analyzing successful initiatives (Manville et al., 2014).

Thus, referring to the available literature for a detailed knowledge of definitions and characteristics of a smart city, we will focus on some key points, allowing a better understanding of how such a widespread concept may contribute to achieve a climate-sensitive urban development.

1.2 THE SMART CITY BETWEEN CONFLICTING APPROACHES AND GOALS: LOOKING FOR NEW DIRECTIONS

One of the most controversial issues in current literature on the smart city is the role of Information and Communication Technologies (ICT) in urban development; it can be synthesized into the well-known and largely debated dichotomy between a techno-centered approach, focused on the potentialities of "hardware" infrastructure (Cairney and Speak, 2000; Washburn and Sindhu, 2010), and a human/social centered approach, which emphasizes the importance of human and social capital (Partridge, 2004; Glaeser and Berry, 2006) to the smart city.

The first approach, so far largely prevailing, has been largely nurtured by multinational companies, leaders in the sector of ICT manufacturing. Nevertheless, in recent years many criticisms of the technological determinism that underlies this approach have been raised, so that the vice president of CISCO pointed out that, despite that "we are crossing the threshold to put internet-based tools to work in cities .... technological devices are merely tools that can make our life better only if they are put in the hands of users who understand and can make the most of them" (Elfkrink, 2012).

The second approach does not ignore the role of technology as a crucial "enabling tool," but points out human and social capital as the main levers for smart development. In this sense, some scholars have emphasized the importance of smart inhabitants, in terms of educational level (Lombardi et al., 2012; Shapiro, 2006); others have emphasized the importance of a highly skilled labor force, and namely a "creative class" (Florida, 2003; Caragliu and Nijkamp, 2008), in terms of employees in "creative" sectors (science, engineering, design, multimedia industry, etc.). Following this line of thought, the label smart city has been recently converted into "human smart city" (Concilio et al., 2015). The new label clearly emphasizes the supporting role of technology to achieve a more effective and, above all, a more equitable urban system, also through participatory planning and design processes (Schuler, 2016). The proponents of the human smart city approach stress not only the potential of ICTs in strengthening the human dimension of urban smartness but clearly note the key role of ICTs in reshaping current cities, by underlining that new technologies cannot be merely "fit into the existing spaces, but actually modify the physical substrate and remodel the city by changing the ways in which people produce it" (Concilio and Rizzo, 2016). Thus, it is possible to argue that the human smart city perspective envisages a radical change of current urban development patterns that, starting from human and social capital, also affects physical, functional, and organizational aspects of cities' development.

Another key issue in current debate on smart cities concerns the main goals to which the increasing use of ICTs has to be addressed. Numerous scholars have focused, indeed, on ICTs as key tools for making cities more and more instrumented and interconnected, for better linking cities' infrastructures and services and, in so doing, for ensuring higher quality services to citizens and improving overall cities' efficiency (Marsa-Maestre et al., 2008; Naphade et al., 2011). In contrast, other scholars outline that a smart city "has to be based on something more than the use of information and communication technologies" (Holland, 2008), since technology is not an end in itself but a means "to reinventing cities for a new economy and society with clear and compelling community benefit" (Eger, 2009).

It is evident that the debate among the supporters of an efficiency perspective and those, more radical, envisioning a substantial change in paradigms on which current urban development patterns are based, is closely linked to, and for some aspects reflects, the previously discussed dichotomy between techno-centered and human-centered approaches.

In light of the famous quote from the Roman philosopher Seneca, "if a man does not know to what port he is steering, no wind is favorable to him" (Epistolae LXXI, 3), it is possible to argue that ICTs represent a favorable wind as long as their use will be explicitly addressed to change or innovate current urban development patterns (in physical, functional, economic, and social terms), given the unsustainable conditions they have led to. The relevance of "direction" was also emphasized some years ago by Meadows (1994) in her seminal presentation to the International Society for Ecological Economics. She argued, in fact, that "even if information, models, and implementation could be perfect in every way, how far can they guide us, if we know what direction we want to move away from, but not what direction we want to go toward?"

Hence, based on the above, the idea of using ICTs for optimizing existing services and infrastructures, by improving cities' efficiency, seems to drive toward a limited (albeit desirable) goal, without envisioning new urban development patterns. In contrast, the smart city concept might present significant opportunities to reshape the way in which cities grow, the relationships between citizens and institutions as well between city and nature, leading to modify current sociotechnical and socioeconomic aspects of growth (Zygiaris, 2013).

1.3 SMART CITY INITIATIVES: FROM SECTORAL TOWARD INTEGRATED PERSPECTIVES

The smart city model developed by Giffinger et al. (2007), as introduced above, represented a reference point not only for scholars but also for the numerous initiatives and projects addressed to improve urban smartness. Thus, smart city initiatives and projects have been initially marked by a sectoral perspective, addressing one or more of the smart cities' operational domains (energy, mobility, etc.), without a comprehensive vision of urban development.

The prevailing sectoral approach also characterized the "Smart Cities and Communities Initiative", launched in 2011 by the European Union (EU) and addressed to support European cities in achieving the targets established by the EU Strategy 2020: 20% reduction of the greenhouse gas emissions (compared to the 1990 levels); 20% of energy from renewable sources; and 20% increase in energy efficiency. The Initiative aimed at funding projects focused on two areas of interest, energy and transport, although the projects could also focus on only one of the two areas.

A severe criticism of the sectoral approach to the smart city has been provided by the Manifesto for Smarter Cities (Kanter and Litow, 2009). The authors noted that in complex urban systems no subsystem can be effective when operating in isolation, and that making each subsystem smarter—by enhancing the performance of individual sectors (from transport to energy, from constructions to urban safety, etc.)—might be insufficient for building up a smart city.

In July 2012, the European Commission launched "The Smart Cities and Communities European Innovation Partnership," aimed at supporting integrated projects in the sectors of energy, transport, and ICTs in urban areas. It is worth noting that only projects capable of integrating the three areas of concern, by creating synergies among them, were considered for funding. One year later, in 2013, the EU adopted the "Strategic Implementation Plan of the European Innovation Partnership for Smart Cities and Communities" (EU, 2013). The background document to the Strategic Implementation Plan clearly emphasized the need for holistic approaches, overcoming the 'silo' mentality and integrating solutions across different sectors (ICTs, energy production, distribution and use, transport and mobility). Moreover, besides promoting cross-sectoral actions, they also emphasized the need for integrating different actors across the innovation chain and engaging citizens in planning decisions at an early stage.

This approach seems better suited to the idea of smart city "as an organic whole" (Kanter and Litow, 2009) and to the emerging and widely agreed upon idea that while the "siloed" city model can be considered as one of the main barriers to the building up of smart city, the latter requires a system-wide view as well as integrated and cross-sectoral approaches (DeKeles, 2015).

The need for embedding smart city initiatives into a comprehensive city vision was also stressed by a 2014 study titled "Mapping Smart City in EU." This study was commissioned by the European Parliament's Industry, Research and Energy Committee in order to provide a picture of the main factors contributing to the success of smart city initiatives and to formulate recommendations for future initiatives. Two main success factors for smart initiatives were identified: (1) with reference to the Giffinger's model, successful initiatives are those capable of covering all of the six characteristics, providing a holistic and integrated approach to the building up of a smart city, and (2) to promote a wide stakeholders' engagement. The study also highlighted pros and cons of top-down and bottom-up approaches, remarking that, whereas a top-down approach promotes a high degree of coordination, a bottom-up approach provides more opportunities for direct people engagement (Manville et al., 2014).

1.4 THE GOVERNANCE OF THE SMART CITY: FROM TOP-DOWN INITIATIVES TOWARD COLLABORATIVE DECISION-MAKING PROCESSES

Even before the study of Manville et al. (2014), people involvement has been considered as a critical factor for successful smart city initiatives, and numerous scholars have stressed that building up a smart city requires the strengthening of the cities' capacity not only to reduce energy consumption or to provide more effective services but, above all, to empower human capital and ensure wide and effective participation of a wide range of stakeholders in urban development processes.

Nam and Pardo (2011), for example, clearly presented the key role of stakeholders' engagement: "successful smart city can be built from top down or bottom up approaches, but active involvement from every sector of the community is essential." Also Batty et al. (2012) emphasized that "although public participation has represented a longstanding tradition in institutionalized planning ..., the emergence of the digital world has turned the activity on its head."

Hence, due to the growing awareness that a smart city "cannot be achieved solely through physical hard planning" (Albrechts, 2016), governance issues have gained increased attention in the debate on smart cities both in academic and in political arenas (Meijer and Bolivar, 2016).

According to Birkmann et al. (2010) it is possible to distinguish between the concept of governance, which "describes all ways in which individuals and institutions exercise authority and manage common affairs at the interface of the public, civil society and private sector," and that one of "good governance" introduced by the United Nations Development Programme (UNDP) in 1997 and related to "participatory, consensus oriented, accountable, transparent, responsive, effective and efficient, equitable and inclusive" processes.

The Discussion Paper on Governance for Sustainable Development issued in 2014 has clearly stated that, although "the Millennium Development Goals (MDGs) did not include a goal or targets on governance," this concept is crucial to sustainable and equitable development. "Governance ... provides," indeed, "the mechanisms through which collaboration can be generated across sectors. It also addresses some of the fundamental obstacles to sustainable development, including exclusion and inequality" (UNDP, 2014).

Due to the key role played by ICTs in enhancing connectivity across different sectors as well as among different urban stakeholders, "smart governance" has been identified as one of the six characteristics or dimensions of a smart city—although more and more widely recognized as a cross-cutting issue in the building up of a smart city—comprising different aspects as public participation in decision making, services for citizens, as well as the transparent functioning of public administration (Giffinger et al., 2007).

Nevertheless, two different schools of thought appear in the debate on smart governance. The first one emphasizes the idea of smart governance as a more effective techno-practice supported by ICTs; the second one stresses the idea that ICTs should be addressed to transform and reshape current decision-making processes.

In detail, the supporters of the idea of smart governance as a "techno-practice" emphasize that the availability of a large number of innovative tools (sensors for collecting real-time data and information; technical platforms as devices for better linking different domains, etc.) may guarantee informed and real-time updated decision-making processes and, in doing so, might increase their effectiveness. Based on this, numerous top-down initiatives have been developed in order to optimize urban functions and services through ICTs in existing cities and, to a very extreme, this line of thought has resulted into newly planned and designed from-scratch cities (Masdar, Songdo, etc.), although widely criticized and largely unsuccessful.

In contrast, some scholars refer to smart governance as "innovative ways of decision-making, innovative administration or even innovative forms of collaboration" (Meijer and Bolivar, 2016). In this way, ICTs are mostly interpreted as facilitating tools for creating new collaborative environments (Komninos, 2011). These environments could be very relevant both for ensuring a better linkage among multiple levels, bodies, and operational tools currently in charge of different aspects of urban and territorial development and for guaranteeing more effective citizen engagement in decision-making processes, although they require, above all, innovation-oriented institutions and virtual collaborative spaces.

Batty et al. (2012) recognized also the significant potential of ICTs in strengthening bottom-up participatory approaches, allowing "informed citizens to engage with experts from many domains in generating scenarios for improving the quality of urban life and urban performance."

In this way, recent experiences in the field of smart city in Europe are exploring bottom-up participatory approaches, paving the way to "pro-active and open-minded governance structures, with all actors involved" (Kourtit et al., 2012) and, in doing so, they are driving toward "community-based models of governance"

(Meijer and Bolivar, 2016). The new governance structures are mainly addressed to create collaborative environments, often based on open platforms that facilitate collaborative interactions and processes, in which different stakeholders, comprising citizens, may interact and work together for codeveloping innovative solutions to specific challenges. In these contexts, the innovation potential is mostly related to the possibility to integrate different sources and types of knowledge—scientific/expert knowledge, developed by different actors, on different geographical scales and in different domains (Galderisi, 2016), and experiential/local knowledge (Albrechts, 2016), which is considered one of the key challenges for adaptive urban governance (Birkmann et al., 2010).

1.5 INFORMATION, KNOWLEDGE, AND LEARNING FOR ADAPTIVE CITIES

In the collaborative environments briefly depicted herein, knowledge and learning arise as key issues for increasing urban smartness. According to Komninos (2011), two relevant features of a smart city can include, on the one hand, the application of a wide range of electronic and digital technologies to create a cyber, digital, wired, informational, or knowledge-based city, and on the other hand, bringing ICT and people together to enhance innovation and learning.

The debate on smart cities has long been focused on the role of information flows. IBM (2010) defined cities as complex networks of interconnected systems, constantly creating new data to be used for monitoring, measuring, and managing urban life. Networks of sensors, wireless devices, and data centers have long been considered as the backbone of smart cities, allowing local authorities to provide urban services in a faster and more efficient manner. Moreover, the potential of ICTs to collect (in real time and from a variety of different sources), store, and manage larger and larger amounts of information and data, as a mean to rationalize planning and management of cities, has been largely emphasized (Townsend, 2013).

Thus, the availability of a large amount and real-time updated data and information related to different aspects of city functioning represents a key feature of a smart city. Nevertheless, the question is if the large amount of available information is capable of guaranteeing better knowledge of urban phenomena and, even more, improving interactions among different stakeholders by engaging them in decision-making processes or supporting innovative solutions to existing problems? As argued by Murgante and Borruso (2015), the larger and larger amount of available data often provides structured and unmanageable information, which does not allow either to integrate different data and information or to effectively engage citizens and other stakeholders in decision-making processes. The recent debate on big data has raised, indeed, numerous doubts related, for example, to the difficulty of integrating data arising from different sources (Assuncao et al., 2013) or to its effectiveness in supporting better decision-making (Shah et al., 2012).

Thus, again, the adoption of a narrow focus to the smart city, exclusively focused on smart technologies and tools, is likely to result in the paradox that the greater and greater amount of available information is ineffective in providing innovative solutions to current problems (Norton et al., 2015).

In contrast, ICTs might play a key role when framed as a wider approach, mainly focused on smart ways to use technology. As remarked by Schuler (2016), "civic intelligence requires that people perceive ('collecting') problems, talk about them ('communicating'), and interpret information ('crunching'), but these chores cannot be accomplished solely with software, no matter how smart it may be."

Thus, smart technologies may effectively contribute to enhancing connectivity across different knowledge domains, but only if we progress from the still-prevailing silo approaches to knowledge and policies (Davoudi and Cowie, 2016) toward systemic-oriented knowledge (Williams and Hardison, 2013; Galderisi, 2016). Such a shift represents a crucial step toward a better understanding of urban phenomena as well as toward more effective policies. As observed by Loevinsohn et al. (2014), indeed, the "disconnect between the different scientific communities and related knowledge and practice hampers comprehensive diagnosis of the problems at stake and the mounting of more effective actions to address it." Moreover, ICTs should support new forms of stakeholders' engagement, by shifting from informative data communication toward manageable (open and interoperable) data and information, enabling different stakeholders to interact to codevelop innovative and shared solutions to the problems at hand.

The potential of ICTs is closely related to another issue that has also been greatly discussed in the recent literature—uncertainty. Nowadays cities are largely interpreted as complex, dynamic, and self-organizing systems, continuously changing under the pressure of heterogeneous perturbing factors due to internal or external processes: these systems are characterized by evolutionary paths that are difficult to foresee in advance. Thus, due to the relevance currently attributed to uncertainty—typical of urban systems and emphasized by a changing climate that is bringing new uncertainties to the table (Head, 2014)—ICTs may contribute, on the one hand, to shed light on interdependencies, by integrating and combining existing knowledge developed by different actors, over different geographical scales and in different domains; on the other hand, they may contribute to better monitoring temporal and spatial dynamics of urban systems, by allowing the establishment of continuous learning processes, capable of enhancing cities' capacities to adapt to changing conditions and to learn from failures, considering that "the current way of doing things is not necessarily the best way" (Albrechts, 2016).

1.6 CONCLUDING REMARKS

Summing up, what is the potential of the smart city metaphor in enhancing cities' capacities to deal with climate issues?

So far, according to the still-prevailing techno-centered and sectoral approach, smart cities initiatives have mainly addressed mitigation issues; since 2011, indeed, the Smart Cities and Communities Initiative has devoted great consideration to the widespread use of ICTs as key tools for achieving the targets established by the EU Strategy 2020. Hence, these initiatives have emphasized the potential for ICTs to improve urban energy performance, by ensuring a more efficient use of energy resources and the reduction of current levels of energy consumption (Kramers et al., 2014).

Nevertheless, even though current initiatives have largely contributed to the achievement of remarkable results in terms of climate change mitigation, at least in Europe, smart city initiatives could be directed toward more ambitious objectives compared to those they have pursued so far, providing significant opportunities to reshape existing cities in their multiple dimensions (physical, functional, organizational, economic, social), enhancing their capacities to deal with current and future challenges, comprising climate change.

First of all, ICTs may enable cities to progress from current silo approaches toward a systemic-oriented knowledge, ensuring a better understanding of urban phenomena and, meanwhile, more effective strategies to cope with the interconnected challenges threatening cities' development. Moreover, the most recent evolution of the Triple/Quadruple Helix approaches—which have been considered as accelerators of smart city solutions, by using ICTs to facilitate interactions and exchange among institutions, universities, industries, and media—has led to place the natural environment in the spotlight of smart city initiatives, by shifting from a Quadruple toward a Quintuple Helix model, with the fifth helix representing the natural environment (Carayannis et al., 2012).

The Quintuple Helix model relates knowledge and innovation to the natural environment, serving as a conceptual framework for driving smart cities initiatives toward wider sustainability goals. This model indeed promotes the building up of collaborative environments capable of tackling—based on a multisectoral and multistakeholder perspective—the heterogeneous environmental challenges threatening cities' development, including climate change, and framing them in the wider process of sustainable development.

Furthermore, in these collaborative environments, ICTs might be oriented to ensure better linkages among the multiple bodies and operational tools currently in charge of different aspects of urban and territorial development and to guarantee effective stakeholder engagement in decision-making processes, in addition to strengthening bottom-up participatory approaches.

Finally, ICTs may effectively support the establishment of continuous learning processes, enhancing cities' capacities to constantly adapt to changing conditions. In other words, smart city initiatives might: enhance networking among different stakeholders; guarantee constant monitoring of conditions and changes in urban systems; improve capacities to learn from past events and failures. In addition, ICTs may significantly increase learning capacity, crucial for improving people and institutions' awareness about climate-related issues, for better anticipating future events and for ensuring effective management of urban systems over time in the face of climate challenges.

References

Albino, V., Berardi, U., Dangelico, R.M., 2015. Smart cities: definitions, dimensions, performance, and initiatives. Journal of Urban Technology 22 (1), 3–21. https://doi.org/10.1080/10630732.2014.942092.

Albrechts, L., 2016. Strategic planning as governance of long-lasting transformative practices. In: Concilio, G., Rizzo, F. (Eds.), Human Smart Cities, Rethinking the Interplay Between Design and Planning. Springer International Publishing, Switzerland, pp. 3–20.

Anthopoulos, L.G., 2015. Understanding the smart city domain: a literature review. In: Rodríguez-Bolívar, M.P. (Ed.), Transforming City Governments for Successful Smart Cities, Public Administration and Information Technology, vol. 8. https://doi.org/10.1007/978-3-319-03167-5_2.

REFERENCES

Assuncao, M.D., Calheiros, R.N., Bianchi, S., Netto, M.A.S., Buyya, R., 2013. Big Data Computing and Clouds: Challenges, Solutions, and Future Directions. Technical Report, CLOUDS-TR-2013-1-Cloud Computing and Distributed Systems Laboratory. The University of Melbourne, Parkville, Victoria.

Batty, M., Axhausen, K.W., Giannotti, F., Pozdnoukhov, A., Bazzani, A., Wachowicz, M., Ouzounis, G., Portugali, Y., 2012. Smart cities of the future. The European Physical Journal Special Topics 214, 481–518. https://doi.org/10.1140/epjst/e2012-01703-3.

Birkmann, J., Garschagen, M., Kraas, F., Quang, N., 2010. Adaptive urban governance: new challenges for the second generation of urban adaptation strategies to climate change. Sustainability Science 5 (2), 185–206. https://doi.org/10.1007/s11625-010-0111-3.

Cairney, T., Speak, G., 2000. Developing a "Smart City": Understanding Information Technology Capacity and Establishing an Agenda for Change. Available at: http://trevorcairney.com/wp-content/uploads/2012/11/IT_Audit.pdf.

Caragliu, A., Nijkamp, P., 2008. The Impact of Regional Absorptive Capacity on Spatial Knowledge Spillovers. Tinbergen Institute Discussion Papers, 08–119/3. Tinbergen, Amsterdam. Available at: https://www.researchgate.net/publication/23775434_The_Impact_of_Regional_Absorptive_Capacity_on_Spatial_Knowledge_Spillovers.

Carayannis, E.G., Barth, T.D., Campbell, D.F.J., 2012. The Quintuple Helix innovation model: global warming as a challenge and driver for innovation. Journal of Innovation and Entrepreneurship 1 (2). https://doi.org/10.1186/2192-5372-1-2.

Concilio, G., Rizzo, F., 2016. Preface: experimenting between design and planning. In: Concilio, G., Rizzo, F. (Eds.), Human Smart Cities, Rethinking the Interplay between Design and Planning. Springer International Publishing, Switzerland, pp. V–X.

Concilio, G., Marsh, J., Molinari, F., Rizzo, F., 2015. Human Smart Cities. A new vision for redesigning urban community and citizen's life. In: Skulimowski, A.M.J., Kacprzyk, J. (Eds.), Knowledge, Information and Creativity Support Systems: Recent Trends, Advances and Solutions. Selected Papers From KICSS'2013-8th International Conference on Knowledge, Information, and Creativity Support Systems, November 7–9, 2013, Kraków, Poland. Springer International Publishing, pp. 269–278. https://doi.org/10.1007/978-3-319-19090-7_21.

Davoudi, S., Cowie, P., 2016. Guiding principles of good territorial governance. In: Schmitt, P., Van Welle, L. (Eds.), Territorial Governance Across Europe. Pathways, Practices and Prospects. Routledge, NY.

DeKeles, J., 2015. Smart City Readiness Guide. Available at: http://readinessguide.smartcitiescouncil.com.

Eger, J.M., 2009. Smart growth, smart cities, and the crisis at the pump. A worldwide phenomenon. I-WAYS-The Journal of E-Government Policy and Regulation 32 (1), 47–53.

Elfkrink, W., 2012. Foreword. In: Campbell, T. (Ed.), Beyond Smart City: How Cities Network, Learn and Innovate. Earthscan, NY.

EU, 2013. European Innovation Partnership on Smart Cities and Communities. Strategic Implementation Plan. Available at: http://ec.europa.eu/eip/smartcities/files/sip_final_en.pdf.

Florida, R., 2003. Cities and the creative class. City and Communities 2 (1), 3–19. Available at: http://www.creativeclass.com/rfcgdb/articles/4%20Cities%20and%20the%20Creative%20Class.pdf.

Galderisi, A., 2016. The nexus approach to disaster risk reduction, climate adaptation and ecosystem management: new paths for a sustainable and resilient urban development. In: Colucci, A., Magoni, F., Menoni, S. (Eds.), Peri-Urban Areas and Food-Energy-Water Nexus. Sustainability and Resilience Strategies in the Age of Climate Change. Springer Tracts in Civil Engineering. https://doi.org/10.1007/978-3-319-41022-7_2.

Gibson, D.V., Kozmetsky, G., Smilor, R.W. (Eds.), 1992. The Technopolis Phenomenon: Smart Cities, Fast Systems, Global Networks. Rowman & Littlefield, New York.

Giffinger, R., Fertner, C., Kramar, H., Kalasek, R., Pichler-Milanovic, N., Meijers, E., 2007. Smart Cities. Ranking of European Medium-Sized Cities. Final Report. Centre of Regional Science, Vienna UT. Available at: http://www.smart-cities.eu/download/smart_cities_final_report.pdf.

Glaeser, E.L., Berry, C.R., 2006. Why are smart places getting smarter? Policy Briefs 2. Available at: https://www.hks.harvard.edu/content/download/68631/1247334/version/1/file/brief_divergence.pdf.

Head, B.W., 2014. Evidence, uncertainty, and wicked problems in climate change decision making in Australia. Environment and Planning C: Government and Policy 32 (4), 663–679. https://doi.org/10.1068/c1240.

Holland, R.G., 2008. Will the real smart city please stand up? City 12 (3), 303–320. https://doi.org/10.1080/13604810802479126.

IBM Global Business Services, 2010. Smarter Cities Assessment. Somers, NY. Available at: http://www-935.ibm.com/services/us/gbs/bus/html/ibv-smarter-cities-assessment.html.

Kanter, R.M., Litow, S.S., 2009. Informed and Interconnected: a Manifesto for Smarter Cities. Working Paper 09-141. Harvard Business School. Available at: http://www.hbs.edu/faculty/Publication%20Files/09-141.pdf.

Komninos, N., 2011. Intelligent cities: variable geometries of spatial intelligence. Intelligent Buildings International 3 (3), 172–188. https://doi.org/10.1080/17508975.2011.579339.

Kourtit, K., Nijkamp, P., 2012. Smart Cities in the Innovation Age. Innovation: The European Journal of Social Science Research 25 (2), 93–95. https://doi.org/10.1080/13511610.2012.660331.

Kramers, A., Höjer, M., Lövehagen, N., Wangel, J., 2014. Smart sustainable cities. Exploring ICT solutions for reduced energy use in cities. Environmental Modelling & Software 56, 52–62. https://doi.org/10.1016/j.envsoft.2013.12.019.

Loevinsohn, M., Mehta, L., Cuming, K., Nicol, A., Cumming, O., Ensink, J.H.J., 2014. The cost of a knowledge silo: a systematic re-review of water, sanitation and hygiene interventions. Health Policy and Planning 1–15. https://doi.org/10.1093/heapol/czu039.

Lombardi, P., Giordano, S., Farouh, H., Yousef, W., 2012. Modelling the smart city performance. Innovation: The European Journal of Social Science Research 25 (2), 137–149. https://doi.org/10.1080/13511610.2012.660325.

Manville, C., Cochrane, G., Cave, J., et al., 2014. Mapping Smart Cities in the EU. Available at: http://www.europarl.europa.eu/studies.

Marsa-Maestre, I., Lopez-Carmona, M.A., Velasco, J.R., Navarro, A., 2008. Mobile agents for service personalization in smart environments. Journal of Networks 3 (5), 30–41. https://doi.org/10.4304/jnw.3.5.30-41.

Meadows, D., 1994. Envisioning a Sustainable World. Third Biennial Meeting of the International Society for Ecological Economics, October 24–28, San Jose, Costa Rica. Available at: http://donellameadows.org/archives/envisioning-a-sustainable-world/.

Meijer, A., Bolivar, M.P.R., 2016. Governing the smart city: a review of the literature on smart urban governance. International Review of Administrative Sciences Vol. 82 (2), 392–408. https://doi.org/10.1177/0020852314564308.

Mosannenzadeh, F., Vettorato, D., 2014. Defining smart city. A conceptual framework based on keyword analysis. TeMA. Journal of Land Use, Mobility and Environment 2014, 683–694. https://doi.org/10.6092/1970-9870/2523. Special Issue INPUT.

Murgante, B., Borruso, G., 2015. Smart cities in a smart world. In: Rassia, S.T., Pardalos, P.M. (Eds.), Future City Architecture for Optimal Living. Springer, Basel, Switzerland, pp. 13–35.

Nam, T., Pardo, T.A., 2011. Conceptualizing smart city with dimensions of technology, people, and institutions. In: Proc. 12th Conference on Digital Government Research, College Park, MD, June 12–15. Available at: https://www.ctg.albany.edu/publications/journals/dgo_2011_smartcity/dgo_2011_smartcity.pdf.

Naphade, M., Banavar, G., Harrison, C., Paraszczak, J., Morris, R., 2011. Smarter cities and their innovation challenges. Computer 44 (6), 32–39. https://doi.org/10.1109/MC.2011.187.

Norton, J., Atun, F., Dandoulaki, M., 2015. Exploring issues limiting the use of knowledge in disaster risk reduction. TeMA. Journal of Land Use, Mobility and Environment 2015, 135–154. https://doi.org/10.6092/1970-9870/3032. Special Issue ECCA.

Partridge, H., 2004. Developing a Human Perspective to the Digital Divide in the Smart City. Available at: http://eprints.qut.edu.au/1299/.

Schuler, D., 2016. Smart cities + smart citizens = civic intelligence? In: Concilio, G., Rizzo, F. (Eds.), Human Smart Cities, Rethinking the Interplay between Design and Planning. Springer International Publishing, Switzerland, pp. 41–60.

Shah, S., Horne, A., Capellá, J., 2012. Good data won't guarantee good decisions. Harvard Business Review. Available at: https://hbr.org/2012/04/good-data-wont-guarantee-good-decisions.

Shapiro, J.M., 2006. Smart cities: quality of life, productivity, and the growth effects of human capital. Review of Economics & Statistics 88 (2), 324–335.

Townsend, A.M., 2013. Smart Cities: Big Data, Civic Hackers, and the Quest for a New Utopia. W. W. Norton & Company, New York.

UNDP (United Nations Development Programme), 1997. Governance for Sustainable Human Development. A UNDP Policy Document. United Nations Development Programme, New York.

UNDP (United Nations Development Programme), 2014. Discussion Paper Governance for Sustainable Development Integrating Governance in the Post-2015 Development Framework. Available at: http://www.undp.org/content/undp/en/home/librarypage/democratic-governance/discussion-paper—-governance-for-sustainable-development.html.

Washburn, D., Sindhu, U., 2010. Helping CIOs Understand "Smart City" Initiatives. Forrester Research. Available at: https://www.forrester.com/report/Helping+CIOs+Understand+Smart+City+Initiatives/-/E-RES55590.

Williams, T., Hardison, P., 2013. Culture, law, risk and governance: contexts of traditional knowledge in climate change adaptation. Climatic Change 120 (3), 531–544. https://doi.org/10.1007/s10584-013-0850-0.

Zygiaris, S., 2013. Smart city reference model: assisting planners to conceptualize the building of smart city innovation ecosystems. Journal of the Knowledge Economy 4 (2), 217–231.

Further Reading

Giffinger, R., Gudrun, H., 2010. Smart cities ranking: an effective instrument for the positioning of the cities? ACE: Architecture, City and Environment 4 (12), 7–26. Available at: http://upcommons.upc.edu/handle/2099/8550.

Papa, R., Galderisi, A., Vigo Majello, M.C., Saretta, E., 2015. Smart and resilient cities. A systemic approach for developing cross-sectoral strategies in the face of climate change. TeMA. Journal of Land Use, Mobility and Environment 8 (1), 19–49. https://doi.org/10.6092/1970-9870/2883.

CHAPTER 2

The Resilient City Metaphor to Enhance Cities' Capabilities to Tackle Complexities and Uncertainties Arising From Current and Future Climate Scenarios

Adriana Galderisi
University of Campania Luigi Vanvitelli, Aversa, Italy

2.1 INTRODUCING THE RESILIENT CITY METAPHOR

The resilient city metaphor is nowadays widely used to depict a city that is capable of withstanding, absorbing, and recovering from sudden events and chronic stresses, caused by heterogeneous pressure factors (resources scarcity, economic crises, natural hazards, climate change, etc.). This metaphor is grounded in a widely debated concept—resilience—that has been used with different meanings since ancient times and intensely investigated and conceptualized starting from the 19th century, according to different disciplinary perspectives, from mechanics and physics to psychology, from ecology to economy, sometimes with controversial outcomes (Alexander, 2013).

Along the evolutionary path of the resilience concept, both objects and objectives of resilience studies have progressively enlarged. The focus has shifted from single elements (materials, individuals) to systems (natural, social) and, more recently, to coupled systems (socioecological; socioecological-technical). Meanwhile, the initial goal of resilience studies moved from the idea of improving elements and systems' capacity to bounce back, by recovering the previous equilibrium state after a crisis, toward a "bounce forward" perspective (Manyena et al., 2011), which includes the strengthening of systems' essential structures and functions and the improvement of their ability to anticipate, in order to better drive complex adaptive systems towards new equilibrium states.

The resilient city metaphor has become prominent in the context of urban studies in the last decade, and it is also heavily emphasized in all of the latest documents on both sustainable development (UN, 2012, 2015a) and disaster risk reduction (DRR) (UN, 2015b). It has also been widely promoted among practitioners and decision makers by different international campaigns. The prominence gained by this metaphor finds its roots in the growing complexities and uncertainties arising from the numerous and interconnected challenges threatening cities (urban population growth; urban development patterns; consumption and degradation of natural resources; natural, human-induced, and coupled hazards; and changes in climate features and related impacts). Thus, resilience has been more and more widely interpreted as a relevant approach for better understanding and managing the interwoven systems of humans and nature as well for empowering cities to better deal with current and emerging challenges and, above all, with the increasing impacts of climate change. Nevertheless, some scholars have raised the idea that "much of what has been recently labeled 'resilience' is 'old wine in new bottles'" (Weichselgartner and Kelman, 2015), that the subject is still vague, has different conceptualizations, and that resilience practices often have been developed without fully accounting for theoretical debate.

Therefore, in the following discussion we will focus on what is new in resilience thinking and how it is informing or could inform resilience-building processes at city scale.

2.2 TRACING THE EVOLUTION OF RESILIENCE ACROSS DISCIPLINARY BOUNDARIES: EMERGING PERSPECTIVES

The long and articulated evolution path of the resilience concept has been at the core of numerous research works in the last decade. Although most scholars trace the roots of the concept back to Holling's studies in the field of ecology (Berkes, 2007; Djalante et al., 2011), the interesting contribution on the etymology of resilience provided by Alexander (2013) allows us to clearly understand the long history of this term and the multiple and sometimes contradictory meanings it has assumed starting from ancient times.

Referring to these studies for in-depth analyses of the evolution path of the resilience concept through different disciplinary fields, we would like to stress here the changes—in objects and objectives—of this concept as it moved across different disciplines (Fig. 2.1). These changes will be briefly outlined according to a disciplinary rather than a temporal perspective and bearing in mind that the transition from one disciplinary field to another has been characterized by interesting and fruitful cross-fertilization, but has also been limited by the difficulties and ambiguities arising from transferring concepts from one field to another, or from one object/system to another.

First of all, it is interesting to notice that both in mechanics and in psychology the conceptualizations of resilience primarily referred to individual elements and were largely based on the common notion of "bouncing-back."

In mechanics, the term was used, indeed, starting from the mid-19th century to describe the capacity of a physical element (e.g., a material) to recover its original state after a perturbation, being resistant to the external pressure or capable of absorbing it through its elasticity, without breaking or being deformed (Gordon, 1978). Also in psychology, the term has been used to describe the process, and the numerous factors influencing it, allowing people to "bounce-back from adversity and go on with their lives" (Dyer and McGuinness, 1996) as well to describe the "recovery trajectory that returns to baseline functioning following a challenge" (Butler et al., 2007).

The well-known and largely quoted work of the ecologist Holling (1973) first brought this concept to prominence into the ecological domain. Holling's contribution can be synthesized into three main steps. Firstly, his work contributed to move the focus of resilience from elements to systems: he defined, indeed, resilience as "a measure of the persistence of systems and of their ability to absorb change and disturbance and still maintain the same relationships between populations or state variables" (Holling, 1973). Secondly, he used "the term resilience to characterize dynamic equilibrium, including that which can exist in several different state spaces" (Alexander, 2013). Thirdly, he clearly distinguished "engineering" from "ecological" approach to resilience, by emphasizing that the former "concentrates on stability near an equilibrium steady state, where resistance to disturbance and speed of

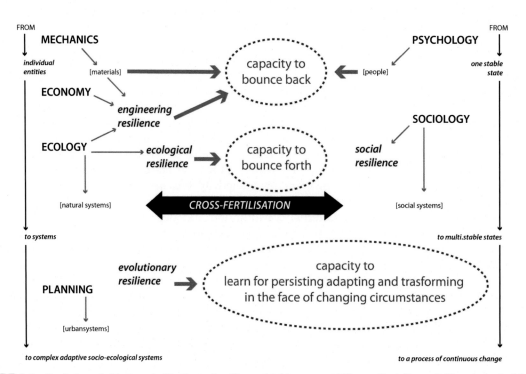

FIGURE 2.1 Evolution of objects and objectives of resilience thinking across different disciplinary fields. *Author's elaboration.*

return to the equilibrium area used to measure the property," while the latter "emphasizes conditions far from any equilibrium steady state, where instabilities can flip a system into (…) another stability domain" (Holling, 1996).

Briefly, Holling clearly underlined the need to shift from the idea that only one stable state exists to the acknowledgment of multiple stable states, stressing that in the face of a disturbance, a system has a twofold possibility: to absorb changes, below a given magnitude of disturbance; and to change its structure, moving into a different, not necessarily better than the previous one, stability domain, when the magnitude of disturbance overcomes a given threshold.

The moving of the resilience concept into the sociology domain started in the late 1990s and moved its focus from people to communities. The approach to resilience in this domain significantly benefited from the conceptualizations of ecological resilience, although the difficulties and criticalities arising when translating resilience from one field to another, and above all from natural to social systems, have been largely emphasized (Davoudi et al., 2012). Social resilience has been generally referred to as "the ability of groups or communities to cope with external stresses and disturbances as a result of social, political and environmental change" (Adger, 2000), and community resilience has been widely interpreted as a dynamic process, based on a "continual learning" (Cutter et al., 2008) that is typical of human systems.

However, the cross-fertilization between ecological and social domains led to further shift the focus of resilience studies from natural and/or social systems to coupled socioecological systems (Berkes and Folke, 1998). In detail, the intertwinement with research studies on complex adaptive systems—capable of learning from experience, process information, adapt, and even transform themselves in the face of changes—allowed to better tailor the resilience concept to the behavior of these systems and progressively moved this concept to embrace a bouncing-forward perspective, focused on processes and dynamics enabling a system to adapt or move into different states in the face of internal or external pressures and including the idea of anticipation and improvement of systems' essential structures and functions.

The concept of resilience was further expanded by some scholars who, based on the metaphor of "panarchy"—introduced by Gunderson and Holling (2001) and explaining the adaptive nature and the evolutionary dynamics of complex adaptive systems—provided an interpretation of the term as a "dynamic interplay of persistence, adaptability and transformability across multiple scales" (Folke et al., 2010). In detail, persistence is defined as the system's ability to withstand an impact, preserving its own characteristics and structure, except for a temporary departure from the ordinary functioning conditions. Adaptability refers to the capacity of a socioecological system of learning, combining experience and knowledge, in order to "adjust its responses to changing external drivers and internal processes, and continue developing within the current stability domain" (Folke et al., 2010). Transformability is defined as "the capacity to create a fundamentally new system when ecological, economic, or social structures make the existing system untenable" (Walker et al., 2004; Folke et al., 2010).

Resilience thinking has recorded further advances in the last decade, when the concept entered the urban studies domain. Despite the numerous contributions on urban resilience and the heterogeneity of their outcomes, the most promising and also the most quoted approach to the resilience concept in this field is the one introduced by Simin Davoudi et al. (2012). It refers to an "evolutionary resilience" that, focusing on cities as complex adaptive systems composed of physical, environmental, social, economic, and cultural resources as well as technical components (Chelleri et al., 2012), overcomes the engineering and ecological perspectives by embracing the idea introduced by Folke et al. (2010) of resilience as "the dynamic interplay between persistence, adaptability and transformability across multiple scales and timeframes," adding a fourth component—preparedness—based on the learning capacity. This fourth component reflects "the intentionality of human action and intervention," (Davoudi et al., 2013) typical of social systems, reflecting the key role of social capital and institutions in the building up of resilient cities.

The concept of evolutionary resilience, "understood not as a fixed asset, but as a continually changing process; not as a being but as a becoming" (Davoudi et al., 2012), seems currently to be the most effective to tackle complexities and uncertainties related to the future development of urban systems in the face of external and internal disturbances.

2.3 THE RESILIENT CITY INITIATIVES: STRENGTHS AND WEAKNESSES

Although the theoretical debate on resilience can be traced back to ancient times and has been largely developed starting from the 1970s, it is only in the last decade, and namely after Hurricane Sandy battered the east coast of the United States in 2012, that the metaphor of the resilient city gained prominence, thanks to a growing number of initiatives aimed at supporting resilience building at city scale.

These initiatives, pushed also by the importance attributed to cities' resilience in all the latest documents on sustainable development and DRR, aim to enhance cities' capacity to deal with current and emerging environmental, social, and economic challenges, with a prevailing focus on the impacts of natural and climate-related hazards.

The on-going initiatives largely differ from each other. They are promoted by different international organizations; pursue different aims; involve cities differing in size as well in geographical, cultural, economic, and social contexts; are based on different principles and guidelines; and adopt different tools to achieve an apparently common goal, that is, the building up of a resilient city.

In the following, we will focus on two of the most important international initiatives, in terms of number of involved cities, aimed at enhancing urban resilience: the "Making Cities Resilient" campaign, launched in 2010 by the United Nations International Strategy for Disaster Reduction (UNISDR); and the "100 Resilient Cities" initiative, promoted in 2013 by the Rockefeller Foundation. After a brief presentation of their goals and adopted methodologies, we will focus here on their main strengths and weaknesses and, above all, on the linkages between the main principles/conceptual frameworks informing these initiatives and the still on-going theoretical debate on resilience.[1]

These two selected initiatives are both promoted by large international organizations: (1) the UNISDR, established in 1999 as part of the United Nations Secretariat, mainly coordinates the international efforts in DRR; and (2) the Rockefeller Foundation, a private foundation established in 1913, whose declared mission is to promote the well-being of humanity throughout the world, by supporting the development of inclusive economies and the building up of resilient cities. The different missions of the two organizations have strongly shaped the aims of their initiatives. The Making Cities Resilient campaign has been developed according to the goals and priorities outlined by the Hyogo Framework for Action 2005–15, and by the Sendai Framework 2015–30. Both these documents aimed at reducing disaster losses while contributing to mainstream DRR and climate adaptation within the wider objectives of sustainability. Hence, this campaign aims at guiding cities to better deal with a main challenge: increase their resilience in the face of different natural and man-made disasters, also taking into account the less-frequent, large-scale events. Climate change and climate-related extremes are also considered, since they increase a city's overall exposure to hazards and risks.

The goals of the 100 Resilient Cities initiative are significantly broader, mirroring the wider goals of the Rockefeller Foundation itself. The initiative was designed to enhance cities' resilience in the face of a broad spectrum of stresses and shocks, ranging from migrations to water shortages, from earthquakes to terroristic attacks, taking into account that each city has to deal with multiple challenges, sometimes interlinked. Therefore, engaged cities are expected to improve their performances "in the face of multiple hazards, rather than preventing or mitigating the loss of assets due to specific events" (The Rockefeller Foundation/ARUP, 2015).

However, regardless of the extent of their aims, which idea of resilience do these two initiatives refer to? This is not a trivial question in light of the multiple and heterogeneous approaches and definitions of resilience previously discussed.

The Making Cities Resilient campaign is grounded in the definition of resilience provided by the UNISDR: "The ability of a system, community or society exposed to hazards to resist, absorb, accommodate to and recover from the effects of a hazard in a timely and efficient manner, including through the preservation and restoration of its essential basic structures and functions" (UNISDR, 2009). This definition mirrors the most widespread approach to resilience in the disaster field and embraces an engineering perspective that, as remarked by Holling (1996), refers to the capacity of a system to return to a previous equilibrium steady state, to "bounce back" after disturbances, and can be measured according to systems' resistance to disturbance and speed of return to the equilibrium area.

The 100 Resilient Cities initiative defines urban resilience as "the capacity of individuals, communities, institutions, businesses, and systems within a city to survive, adapt, and grow no matter what kinds of chronic stresses and acute shocks they experience."[2] Despite the broader adopted definition, which heavily emphasizes the capacity of individuals, communities, and systems to withstand, adapt, and also to continue to grow in the face of shocks and stresses, the numerous documents available on the initiative's website emphasize the aim to support cities in developing their capacity to recover quickly and effectively when crises arise.[3]

---

[1] http://www.unisdr.org/campaign/resilientcities/home/index; http://www.100resilientcities.org/about-us/.

[2] http://www.100resilientcities.org/resources/.

[3] https://www.rockefellerfoundation.org/blog/valuing-resilience-dividend/.

The different aims and the adopted approaches to resilience also affect the choice of the guiding principles provided by these initiatives. The Making Cities Resilient campaign provides local governments with principles and procedures aimed at guiding them through the process of resilience building. The main principles guiding this campaign are synthesized in the "Ten Essentials," whose first version released in 2010 was slightly revised in 2016, following the Sendai Framework. These principles provide the "golden rules" for enhancing cities' resilience by improving, for example, the organizational institutional structure or by investing in risk knowledge, taking into account both current and future risk scenarios or, even, by carrying out risk-informed urban planning tools. The very general targets provided by the Ten Essentials represent the starting point of the so-called "Cycle of Resilience Building," a cyclical process that, starting from the preparatory step, goes through the assessment of local disaster resilience as a base for outlining and implementing the DRR Action Plan and monitoring its achievements (UNISDR, 2017). The assessment phase can be carried out by using the Disaster Resilience Scorecard,[4] which includes for each principle a set of questions and comments, aimed at guiding local governments in evaluating current disaster resilience, monitoring and reviewing progress toward the effective implementation of the Sendai Framework. The Scorecard is structured as a two-level process (including a preliminary and a detailed assessment), involving multiple stakeholders (e.g., public bodies, private businesses, community groups, academic institutions, etc.). The provided assessment tools, although mainly qualitative, set a clear pathway for defining the current baseline, outlining future goals and monitoring achievements, raising the awareness of all the involved stakeholders on the key actions to be carried out for increasing cities' resilience in the face of multiple hazard factors. In some cases, quick assessment tools are also provided (e.g., the Quick Risk Estimation Tool).

The 100 Resilient Cities initiative is also based on a set of tools aimed at supporting selected cities in the building up of a resilience strategy. These principles have been structured into the "City Resilience Framework" (CRF), set up by the Rockefeller Foundation in cooperation with the global design firm Arup, with the aim of understanding and measuring cities' resilience. The CRF is a circular model structured into different rings and sectors. It identifies four key sectors (Health and Well-being; Economy and Society; Infrastructure and Environment; and Leadership and Strategy) and 12 key goals (three for each dimension) that cities should achieve for improving their resilience. Then, a set of 52 indicators, related to the different sectors and key goals, as well the most adequate qualitative or quantitative metrics for their measurement and assessment, are provided. The set of indicators is related to seven characteristics that a resilient system should strengthen to effectively withstand, respond to, and readily adapt to shocks and stresses: inclusiveness, integration, reflectiveness, resourcefulness, robustness, redundancy, and flexibility (The Rockefeller Foundation/ARUP, 2015).

Although largely different in their foci and adopted methodologies, the selected initiatives have some unquestionable merits as well some weaknesses.

First of all, the involvement of thousands of local governments all over the world is undoubtedly a merit of these initiatives. They have significantly contributed, indeed, to bring urban resilience issues out of the theoretical debate, confined to some disciplinary boundaries, into a priority goal of the cities' political agenda, meanwhile raising community and decision makers' awareness.

Moreover, despite the differences in the adopted definitions and tools, both of them provide local governments with useful tools for understanding, assessing, and improving their capacity to cope with different stresses and shocks, by supporting them in establishing a baseline to understand current gaps, to establish and prioritize their goals and actions, and to monitor their progress through a cyclical process. As well, both of them require the engagement of multiple stakeholders that is relevant to strengthen collaboration among different sectors of local governments as well as among different institutional bodies, businesses, and research units. Such collaboration may enable cities to progress from current silo approaches toward a systemic-oriented knowledge (Williams and Hardison, 2013), considered a prerequisite for promoting cross-sectoral strategies to better cope with the interconnected challenges threatening cities' future development (Galderisi, 2017).

Furthermore, both of them seem to be based more on a "formative" than on a "summative" approach to resilience assessment; the former is generally "used to inform decisions about improvement, to evaluate a project or a program development and to learn about incremental changes" (Turner et al., 2014), and the latter is generally used to evaluate the "overall effectiveness of an initiative and informs decisions about whether to continue or end" (Turner et al., 2014) after the completion of the initiative itself. Therefore, the "formative" approach allows not only to better grasp the "dynamic nature of resilience" (Sharifi, 2016) but also to guarantee "continual learning" along the process of resilience capacity building.

---

[4] http://www.unisdr.org/campaign/resilientcities/home/toolkitblkitem/?id=4.

Another important aspect is that these initiatives make available a large array of knowledge on challenges that cities all over the world are coping with, as well as on their visions and goals for the future.

However, the two briefly described initiatives also present some weaknesses. Leaving aside the differences and validity of the selected resilience assessment tools, among the numerous currently available and largely discussed in recent literature (see Cutter, 2016; Sharifi, 2016), some points deserve to be highlighted here.

A first point refers to the governance of the resilience building processes: the latter require, indeed, active participation from governments, citizens, scientists, and private sectors (Borquez et al., 2016). In the examined initiatives, the strong role of international organizations in advocating for local actors raises the reasonable doubt that, although these initiatives surely contribute to increase local awareness about resilience issues, they do not necessarily lead to an empowerment of local institutions and, especially, of local communities. Effective engagement of the diverse urban stakeholders, including scientists, politicians, businesses, media, and citizens, in all the phases of the process, is indeed generally time-consuming, and this choice cannot be fully compatible with the short-term horizon of policy makers and funding agencies.

Moreover, it is questionable if the provided tools, based on literature and external experts' inputs, are usable in different contexts all over the world: these contexts are, indeed, heterogeneous in size (from metropolitan areas or provinces to small cities); show different geographical, cultural, social, political, and economic features; and deal with different challenges. Hence, it's worth wondering if these tools are flexible enough to be tailored to the peculiarities of different cities, allowing them to remove or add principles or criteria, according to their needs. Besides, the "customization" of these tools to the different cities should be developed through a transdisciplinary process of coevaluation and codesign, capable of combining the scientific/expert knowledge and the experiential knowledge of local stakeholders. Unfortunately, based on the available documents, it is not clear if the "customization" of these tools to the specific contexts is currently developed through a participatory and transparent process, which is considered a key challenge for promoting adaptive urban governance, enabling cities to better deal with current and future challenges.

Furthermore, the weak link between the theoretical debate on resilience and these initiatives causes them to often overlook the importance of cross-scale interactions—crucial to an effective understanding of resilience—focusing on cities as isolated entities (Sharifi, 2016). It is worth noting also that, as previously stressed, resilience is nowadays widely interpreted as a "dynamic interplay of persistence, adaptability and transformability across multiple scales" (Folke et al., 2010), while most current practices tend to focus on one or two of these components, namely on persistence and adaptability.

Finally, whereas in scientific literature the key dimensions of a resilient system are intended as dynamically interacting over time and across space, and resilience is often interpreted as a set of interconnected properties/capacities (Norris et al., 2008; Papa et al., 2015), both the analyzed tools focus on separate principles, goals, and sectors; such an approach leads to undervalue the potential synergies and trade-offs among different goals and actions.

In brief, these initiatives, despite their importance in raising local awareness on resilience issues and in advancing toward the "operationalizing" of the resilience building at city-scale, seem to benefit only partially from the significant evolution of the resilience thinking as a result of the integration and cross-fertilization among different disciplines, mirroring a persistent gap between theory and practice.

To fill this gap, a key role should be played by urban planners, both due to their main focus on cities as systems of interconnected systems, which is crucial to better understand synergies, potential conflicts and trade-offs among different resilience goals, and to the peculiar nature of planning, defined as "a professional practice that specifically seeks to connect forms of knowledge with forms of action in the public domain" (Friedmann, 1993).

2.4 CONCLUDING REMARKS

Although the theoretical debate on resilience can be traced back to ancient times and is well documented in different disciplinary fields, this concept is still struggling to find effective integration, capable of overcoming disciplinary boundaries; so far, it has been largely interpreted as a fashionable umbrella concept, "an all-encompassing, multi-interpretable idiom" (Weichselgartner and Kelman, 2015), capable of attracting wide interest and funds but difficult to translate into operational terms.

The resilient city metaphor has gained prominence only in the last decade as a consequence of relevant urban disasters (e.g., Hurricanes Katrina and Sandy) and pushed by important international organizations (UNISDR, ICLEI, etc.) through campaigns aimed at improving cities' capacity to cope with existing and emerging environmental and social challenges and, above all, with the growing impacts of climate change; most of the on-going international initiatives are focused, indeed, on urban climate resilience. Unfortunately, the resilient city metaphor, as it has been used so far, does not fully mirror the theoretical debate on resilience, since most of the current experiences have been developed according to a "bounce-back" perspective, which clearly informs, for example, the UNISDR's Making Cities Resilient campaign. This perspective leads to stress on "climate adaptation" strategies, mainly addressed to increase cities' capacities to withstand, absorb, and recover from climate impacts.

Nevertheless, the theoretical debate on resilience could bring out new perspectives, paving the way for resilience-based climate strategies capable of overcoming a optimization-based approach, proper tothe engineering resilience, emphasizing on the opposite the capacity of living systems to continuously adapt or change, by "inventing new practices in front of novel problems" (Grøtan, 2014), as the ones posed by climate change.

Based on the idea of "evolutionary resilience" introduced by Davoudi et al. (2012), cities should develop proactive and site-tailored climate strategies, based on local communities' capacities for active learning, and addressed to improve their capacity to cope in the short term with climate impacts (persistence), to continuously adapt in the face of changing conditions through incremental adjustments (adaptability), and to innovate in the long term (transformability), by introducing fundamental changes within and across urban systems.

This idea better fits with the recently emerging idea of "transformational adaptation" (Lonsdale et al., 2015), mirroring the growing awareness that current climate strategies, based on a coping or an incremental approach, could be insufficient or inadequate if global mitigation strategies, aimed at counteracting the anthropogenic causes of climate change, fail. Thus, transformational adaptation seeks to address "underlying failures of development, including the increase in greenhouse gas emissions, by linking adaptation, mitigation and sustainable development" (EEA, 2016). However, these fundamental changes require significant "shifts in perception and meaning, social network configurations (…) and associated organizational and institutional arrangements" (Folke et al., 2010). They require, indeed, a long-term vision, an integrated approach to urban development—capable of better linking mitigation and adaptation and of embedding these goals into the broader one of sustainable development— and appropriate governance mechanisms—capable of engaging a wide range of stakeholders and promoting collaboration and cooperation among different municipal departments and different governmental levels.

References

Adger, W.N., 2000. Social and ecological resilience: are they related? Progress in Human Geography 24 (3), 347–364. https://doi.org/10.1191/030913200701540465.

Alexander, D.E., 2013. Resilience and disaster risk reduction: an etymological journey. Natural Hazards and Earth System Science 13, 2707–2716. https://doi.org/10.5194/nhess-13-2707-2013.

Berkes, F., 2007. Understanding uncertainty and reducing vulnerability: lessons from resilience thinking. Natural Hazards 41, 283–295. https://doi.org/10.1007/s11069-006-9036-7.

Berkes, F., Folke, C. (Eds.), 1998. Linking Social and Ecological Systems: Management Practices and Social Mechanisms for Building Resilience. Cambridge University Press, Cambridge, UK.

Borquez, R., Aldunce, P., Adler, C., 2016. Resilience to climate change: from theory to practice through co-production of knowledge in Chile. Sustainability Science 12, 163–176. https://doi.org/10.1007/s11625-016-0400-6.

Butler, L., Morland, L., Leskin, G., 2007. Psychological resilience in the face of terrorism. In: Bongar, B., Brown, L., Beutler, L., Breckenridge, J., Zimbardo, P. (Eds.), Psychology of Terrorism. Oxford University Press, NY.

Chelleri, L., Kunath, A., Minucci, G., Olazabal, M., Waters, J.J., Yumalogava, L., 2012. Multidisciplinary Perspective on Urban Resilience. Workshop report. BC3, Basque Centre for Climate Change, ISBN 978-84-695-6025-9. Available at: https://www.researchgate.net/publication/235223379_Multidisciplinary_perspectives_on_Urban_Resilience.

Cutter, S.L., 2016. The landscape of disaster resilience indicators in the USA. Natural Hazards 80, 741. https://doi.org/10.1007/s11069-015-1993-2.

Cutter, S.L., Barnes, L., Berry, M., Burton, C., Evans, E., Tate, E., Webb, J., 2008. A place-based model for understanding community resilience to natural disasters. Global Environmental Change 18, 598–606. https://doi.org/10.1016/j.gloenvcha.2008.07.013.

Davoudi, S., Shaw, K., Haider, J.L., Quinlan, A.E., Peterson, G.D., Wilkinson, C., Fünfgeld, H., McEvoy, D., Porter, L., Davoudi, S., 2012. Resilience: a bridging concept or a dead end? "Reframing" resilience: challenges for planning theory and practice interacting traps: resilience assessment of a pasture management system in Northern Afghanistan urban resilience: what does it mean in planning practice? Resilience as a useful concept for climate change adaptation? The politics of resilience for planning: a cautionary note. Planning Theory & Practice 13 (2), 299–333. https://doi.org/10.1080/14649357.2012.677124.

Davoudi, S., Brooks, E., Mehmood, A., 2013. Evolutionary resilience and strategies for climate adaptation. Planning Practice & Research 28 (3), 307–322. https://doi.org/10.1080/02697459.2013.787695.

Djalante, R., Holley, C., Thomalla, F., 2011. Adaptive governance and managing resilience to natural hazards. International Journal of Disaster Risk Science 2, 1–14. https://doi.org/10.1007/s13753-011-0015-6.

Dyer, J.G., McGuinness, T.M., 1996. Resilience: analysis of the concept. Archives of Psychiatric Nursing 10 (5), 276–282. https://doi.org/10.1016/S0883-9417(96)80036-7.

EEA, 2016. Urban Adaptation to Climate Change in Europe 2016: Transforming Cities in Changing Climate. EEA Report 12/2016. Available at: https://www.eea.europa.eu/publications/urban-adaptation-2016.

Folke, C., Carpenter, S.R., Walker, B., Scheffer, M., Chapin, T., Rockstrom, J., 2010. Resilience thinking: integrating resilience, adaptability and transformability. Ecology and Society 15 (4), 20. Available at: http://www.ecologyandsociety.org/vol15/iss4/art20/.

Friedmann, J., 1993. Toward a non-euclidean mode of planning. Journal of the American Planning Association 59, 482–485. https://doi.org/10.1080/01944369308975902.

Galderisi, A., 2017. The nexus approach to disaster risk reduction, climate adaptation and ecosystem management: new paths for a sustainable and resilient urban development. In: Colucci, A., Magoni, F., Menoni, S. (Eds.), Peri-Urban Areas and Food-Energy-Water Nexus. Sustainability and Resilience Strategies in the Age of Climate Change, Springer Tracts in Civil Engineering. https://doi.org/10.1007/978-3-319-41022-7_2.

Gordon, J., 1978. *Structures*. Harmondsworth. Penguin Books, UK.

Grøtan, T.O., 2014. Hunting high and low for resilience: sensitization from the contextual shadows of compliance. In: Steenbergen, R.D., van Gelder, P.H., Miraglia, S., Vrouwenvelder, A.C. (Eds.), Safety, Reliability and Risk Analysis: Beyond the Horizon. Taylor & Francis Group, London, pp. 327–335.

Gunderson, L., Holling, C.S. (Eds.), 2001. Panarchy: Understanding Transformations in Human and Natural Systems. Island Press, Washington DC, USA.

Holling, C.S., 1973. Resilience and stability of ecological systems. Annual Review of Ecology and Systematics 4, 1–23. https://doi.org/10.1146/annurev.es.04.110173.000245.

Holling, C.S., 1996. Engineering resilience versus ecological resilience. In: Schulze, P. (Ed.), Engineering with Ecological Constrains. National Academy, Washington DC, USA.

Lonsdale, K., Pringle, P., Turner, B., 2015. Transformative Adaptation: What it Is, Why it Matters & what Is needed. UK Climate Impacts Programme. University of Oxford, Oxford, UK, ISBN 978-1-906360-11-5. Available at: http://www.ukcip.org.uk/wp-content/PDFs/UKCIP-transformational-adaptation-final.pdf.

Manyena, S.B., O'Brien, G., O'Keefe, P., et al., 2011. Disaster resilience: a bounce back or bounce forward ability? Local Environment 16 (5), 417–424. https://doi.org/10.1080/13549839.2011.583049.

Norris, F.H., Stevens, S.P., Pfefferbaum, B., Wyche, K., Pfefferbaum, R., 2008. Community resilience as a metaphor, theory, set of capacities, and strategy for disaster readiness. American Journal of Community Psychology 41, 127–150. https://doi.org/10.1007/s10464-007-9156-6.

Papa, R., Galderisi, A., Vigo Majello, M.C., Saretta, E., 2015. Smart and resilient cities. A systemic approach for developing cross-sectoral strategies in the face of climate change. TeMA. Journal of Land Use, Mobility and Environment 8 (1), 19–49. https://doi.org/10.6092/1970-9870/2883.

Sharifi, A., 2016. A critical review of selected tools for assessing community resilience. Ecological Indicators 69, 629–647. https://doi.org/10.1016/j.ecolind.2016.05.023.

The Rockefeller Foundation/ARUP, 2015. City Resilience Index. Understanding and Measuring City Resilience. Available at: https://assets.rockefellerfoundation.org/app/uploads/20160201132303/CRI-Revised-Booklet1.pdf.

Turner, S., Moloney, S., Glover, A., Fünfgeld, H., 2014. A Review of the Monitoring and Evaluation Literature for Climate Change Adaptation. Centre for Urban Research, RMIT University, Melbourne. Available at: https://gallery.mailchimp.com/b38874b25e686137780eb836e/files/M_E_Lit_Review.pdf.

UN, 2012. The Future We Want. Outcome Document of the United Nations Conference on Sustainable Development. Available at: https://sustainabledevelopment.un.org/content/documents/733FutureWeWant.pdf.

UN, 2015a. Transforming Our World. The 2030 Agenda for Sustainable Development. Available at: https://sustainabledevelopment.un.org/content/documents/21252030%20Agenda%20for%20Sustainable%20Development%20web.pdf.

UN, 2015b. Sendai Framework for Disaster Risk Reduction 2015-2030. Available at: http://www.preventionweb.net/files/43291_sendaiframework fordrren.pdf.

UNISDR, 2009. Terminology on Disaster Risk Reduction. Available at: http://www.unisdr.org/files/7817_UNISDRTerminologyEnglish.pdf.

UNISDR, 2017. How to Make Cities More Resilient – A Handbook for Mayors and Local Government Leaders. Geneva, Switzerland. Available at: http://www.unisdr.org/campaign/resilientcities/assets/documents/guidelines/Handbook%20for%20local%20government%20leaders%20[2017%20Edition].pdf.

Walker, B., Holling, C.S., Carpenter, S.R., Kinzig, A., 2004. Resilience, adaptability and transformability in social-ecological systems. Ecology and Society 9 (2), 5. Available at: http://www.ecologyandsociety.org/vol9/iss2/art5/.

Weichselgartner, J., Kelman, I., 2015. Geographies of resilience: challenges and opportunities of a descriptive concept. Progress in Human Geography 39 (3), 249–267. https://doi.org/10.1177/0309132513518834.

Williams, T., Hardison, P., 2013. Culture law, risk and governance: contexts of traditional knowledge in climate change adaptation. Climatic Change 120 (3), 531–544. https://doi.org/10.1007/s10584-013-0850-0.

CHAPTER 3

The Transition Approach in Urban Innovations: Local Responses to Climate Change

Angela Colucci
Co.O.Pe.Ra.Te. ldt., Pavia, Italy

3.1 SOCIETAL TRANSITION MODELS

In chemical and physical disciplinary fields the term *transition* refers to the "phase of transition" of substances in changing state from solid to liquid to gas. The concept, with the meaning of chaotic and nonlinear process of states changing, has been applied to a wide variety of different types of systems in order to describe the shifting between different states. Models of nonlinear transition—also called punctuated equilibrium—have been applied in ecology, psychology, technology studies, economics, and demography (Gersick, 1991). "In the field of social development, for example, the transition concept was empirically founded and validated on the demographic transition. What is new, however, is the use of the concept of transitions to describe broad social, ecological and economic changes and to explain their mutual connection" (Martens and Rotmans, 2005, p. 1136).

Since the last decades of the 20th century the concept of transition has been associated with complex systems modification. Transitions of societal systems are processes of structural modification of social components of systems implying alteration in culture, in structure, and in behaviors of societal systems. Societal transition as modifications involving the entire societal system and resulting from interacting changes in all societal domains (e.g., economy, ecology, institutions, technology, etc.) is a long-period process. In contrast, modification processes affecting single domain or single aspect of complex systems can be faster.

In the transition research field these processes are studied from different perspectives producing different transition models or modeling the societal transitions from different points of observation. Some models observe the sociotechnical transitions of complex systems describing how innovations can influence the transition of complex systems; an example are models related to "strategic niche management" (Schot and Rip, 1997; Kemp et al., 1998, 2001; Geels, 2002; Berkhout et al., 2004). Other research focuses on the dynamics of adaptation in complex systems to innovation and transitions phenomena, developing models describing complex systems dynamics and focusing on the governance of transition processes (Rotmans et al., 2001; De Haan, 2006; Loorbach, 2007, 2010). The proposed transition models are based on complex and adaptive systems theories that give a theoretical framework to understand the peculiarities of societal transitions dynamics including multidimensional aspects (in time and spatial scale) and interactions among partial societal changes (transitions related to single domains) and transition process regarding the overall societal systems. The sociotechnical transition model (Geels, 2002, 2005) describes how societal transitions happen and how innovation can modify the complex systems focusing on historical perspectives (analyzing how transitions appended in the past to understand transitions dynamics). The transition management model (Rotmans et al., 2001) was developed to understand the transition mechanisms or transition pathways and develop a theoretical framework supporting complex adaptive systems transitions toward sustainability (Fig. 3.1).

3.1.1 Governance of Transition Management

The transition management focus is the governance of transition processes and how, in relation to established (shared) long-term sustainability strategic visions, innovative transition initiatives (microlevel bottom-up initiatives)

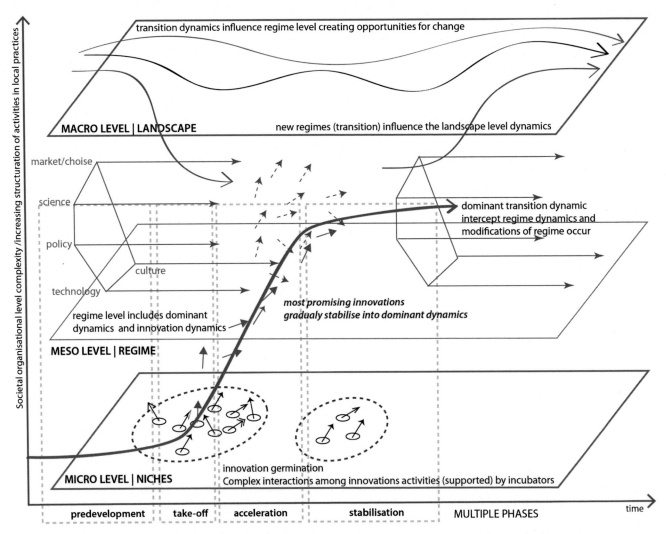

FIGURE 3.1 A scheme summarizing the two proposed models of sociotechnical transitions (Geels, 2002, 2005) and transitions management (Rotmans et al., 2001). *Author's elaboration based on Geels, F.W., 2002. Technological transitions as evolutionary reconfiguration processes: a multilevel perspective and a case study. Research Policy 31 (8), 1257–1274. https://doi.org/10.1016/S0048-7333(02)00062-8; Rotmans, J., Kemp, R., Van Asselt, M., 2001. More evolution than revolution: transition management in public policy. Foresight 3 (1), 15–31. https://doi.org/10.1108/14636680110803003; Loorbach, D., 2010. Transition management for sustainable development: a prescriptive, complexity based governance framework. Governance 23 (1), 161–183. https://doi.org/10.1111/j.1468-0491.2009.01471.x.*

can influence modifications of mesolevel toward more sustainable "states." The governance model proposes a radical shift from traditional policy models based on decision making (goals and priorities) and implementation (actions) that are oriented to the achievement of a set of targeted goals in a defined period of time to adaptive governance models. These emphasize the role of key actors' involvement in defining multiple visions articulated in alternative (future) scenarios in order to reach a common set of defined goals. At the same time, opportunities and obstacles to transformations could emerge from experiments developing technological and social innovations. Experimentation plays a crucial role in identifying possible future development trajectories. The transition (adaptive) model is based on a learning and experimental approach where the innovation/transition processes are constantly monitored in order to adjust and reorient the set of defined goals and strategies in relation to feedback emerging from innovation practices. Strong connections between long- and medium-term visioning (mesolevel regime and macrolevel regime) characterize this transition model, a model that is able to connect long-term vision and short-term action through circular and iterative process (Loorbach, 2010). In the transition (governance) arena of developed models, four different types of governance activities (Loorbach, 2007, 2010) supporting societal transitions are identified: strategic activities focusing on system transition in a long-term period (30 years), tactical activities focusing on subsystem transition in medium-term (5–15 years), operational activities focusing on practices (project and actions) in short-term period, and reflexive activities. Reflexive activities refer to the continuous monitoring as a vital part of the search-and-learning process of transitions (Taanman et al., 2008). Monitoring activities are distinguished between the monitoring of the transition process itself and the transition management process. Monitoring activities are articulated in order to evaluate single projects of actors involved and the level of their general engagement (role in the transition process and the level and characteristics of the alliance/network), to monitor transition experiments (in particular, in relation to social and institutional learning level) and to evaluate the large transition process (rate of progress, barriers, and opportunities).

Rotmans, Kemp, and others (Rotmans et al., 2001; Rotmans and Loorbach, 2009) applied the transition concept to complex systems in relation to sustainable development (governance and polices) activating transitions processes of structural change in specific societal domains such as energy supply, housing, mobility, agriculture, health care, and so on (Rotmans et al., 2001; Geels, 2002). In the framework of the polices launched in the Netherlands since the early 2000s, research was developed in order to understand the dynamics of social and technological innovations supporting the complex systems transition toward more sustainable conditions (Martens and Rotmans, 2005; Rotmans et al., 2001). In particular, this research developed models and tools supporting the activation and the acceleration of transition processes toward specific environmental issues like energy polices, circular waste management, low-carbon strategies, health care and welfare transition, and so on. For instance, the Energy Transition Program (www.senternovem.nl/energytransition) of the Netherlands is based on transition management models aiming to accelerate and orient the transition of the Dutch energy supply system toward more sustainable visions. The "energy transition" (as is usually labeled) was launched in 2001, and it is strictly and explicitly based on transition management principles (principles of collaborative policy making, long-term planning, and innovative environmental policies) as formulated by Rotmans et al. (2001). Other examples (not exhaustive) of programs and projects based on transition management models are:

- on energy topics (topic most explored): Dutch Energy Transition program, Parkstad Limburg energy transition, Energietransitie, Urgenda - Energy Transition;
- on health care topics: Transitie Programma Langdurige Zorg;
- on waste management and circular economies: Plan C (http://www.plan-c.eu/en/about-us)
- on sustainable housing and building: *DuWoBo project*
- and a more general program: Fourth National Environmental Policy plan of the Netherlands (*NMP-4*).

In general, in literature and research on transition, actions and polices are oriented to environmental issues (sustainability), often focusing on urban and metropolitan contexts. Consideration of urban context can be identified in urban condition modifications. The major impacts (in terms of energy consumption and emissions, for example) are generated by urban contexts. At the same time, in urban contexts it is possible to find a richer set of actors including actors directly connected (competencies and role) in managing key infrastructures and providing innovation (for example, in local transport, waste and water systems), and the involvement of local communities can be easier. Cities plays a central role in implementing transformations and activating transition processes toward sustainability and climate action.

3.2 TRANSITION INITIATIVES

In the last few decades, academic and disciplinary debates and research (societal transition models are examples) have increasingly paid attention to transition initiatives as spontaneous intervention and grassroots innovations and to their growth, pervasiveness (both in relation to geographical diffusion and issues of interest and action), and spread. Differentiated initiatives activated and managed by local groups acting on both physical dimensions (for instance, on urban public space) and on immaterial dimensions (for instance, on environmental issues or social services and facilities) of commons became a relevant worldwide phenomenon. Also in planning, the debate about the role of urban design and urban polices became more central (Banerjee, 2001). In planning and urban polices, recent research and projects have proposed renovated approaches to urban transformation, integrating and valorizing energies and innovations emerging from spontaneous initiatives and from stronger participative and inclusive codesign processes. For example, in the Handmade Urbanism experience (Rosa and Weiland, 2013), the urban design process is based on a few physical interventions on urban environment directly acted and managed by citizens engaged since the decision-making process through coproduction and codesign tools (another example could be the Tactical Urbanism experience where urban designers are also directly actors of urban interventions (Lydon and Garcia, 2015)).

The attention to transition initiatives from academic, disciplinary, and institutional debate is also demonstrated by the US pavilion presented in at the 2012 Biennale di Venezia event, where "Spontaneous Interventions: Design Actions for the Common Good" celebrated and presented initiatives activated by citizens that are able to convert the critical urgencies into new opportunities for urban regeneration and public life improvement (http://www.spontaneousinterventions.org/). Practices presented in the US project show how alliances and collaboration between "bottom-up" and "top-down" can enhance urban regeneration processes toward improved urban context and public life. Collected good practices demonstrate how citizen-led initiatives create innovative and creative solutions and manage conscious processes of transformation toward sustainability in the long-term perspective.

In the literature debate transition initiatives are investigated through a differentiated lens in relation to innovative and creative proposals and solutions (grassroots innovations, social innovation, etc.), in relation to community-led governance processes (coproduction and codesign, participative urban interventions, etc.), in relation to social aspects (common and public services, urban and public polices, etc.) and others. In this discussion, the definition developed in the literature debate on societal transitions in terms of societal processes of fundamental change in culture, structure, and practices (Frantzeskaki et al., 2012; Nevens et al., 2013; Frantzeskaki and Kabisch, 2016) is adopted. If in transition literature also examples of historical transitions without a defined set of long-term goals (Geels and Schot, 2007) have been investigated, discourses related to transition explicitly connected to sustainable development (Geels and Schot, 2007; Grin et al., 2010) are more centered and coherent with the aim of the chapter. Within the Accelerating and Rescaling Transitions to Sustainability project (http://acceleratingtransitions.eu/), transition initiatives are defined as locally based activities, which aim to drive transformative change toward environmental sustainability of existing societal systems in multiple dimensions. This means that a transition initiative is driven by multiple actors that live or work in the city region, and its activities are directly aimed at transforming services, approaches, routines, practices, and/or infrastructures existing within the city region boundaries (Frantzeskaki and Kabisch, 2016).

The transition initiatives umbrella includes a large range of locally based processes led by citizens groups (NGOs), as well as by private sectors and experts and acting on different sets of issues from social innovation to sustainable local behavior and social inclusion that could be addressed toward a general sustainability goal. A peculiarity is the differentiated range of issues that usually result from a balanced mix of local urgencies characterizing the local context (in terms of local complex system) and of priorities of actions (defined by local actors). This process, "balancing" local context urgencies and community priorities of action, can give unexpected solutions and activate a large range of possible interventions. The results of these processes can differ significantly, even if they are characterized by similar initial conditions. Climate change and resilience issues are the focus of a large range of transition initiatives and, also if the levers of local initiatives are connected to social urgencies (for instance, social innovation initiatives), environmental and climate change issues are often integrated in transition initiatives.

Some commonalities characterize the practices and initiatives, describing the transition initiatives as local-based, community-led, crosscutting (acting on several issues) and promising in identifying innovative solutions initiatives, able to cope with urban urgencies and based on inclusive governance processes.

3.3 THE TRANSITION (TOWNS) NETWORK

We now have 10 years' experience of supporting groups bringing Transition to life in over 50 countries, in towns, cities, villages, institutions. We have a pretty clear idea now of what works and what doesn't, and we want to share that with you so you can be as effective as possible as quickly as possible. We have created a lot of resources to support groups doing and being, Transition.

[…]Take it, run with it, do amazing things. *Introduction, The Essential Guide to Doing Transition, 2016.*

The Transition Network originated in 2006 (Hopkins, 2008, 2011) in Totnes, England. It is an international grassroots movement that seeks to deal with climate change, peak oil and proposes a "different" economic and social development model (downsides of current economic model). The Transition Network promotes an energy descent model, a decrease in resources consumption, and local resilience based on local community empowerment with a particular emphasis on local community creativity, awareness, and knowledge. Local chain of production and consumption (relocalization/local based) is a central issue in Transition Town initiatives aimed at the reduction of the dependency on global markets, decreasing transport, and improving local (circular) economies based on sustainable production and transformation. The citizen-led interventions of the Transition Network focus on a consolidated range of issues (the most frequent issues are food production and transformation, energy, and local currencies; Hopkins, 2008).

The Transition Network is structured on Transition Initiatives and on transition national (or regional) hubs. The Transition Network (Totnes Transition Initiative became the center of TN) plays different roles for strengthening the movement: coordination of networking among local Transition Initiatives, development of strategies for transition (including communication strategies), organizing learning modules and practices/information exchange among local Transition Initiatives, producing guidelines, delivering training and differentiated tools in supporting transitioners including consultancy services.

In 10 years of "activity," the Transition Network has developed different supporting instruments for Transition Initiatives and three books have been published: The Transition Handbook (Hopkins, 2008), Transition Initiatives Primer (Brangwyn and Hopkins, 2008), and Transition Companion (Hopkins, 2011).

It is possible to point out an evolution in approaching tools supporting the local initiatives from a model based on "steps of transition" to a more flexible model referring to principles or "essentials." The first transition model proposed in 2008 (based on the Totness experience) was based on 12 "steps to transition" (presented in Transition Handbook) (Hopkins, 2008; Brangwyn and Hopkins, 2008). It offered a practical guide in order to organize and launch a transition initiative. More recently in The Essential Guide to Doing Transition (Transition Network, 2016), the 12-steps model was reviewed (as the entire Transition process that became Transition 2.0); the new approach developed is based on "7 essential ingredients" and enriched by practical suggestions for transition initiatives success. The Transition 2.0 is based on the Head, Heart and Hands principle: "doing Transition successfully is about finding a balance between these: The Head: we act on the basis of the best information and evidence available and apply our collective intelligence to find better ways of living. The Heart: we work with compassion, valuing and paying attention to the emotional, psychological, relational and social aspects of the work we do. The Hands: we turn our vision and ideas into a tangible reality, initiating practical projects and starting to build a new, healthy economy in the place we live. The seven essential ingredients are Learning how to work well together; imagining the future you want to co-create, community involved developing relationships (beyond friends and natural allies), Collaborating with others, Developing inspirational projects, Linking up with other Transitioners, Reflect & celebrate (celebrating the difference you're making)" (Transition Network, 2016).

Recent emphasis is given to the narrative and celebration of transition initiatives and to communication strategies of Transition Network that became more oriented in publishing and sharing videos and storytelling also in training modules and in transition principles presentation. Transition Network established also a system of branding and guidelines for Transition Network brand use.

To be recognized as "official" members of Transition Network, a transition initiative has to comply with a set of criteria such as having attended a training session, having drafted and approved a constitution, composed of at least four or five people, and having demonstrated commitment to network with others, including local actors and authorities (Brangwyn and Hopkins, 2008; Smith, 2011). In Transition Network principles a central role is played not only by environmental (and health) aspects but also by social and community aspects in terms of social interaction and informal community networking. In addition to strategies and interventions related to environmental issues (environmental sustainability, peak oil, climate change mitigation, etc.), it is possible to highlight:

- actions aimed at part of the community taking back areas and regions or their living environment (rebuilding links with housing areas, active participation and involvement of the population, stewardship of periurban and rural areas, etc.) and measures related to community living and environments (human scale);
- strategies addressed to social inclusion (shared and community services, social informal networks, etc.);
- measures related to food supply chains (local agriculture, community market gardens and so on).

Community participation is not only concentrated in strategic vision definition as participation is extended to the whole town as "agent and party" that implements the strategies for achieving shared goals and goals (Hopkins, 2008).

A more detailed narrative of Transition Towns Initiatives and Network is presented in Section 3.4; in this paragraph some aspects connected to the metaphor of transition and how it is applied in the Transition Towns network are highlighted:

- resilience and adaptation: Principles of Transition Network explicitly refer to resilience: "We respect resource limits and create resilience: the urgent need to reduce carbon dioxide emissions, greatly reduce our reliance on fossil fuels and make wise use of precious resources is at the forefront of everything we do" (Principles of Transition Network, Transition Network website, 2017);
- in Transition 2.0 the relationships between local and global levels are more explicit: Transition Network and local initiatives have to deal with long-term and large-scale dynamics and issues acting on local urgencies;

- Transition Network initiatives are based on positive and strategic envisioning approaches;
- global "networking" is assumed as collective reflexive opportunity: the Network is described as a real-life, real-time global social experiment (network as space of exchange of experiences and insights) where it is possible to learn both from successes and from failures and share creative innovations and solutions;
- transition initiatives are based on community involvement and engagement of people activating thematic groups.

3.4 TRANSITION INITIATIVES: INNOVATIONS FOR FACING URBAN CLIMATE CHANGE ISSUES

Main shared principles and concepts emerging from societal transition debate and transition initiatives practices can be identified in order to highlight promising ideas and tools for urban innovations coping with climate change issues and resilience.

The sociotechnological transition and transition management models were developed in order to understand with a holistic approach the processes of transition of complex systems and to develop governance models and possible tools able to support the transition toward more sustainable scenarios. In particular, a transition management model was applied in transition processes related to specific domains as energy and low carbon transition of complex systems.

Transition initiatives are characterized by a large range of issues, goals, and tools and present common aspects. They generally act on "commons" (public spaces or public life): in their interventions—finalized mostly to public spaces and public environments improvement through community-led actions—transition initiatives generate positive impacts on environmental phenomena as climate change mitigation and adaptation issues and resilience (for instance, Depave initiatives, urban gardening, and other transition initiatives acting on urban-greening improvement). Environmental issues are explicitly integrated goals in a large rage of transition initiatives.

Single Transition Towns Initiatives joining the Transition Towns Network have to share a Transition Network framework of values and goals in which the environmental issues are explicitly identified and declared (energy-decreasing model, local resilience, adaptation, reduction of natural resources consumption, improvement of local cycles of productions, transformation and market, and so on); single transition initiatives, characterized by specific local issues, in joining the Transition Network assume resilience and climate change mitigation and adaptation as general principles also for local action. In community engagement processes, Transition Town Network highlights the crucial relevance in organizing training and participative events to improve citizen and local community knowledge and awareness on climate change dynamics, peak oil scenarios, and resilience principles.

3.4.1 Complex Systems Approach

Societal Transition models are based on complex systems theory that emerged since the middle of 20th century in different disciplinary fields (from ecology to sociology, economy and political science) as a universal language to describe complex interactions between different components in complex adaptive systems (Gell-Mann, 1994; Loorbach et al., 2017). Transitions deal with systemic innovations, not only entailing new technologies but also changes in markets, user practices, cultural discourses, policies, and governing institutions (Geels and Schot, 2007). Societal transitions emerge from continuous dynamic interactions among adaptive processes and innovation practices activated in the systems and their subsystems (or components). Adaptive complex system is also a common theoretical framework for several transition initiatives.

3.4.2 Process as Focus of Transition Metaphor

Societal transition models and transition initiatives focus on the process of transition toward a defined set of long-term goals; in addition, goals and issues of transition (contents) and the process are often inseparable (Loorbach, 2010). The transition process is the focus of studies on transition management, of literature on urban innovations, innovation grassroots, and, for instance, discourses and polices emphasizing local responses to climate change. In single transition initiatives, the "process" of transition is often not "explicitly" planned but all the "networks" of transition initiatives have developed frameworks to describe and tools to support local and global processes of transition. Transition network frameworks refer often to sustainability and resilience concepts in their long-term visions.

3.4.3 Governance of Transition Processes

Due to the central role of process in transition approaches and initiatives, the governance of these processes is studied and approached in terms of "instruments" supporting the transition of complex systems toward sustainability and resilience.

Flexibility and adaptability are two attributes that characterize (or that have to characterize) the governance of transition processes both in societal transition models and in transition initiatives. Adaptive governance models and related tools have to be able to adapt themselves to innovation dynamics and modifications of micro-, meso-, and macrolevels. Transition Town Network principles for managing transition initiatives are based on adaptive model and feedback cycles between action implementation and reflexive leering phases (learn by doing).

3.4.4 Institutional-Led and Community-Led Processes

In general, transition initiatives, due to their nature in engaging local communities and providing good practices and experimentation in the field, in recent decades became actors in the institutional urban process aiming to produce climate change polices mitigation and adaptation and urban resilience improvement (polices based on inclusive governance process based on large partnerships). Transition initiatives are launched and generated in urban contexts where, as discussed previously, are at the same time present as critical phenomena and resources for social innovation.

In the last two decades, due to the diffusion of transition initiatives in metropolitan and urban contexts a large range of institution-led polices were involved and engaged in transition initiatives in long-term and strategic processes of transformation on climate change issues but also on other environmental polices (energy, mobility, as well as food policies). The transition initiatives became fundamental actors both in decision-making processes and in policy implementation.

Extensive literature on Transition Towns Networks investigates the "political/apolitical" role that Transition Towns plays in the "local arena" (MacKinnon and Derickson, 2013; Feola and Nunes, 2013; Feola and Butt, 2017; Barnes, 2016). If this specific aspect is not the focus of this book, still it is an interesting aspect to be considered in the perspective of governance of transition processes. Thus, it is interesting to underline how the Transition Networks are activating partnership and cooperation with institutions but only if Transition Towns Initiatives maintain their recognizable autonomy in polices governance processes.

3.4.5 Nonlinear Transition Processes and Seeds of Innovation

In general, societal transition models (and transition initiatives) are based on nonlinear process characterized by coevolution and crosscutting scales and components (Rotmans, 2009). Nonlinear and coevolution processes imply that at the same time large-scale and long-term processes of transition and innovation in technological, economic, ecological, sociocultural, and institutional fields and short-term innovation initiatives could be activated, and that these different processes and initiatives influence and reinforce each other. In these terms the transition initiatives models imply reiterative and cyclical interactions between different scale levels (niche, regime, landscape) and different perspectives along time scale. Supporting germination processes of innovation experiments (single-transition initiatives) is crucial as well as the creation of opportunities in terms of providing a fertile environment (or transition arena) for the germination of seeds of innovation. The agglutination of single-transition initiatives in global or regional network plays this role in creating a fertile environment supporting dispread and isolated local initiatives. Networks developing training and tools, sharing experiences can increase the success of initiatives.

3.4.6 Local-Global Dimensions

The role of networking is crucial in transition initiatives and in particular in Transition Towns Networks. Considering other transition initiatives (for example DEPAVE initiatives, Repair Café network and others) the networking supports the upscaling and gives the possibility to replicate the initiatives in other contexts; based on the first successful experiment(s) a process of "standardization" of the main principles of actions (ethical values, set of "global" goals) and the development of a toolbox (how to start and how to do) are usually identified and shared aiming at knowledge transfer and local action replication in other local contexts. Other citizen groups that share principles and long-term visions (that are already interested or have already launched similar initiatives) can join the network and be supported in transition initiative management. The Transition Network is characterized by peculiarities because since the beginning it was more structured and long-term visions and principles were assumed. Transition Towns demonstrate the crucial role played by the Global Transition Network in giving identity (feeling part of a global movement) and support (training, toolboxes, etc.) in the success of single Transition Towns initiatives.

3.4.7 Transition Tools: Coproduction, Adaptive Learning, and Codesign Tools

Single-transition initiatives could be considered as "seeds" of innovation and be characterized by unavoidable fragility; thus, the networking (of transition initiatives) could be a first step of consolidation that might nurse and allow single (fragile) seeds to germinate and grow. At the same time, the networking is not a sufficient condition for the consolidation and success of single initiatives. Transition Towns Networks list, for example, all the local initiatives that "joined" the Network; indeed the Network does not activate a monitoring system to verify the vitality or the progress of single initiatives around the world. In general the "aggregation" process and the "upscaling" process are crucial phases in transition initiatives that imply both stabilization and diffusion of innovations proposed.

3.4.8 The Networking Level Develops Tools to Support the Success of Single-Transition Initiatives

Transition initiatives and the transition management model developed tools for knowledge coproduction, codesign, and learning and training. Coproduction and codesign are fundamental tools in community-based initiatives in order to share knowledge resources (from experts, citizens, etc.), identify questions and problems afflicting local contexts (and also to spread climate change or other environmental and global phenomena awareness), and define priorities of local actions and initiatives.

In Transition Network, training is one of core principles and an unavoidable step to join the global network for a single-candidate initiative. Training and learning are principles and tools developed and implemented in the transition management model and in transition initiatives. Knowledge coproduction and training is fundamental in the valorization and capitalization of knowledge resources characterizing the local community involved (from more "technological" or "expert" knowledge to operative and action-oriented skills), in order to engage all local community components (all are asked to share their competencies, knowledge, and expertise), and in order to amplify and generate transition in the cultural sphere toward more sustainable individual and social behaviors.

3.4.9 Transition Tools: Monitoring

The aspect of monitoring and assessment that theoretically has to be assumed as a common strategic tool due to the adaptive and coevoluntionary approach of all transition initiatives and the transition literature debate is not interpreted with a univocal approach in all transition initiatives and in the literature debate. The Transition Towns framework principles explicitly include goals that could be quantifiable and assessed (energy decreasing, resources consumption decrease, etc.). In training and for some of global goals the Transition Network activates strategies addressed to evaluate the level of goals achievement and the positive impacts deriving from interventions implementation (for instance, the energy decreasing monitoring group in the Totnes Transition Initiative). Transition Network did not develop a general framework that could be applicable to all transition initiatives in order to evaluate the positive impact (on environment, on community, etc.) and the level of efficacy in goals achievement. Other transition initiatives "networks" in the toolboxes that have been developed include some assessment tools for self-evaluation of practices (checklists) and for explicitly collecting and assessing some quantitative data able to describe the positive impact derived from single initiatives (for instance, an annual "report" monitoring the results reached by DEPAVE initiatives is published assessing the surface depaved and related positive impacts on urban ecosystem). In societal transition models and in management transition models the assessment (as highlighted in previous paragraphs) plays a crucial role in monitoring the transition process from the "transition arena" of networking and actors' engagement assessment to the monitoring of long-term visions and scenario achievement.

References

Banerjee, T., 2001. The future of public space: beyond invented streets and reinvented places. Journal of the American Planning Association 67 (1), 9–24.

Barnes, P., 2016. Transition Initiatives and Confrontational Politics: Guidelines, Opportunities, and Practices. Western Political Science Association, 2016 Annual Meeting, 0–29. Available at: http://wpsa.research.pdx.edu/papers/docs/Transition%20initiatives%20and%20confrontational%20politics%20(Barnes).pdf.

References

Berkhout, F., Smith, A., Stirling, A., 2004. Socio-technical Regimes and Transition Contexts. In: Elzen, B., Geels, F.W., Green, K. (Eds.), System Innovation and the Transition to Sustainability. Edward Elgar, Cheltenham.

Brangwyn, B., Hopkins, R., 2008. Transition Initiatives Primer. Transition Network. Available at: http://transitiontowns.org/TransitionNetwork/TransitionNetwork#primer.

De Haan, J., 2006. How emergence arises. Ecological Complexity 3 (4), 293–301. https://doi.org/10.1016/j.ecocom.2007.02.003.

Feola, G., Butt, A., 2017. The diffusion of grassroots innovations for sustainability in Italy and Great Britain: an exploratory spatial data analysis. The Geographical Journal 183 (1), 16–33. https://doi.org/10.1111/geoj.1215.

Feola, G., Nunes, R., 2013. Failure and Success of Transition Initiatives: A Study of the International Replication of the Transition Movement. Research Note 4. Walker Institute for Climate System Research, University of Reading. Available at: www.walker-institute.ac.uk/publications/research_notes/WalkerInResNote4.pdf.

Frantzeskaki, N., Kabisch, N., 2016. Designing a knowledge co-production operating space for urban environmental governance - lessons from Rotterdam, Netherlands and Berlin, Germany. Environmental Science & Policy 62, 90–98. https://doi.org/10.1016/j.envsci.2016.01.010.

Frantzeskaki, N., Loorbach, D., Meadowcroft, J., 2012. Governing societal transitions to sustainability. International Journal of Sustainable Development 15 (1–2), 19–36. https://doi.org/10.1504/IJSD.2012.044032.

Geels, F.W., 2002. Technological transitions as evolutionary reconfiguration processes: a multi-level perspective and a case-study. Research Policy 31 (8), 1257–1274. https://doi.org/10.1016/S0048-7333(02)00062-8.

Geels, F.W., 2005. Technological Transitions and System Innovations: A Co-evolutionary and Socio-technical Analysis. Edward Elgar Publishing.

Geels, F.W., Schot, J., 2007. Typology of sociotechnical transition pathways. Research Policy 36 (3), 399–417. https://doi.org/10.1016/j.respol.2007.01.003.

Gell-Mann, M., 1994. Complex adaptive systems. In: Cowan, G., Pines, D., Meltzer, D. (Eds.), Complexity: Metaphors, Models, and Reality. Santa Fe Institute Studies in the Sciences of Complexity, Proc. vol. XIX. Addison-Wesley. Available at: http://resolver.caltech.edu/CaltechAUTHORS:20150924-144445402.

Gersick, C.J., 1991. Revolutionary change theories: a multilevel exploration of the punctuated equilibrium paradigm. Academy of Management Review 16 (1), 10–36. https://doi.org/10.5465/AMR.1991.4278988.

Grin, J., Rotmans, J., Schot, J., 2010. Transitions to Sustainable Development: New Directions in the Study of Long Term Transformative Change. Routledge, New York.

Hopkins, R., 2008. The Transition Handbook. From Oil Dependency to Local Resilience. Green Books, Totnes.

Hopkins, R., 2011. The Transition Companion: Making Your Community More Resilient in Uncertain Times. Green Publishing, Chelsea.

Kemp, R., Schot, J., Hoogma, R., 1998. Regime shifts to sustainability through processes of niche formation: the approach of strategic niche management. Technology Analysis & Strategic Management 10 (2), 175–198. https://doi.org/10.1080/09537329808524310.

Kemp, R.P.M., Rip, A., Schot, J., 2001. Constructing transition paths through the management of niches. In: Garud, R., Karnoe, P. (Eds.), Path Dependence and Creation. Lawrence Erlbaum, Mahwa (N.J.) and London.

Loorbach, D., 2007. Transition management. New mode of governance for sustainable development. International Books, Utrecht.

Loorbach, D., 2010. Transition management for sustainable development: a prescriptive, complexity based governance framework. Governance 23 (1), 161–183. https://doi.org/10.1111/j.1468-0491.2009.01471.x.

Loorbach, D., Frantzeskaki, N., Avelino, F., 2017. Sustainability transitions research: transforming science and practice for societal change. Annual Review of Environment and Resources. https://doi.org/10.1146/annurev-environ-102014-021340.

Lydon, M., Garcia, A., 2015. A tactical urbanism how-to. In: Tactical Urbanism. Island Press, Washington.

MacKinnon, D., Derickson, K.D., 2013. From resilience to resourcefulness: a critique of resilience policy and activism. Progress in Human Geography 37 (2), 253–270. https://doi.org/10.1177/0309132512454775.

Martens, P., Rotmans, J., 2005. Transitions in a globalising world. Futures 37 (10), 1133–1144. https://doi.org/10.1016/j.futures.2005.02.010.

Nevens, F., Frantzeskaki, N., Gorissen, L., Loorbach, D., 2013. Urban Transition Labs: co-creating transformative action for sustainable cities. Journal of Cleaner Production 50, 111–122. https://doi.org/10.1016/j.jclepro.2012.12.001.

Rosa, M.L., Weiland, U.E. (Eds.), 2013. Handmade Urbanism: From Community Initiatives to Participatory Models. Jovis Verlag.

Rotmans, J., Loorbach, D., 2009. Complexity and transition management. Journal of Industrial Ecology 13 (2), 184–196. https://doi.org/10.1111/j.1530-9290.2009.00116.x.

Rotmans, J., Kemp, R., Van Asselt, M., 2001. More evolution than revolution: transition management in public policy. Foresight 3 (1), 15–31. https://doi.org/10.1108/14636680110803003.

Schot, J., Rip, A., 1997. The past and future of constructive technology assessment. Technological Forecasting and Social Change 54 (2–3), 251–268. https://doi.org/10.1016/S0040-1625(96)00180-1.

Smith, A., 2011. The transition town network: a review of current evolutions and renaissance. Social Movement Studies 10 (01), 99–105. https://doi.org/10.1080/14742837.2011.545229.

Taanman, M., de Groot, A., Kemp, R., Verspagen, B., 2008. Diffusion paths for micro cogeneration using hydrogen in The Netherlands. Journal of Cleaner Production 16 (1), S124–S132. https://doi.org/10.1016/j.jclepro.2007.10.010.

Transition Network, 2016. The Essential Guide to Doing Transition. Your Guide to Starting Transition in Your Street, Community, Town or Organisation. Transition Network, Totnes.

Websites

ARTS (Accelerating and Rescaling Transitions to Sustainability) Project. http://acceleratingtransitions.eu/.
DEPAVE Volunteer-Driven Organization. Web site: http://depave.org/.
Repair Café Foundation. https://repaircafe.org/en.
Spontaneous Interventions: Design Actions for the Common Good. http://www.spontaneousinterventions.org/.
Transition Network Web site https://transitionnetwork.org.

Further Reading

Adger, W.N., 2000. Social and ecological resilience: are they related? Progress in Human Geography 24 (3), 347–364. https://doi.org/10.1191/030913200701540465.

Bosman, R., Loorbach, D., Frantzeskaki, N., Pistorius, T., 2014. Discursive regime dynamics in the Dutch energy transition. Environmental Innovation and Societal Transitions 13, 45–59. https://doi.org/10.1016/j.eist.2014.07.003.

Gillberg, D., Berglund, Y., Brembeck, H., Stenbäck, O., 2012. Urban Cultures as a Field of Knowledge and Learning. Mistra Urban Futures, Gothenburg. Available at: https://www.mistraurbanfutures.org/en/urban-cultures-field-knowledge-and-learning.

Holling, C.S., Gunderson, L.H., 2002. Resilience and Adaptive Cycles. In: Gunderson, L., Holling, C.S. (Eds.), Panarchy: Understanding Transformations in Human and Natural Systems. Island Press, Washington, pp. 25–62.

Kemp, R., Loorbach, D., Rotmans, J., 2007. Transition management as a model for managing processes of co-evolution towards sustainable development. The International Journal of Sustainable Development & World Ecology 14 (1), 78–91. https://doi.org/10.1080/13504500709469709.

Kauffman, S., Macready, W., 1995. Technological evolution and adaptive organizations: ideas from biology may find applications in economics. Complexity 1 (2), 26–43. https://doi.org/10.1002/cplx.6130010208.

Lerner, J., 2014. Urban Acupuncture. Island Press, Washington.

Lansing, J.S., 2003. Complex adaptive systems. Annual Review of Anthropology 32 (1), 183–204. https://doi.org/10.1146/annurev.anthro.32.061002.093440.

Loorbach, D., Van der Brugge, R., Taanman, M., 2008. Governance in the energy transition: practice of transition management in The Netherlands. International Journal of Environmental Technology and Management 9 (2–3), 294–315. https://doi.org/10.1504/IJETM.2008.019039.

Loorbach, D., Rotmans, J., 2006. Managing transitions for sustainable development. Understanding industrial transformation. In: Olsthoorn, X., Wieczorek, A. (Eds.), Understanding Industrial Transformation, Environment & Policy, vol. 44. Springer, Dordrecht, pp. 187–206. https://doi.org/10.1007/1-4020-4418-6_10.

Veldkamp, A., et al., 2009. Triggering transitions towards sustainable development of the Dutch agricultural sector: Trans Forum's approach. In: Lichtfouse, E., Navarrete, M., Debaeke, P., Véronique, S., Alberola, C. (Eds.), Sustainable Agriculture. Springer, Dordrecht. https://doi.org/10.1007/978-90-481-2666-8_41.

Voss, J.P., Bauknecht, D., Kemp, R. (Eds.), 2006. Reflexive Governance for Sustainable Development. Edward Elgar Publishing, Cheltenham.

CHAPTER 4

Smart, Resilient, and Transition Cities: Commonalities, Peculiarities and Hints for Future Approaches

Adriana Galderisi[1], Angela Colucci[2]
[1]University of Campania Luigi Vanvitelli, Aversa, Italy; [2]Co.O.Pe.Ra.Te. ldt, Pavia, Italy

4.1 THREE URBAN METAPHORS

In the previous chapters, three out of the currently most widespread urban metaphors— smart, resilient, and transition cities—have been presented, emphasizing, according to scientific literature, consolidated and common aspects as well as controversial issues. The main aspects presented in relation to each urban metaphor can be summarized as follows.

The smart city metaphor is widely used to underline the key role of Information and Communications Technologies (ICTs) in urban development; the debate has long been dominated by a techno-centered approach, focused on the potentialities of the "hardware," leaving aside the significant role of human and social capital in the process of building up a smart city. Although this approach still largely prevails, both in theory and in practice, over the last few years a human/social-centered approach has gained relevance. The latter looks at ICTs as crucial "enabling tools," whereas social capital is considered as the main lever for smart development (some scholars emphasize the importance of smart inhabitants while others emphasize the highly skilled labor force in "creative" sectors). In relation to climate strategies, so far initiatives framed within this metaphor have been mostly focused on mitigation issues.

The resilient city metaphor is widely used to describe a city capable to withstand, absorb, and recover from sudden adverse events as well from chronic stresses, caused by heterogeneous pressure factors (resource scarcity, economic crises, natural hazards, climate change, etc.). The resilience concept has been used with different meanings since ancient times and deeply investigated and conceptualized beginning in the 19th century, according to different disciplinary perspectives, from mechanics to psychology, from ecology to economy. The concept has recently entered the urban studies domain leading to an interpretation of resilience as a dynamic interplay between persistence, adaptability, and transformability across multiple scales and time frames, largely based on learning capacity, that reflect "the intentionality of human action and intervention" (Davoudi et al., 2013), typical of social systems. Hence, the key role of social capital and institutions in building resilient cities has been clearly emphasized. In relation to climate issues, so far initiatives framed within this metaphor have been mostly focused on adaptation issues.

Societal transitions are considered as processes of fundamental modification in culture, structures, and practices. Transition models describe systemic innovations in all the components of complex systems: in economic spheres or markets, in social and cultural spheres or behaviors, and in infrastructures, policies, and governing institutions. In general terms, sustainable transition initiatives can be defined as locally based activities addressed to drive transformative change toward environmental sustainability of existing societal systems. The Transition Town Network explicitly adopts transition as a metaphor and most of the numerous local initiatives directly activated by citizens and addressed to deal with climate issues and enhance local resilience refer to this metaphor.

In the face of the increasing importance gained by these metaphors, the first key question is why they rapidly spread over the last decade in different geographical contexts? We have identified at least three key factors:

Changing Urban Conditions—During the 1990s, urban and metropolitan systems have been affected by more and more pressing environmental threats (climate-related hazards, water scarcity, environmental decay, etc.). Nowadays, cities are largely considered as the main "drivers" of these global environmental pressures, and, meanwhile, "contexts" where complex and innovative responses to them can be activated. According to Secchi, metaphors "appear in urban discourse, when urban condition is transformed and shifting; that is, when the urban condition changes. This calls for new ways of description and thus for metaphors" (Secchi, 2014). Hence, in the face of the growing environmental pressures arising from the complex interactions among human activities and natural dynamics, numerous urban metaphors (smart, resilient, green, etc.) have been introduced to outline long-term visions toward similar goals: creating better environmental, social, and economic conditions and empowering cities in the face of complexities and uncertainties arising from the interconnected challenges at stake.

Emergence of Complex Challenges Requiring Integrated Approaches—Urban metaphors allow to outline strategic visions based on crosscutting and multisectoral approaches and involving different disciplinary fields as well different social and organizational components. Over the past two decades, interdisciplinary studies, focusing on cities and on the need for systemic changes toward sustainability and involving a wide range of research areas dealing with urban development, systemic configurations, and system innovation dynamics, have developed complementary theories, concepts and metaphors, showing peculiar features and overlapping areas. Current interconnected environmental, social, and economic challenges, and above all, climate change, require strategies capable of linking different spheres and components of urban complex systems, to outline long-term visions and promote significant cultural and behavioral changes. In the face of current complex challenges, the most widespread urban metaphors support innovative processes and paths toward sustainability, with a specific emphasis on the innovation of existing governance processes. The proliferation of locally based practices, activated and managed by communities and promoting actions to counterbalance climate change, clearly reveals, for example, the demand for new governance models.

Visioning and positive envisioning—The growing use of metaphors is also related to the need for "positive" visions, capable of supporting innovation and defining new urban development models, focusing on strategic and synergic actions rather than on regulative (planning) tools.

Another important issue to be discussed here refers to the commonalities and peculiarities of these metaphors. Comparison among them seems to reveal, indeed, that commonalities and convergences are more numerous than conflicting aspects. In detail, approaches, models, and, above all, solutions framed by these metaphors show a shared and common system of long-term sustainability goals, although practices and initiatives refer to theoretical and operational frameworks not fully aligned.

Hence, in the following in-depth comparison among the three urban metaphors will be provided, with the aim to identify crosscutting aspects and novel inputs toward sustainable urban development. Effective integration between approaches, strategies, and actions arising both from theoretical debate and from practices and initiatives developed under these labels could allow for significantly improving urban policies in the face of climate change and sustainability issues.

4.2 COMMONALITIES AND PECULIARITIES OF THE THREE METAPHORS IN SCIENTIFIC LITERATURE

The three considered urban metaphors arise from theoretical evolution paths developed through different disciplinary fields. Hence, numerous terms related to each metaphor have been differently used and applied in each disciplinary domain (e.g., the term *resilience*). It is only since the 1990s that these metaphors have been explicitly applied to complex urban systems.

In the following discussion we will present commonalities and peculiarities arising from the scientific debate on the three urban metaphors—smart city, resilient city, and transition towns, individually and thoroughly discussed in the previous chapters of this book—with respect to three key themes: goals, approaches and concepts, and governance models.

Regarding the goals, as just mentioned, the considered metaphors explicitly share long-term sustainability goals. Urban resilience is considered as a key goal for sustainable urban development in all recent international documents (UN, 2012, 2017), and the model of Quintuple Helix (Carayannis et al., 2012) provides a conceptual framework for driving smart city initiatives toward sustainability goals.

Furthermore, besides the common goal of increasing human well-being in urban areas, the three metaphors share specific objectives related, for example, to the reduction of energy consumption, to the regeneration of natural resources, and to climate change mitigation and adaptation.

However, a crucial difference between these metaphors arises when we move from the general goals to the ways to achieve them; strategies and solutions are based, indeed, on different models and conceptual frameworks. Whereas the debate on smart cities focuses on the optimization of features and performances of existing urban systems, the debate on resilient cities and transition towns largely emphasizes the need for enhancing capabilities of complex urban systems to deal with internal and external pressures, by adapting or even transforming themselves in the face of changing conditions.

Finally, it is worth noting that the three metaphors are also clearly interlinked. For example, the resilience concept is explicitly quoted as a "conceptual framework" by the Transition Town Movement; the latter aims indeed to inspire, encourage, connect, support, and train communities in creating initiatives capable of enhancing resilience and reducing CO^2 emissions (Hopkins, 2008; Transition Network, 2016). Some authors clearly emphasize that transition initiatives offer a bottom-up approach for local development with emphasis on a shift from sustainability to resilience (Mehmood and Franklin, 2013).

Focusing on the approaches and concepts that the three metaphors are based on, it is worth noting that all of them refer, implicitly or explicitly, to cities as complex adaptive systems, capable of processing information, learning from experience, and adapting in the face of changing conditions.

Based on the complexity theory, all these metaphors focus on dynamics rather than on states. Resilience is clearly described as a dynamic process, "not as a state but as a becoming" (Davoudi et al., 2012). Transition models (namely sociotechnological transition and transition management models) (Geels, 2002; Rotmans and Loorbach, 2009) have been specifically developed to grasp the dynamics of societal changes. ICTs—at the core of the smart city metaphor—are widely interpreted as key tools to monitor temporal and spatial dynamics of urban systems.

Furthermore, these metaphors share the widespread awareness that current society has to deal with global risks, characterized by increasing levels of uncertainty (Beck, 2008); hence, all of them emphasize the need for continual learning processes (Cutter et al., 2008), enabling cities to better cope with uncertainty. The three metaphors assign a key role to a practice-based knowledge too: they provide tools based on a "learning-by-doing" approach and promote adaptive processes based on the continuous monitoring of practices, in order to constantly review and improve undertaken strategies and actions (for instance, training and learning from experiences is a core mission of the Transition Towns Network).

Knowledge and learning represent core concepts in all the three considered urban metaphors. Namely, both smart city and transition town metaphors explicitly focus on knowledge coproduction, emphasizing the need for developing comprehensive knowledge (including academic but also communities' knowledge), capable of overcoming traditional disciplinary barriers as well as current sectoral and fragmented institutional competencies, driving toward innovative transdisciplinary research paths and crosscutting strategies, crucial for dealing with complex phenomena.

Finally, the three metaphors share a focus on governance issues; the scientific debate on these metaphors clearly emphasizes, indeed, the need for adaptive governance models, capable of engaging a wide range of stakeholders and promoting collaboration and cooperation among different municipal departments and different governmental levels. The smart city metaphor introduces smart governance as an "innovative way of decision-making" (Meijer and Bolivar, 2016), with ICTs as facilitating tools for creating new collaborative environments (Komninos, 2011), linking multiple levels, bodies, and operational tools currently in charge of different aspects of urban development and supporting effective citizen engagement in decision-making processes.

Even though adaptive governance is a shared principle, it is worth noting that whereas scientific literature on smart and resilient cities mostly refers to institutional-driven governance models, aimed at engaging a large range of stakeholders, comprising citizens, transition initiatives are directly acted by citizens and based on "bottom-up" governance models, often involving private and institutional stakeholders. In particular, the transition management model (Loorbach, 2010) highlights a governance framework integrating institutional governance (long-term and strategic visioning) with bottom-up intervention practices.

4.3 COMMONALITIES AND PECULIARITIES OF THE THREE METAPHORS IN CURRENT PRACTICES

Practices implementation is one of the main objectives of the three considered urban metaphors. The latter could be considered as "brands" to promote local initiatives, which are generally strongly place-based and, at the same time, linked to international networks that act as platforms for sharing and exchanging knowledge, principles, guidelines, and operational tools.

Nowadays, practices framed within the three urban metaphors are significantly heterogeneous, although sharing numerous commonalities.

As briefly mentioned before, an important difference among current practices is related to the governance processes and to the key actors leading them. Whereas the resilient city initiatives are generally guided by institutional actors, indeed, smart city initiatives are often led by the private sector (e.g., large enterprises or multinational corporations), while transition initiatives (namely transition towns) are directly activated and managed by citizens and community associations. Even though single transition initiatives have carried out projects or programs in partnership with local institutions, the Transition Towns Network clearly underlines that a transition initiative has to be recognized as an independent actor (Hopkins, 2010).

As a common point, it is worth noting that all these metaphors assign a central role to practice-based knowledge, looking at urban innovation as a learning process based on practice-based innovation, such as the development and use of new methods, procedures, products, or services (Ellström, 2010). Moreover, local practices and initiatives related to the three metaphors are generally activated and implemented according to specific methodological frameworks, comprising both general principles and operational tools.

An important difference among the considered metaphors is the role assigned to the monitoring and assessment phases. While practices related to smart and resilient city metaphors are generally based on sets of indicators, used for benchmarking cities at different scales (national, continental, and global levels) and for guiding both the decision-making process and the monitoring of expected goals and outcomes, only a few transition town initiatives have developed monitoring tools in relation to specific action domains. In detail, although in the transition management model "continuous monitoring is a vital part of the search and learning process of transitions" (Loorbach, 2010), monitoring is generally focused on the societal transition process rather than on the specific goals to be achieved. The Transition Towns platform collects and publishes general data and information about transition initiatives all around the world, focusing on the "steps" of the transition process and on the main issues of the different initiatives, without providing performance indicators to evaluate impacts and benefits arising from their implementation.

The availability of monitoring tools is considered as crucial in the scientific debate in order to deal with climate and resilience issues through adaptive and flexible approaches; they not only allow for understanding the benefits emerging from each initiative but also permit dynamic improvement and realignment of strategies and actions. Nevertheless, with respect to the transition towns initiatives, only a few surveys on "level and quality" of transition initiatives' implementation have been developed (Feola and Nunes, 2013) and only few initiatives provide indicators for measuring the achieved benefits (e.g., energy saved, economic savings, etc.). Unfortunately, so far available data refer only to individual initiatives and do not allow for comparison between them.

Another peculiarity of the transition town initiatives is the local/global scale relationship; local and global dimensions are used to describe environmental (and economic) dynamics as well as to create narrative of future alternative scenarios (emphasizing the role of local scale). Single initiatives are deeply locally based and characterized by actions that usually respond to local urgencies, but they are globally connected through the Transition Network. Hence, the latter is a global network of punctual initiatives sharing few principles and values: the Network does not provide a common methodological and operational framework but disseminates a set of principles and approaches inspiring local goals, strategies, and actions.

4.4 CONCLUDING REMARKS

According to our previous discussion, the three considered urban metaphors have numerous commonalities and some peculiarities, mostly related to the adopted methodological and operational tools and to the selected levers for the activation of practices. However, the resilient city metaphor seems to embed both smart and transition practices; both of them, although differently, refer to resilience as an umbrella concept or, even, as an ultimate goal.

Therefore, in Fig. 4.1 commonalities and overlapping areas among the three urban metaphors have been highlighted according to the conceptual framework of "evolutionary resilience" (Davoudi et al., 2013), interpreted as a dynamic interplay between persistence, adaptability, and transformability across multiple scales and time frames and based on the learning capacity, typical of social systems (see Chapter 2). In detail, while the smart city metaphor so far has been mainly focused on the persistence property, guiding strategies and actions addressed

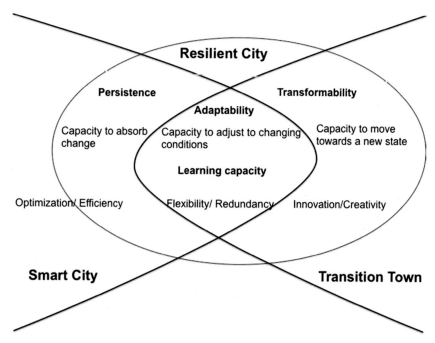

FIGURE 4.1 The scheme shows shared concepts and overlapping areas among the three urban metaphors. *Authors' elaboration.*

to improve urban systems' efficiency in the face of adverse events, transformability is considered as a crucial property both in resilient and in transition metaphors that look at novelty and innovation as crucial features to drive urban systems toward new conditions/states, when current ones are inadequate or unsustainable. However, it is worth stressing that the emerging approaches to the smart city metaphor clearly envisage a radical change of current urban development patterns that, starting from human and social capital, also affect physical, functional, and organizational aspects of cities' development (Concilio and Rizzo, 2016). Finally, adaptability and learning capacity represent the overlapping areas among the three considered metaphors.

Focusing on climate issues, although all three metaphors consider the carrying out of adequate strategies to counterbalance climate change as a key goal, each of them has devoted attention to different aspects and has identified different levers for activating these strategies.

Smart city initiatives, according to the prevailing techno-centered and sectorial approach, have been mainly addressed to mitigation issues; since 2011, indeed, the Smart Cities and Communities Initiative has devoted increased attention to the widespread use of ICTs as key tools for achieving the targets established by the EU Strategy 2020. Hence, these initiatives have emphasized the potential for ICTs to improve urban energy performance, by ensuring a more efficient use of energy resources and the reduction of current levels of energy consumption (Kramers et al., 2014).

Resilient city initiatives, for long widely interpreted according to a bounce-back perspective, have been mainly used to support climate adaptation strategies, aimed at increasing cities' capacities to withstand, absorb, and recover from climate impacts. Only recently, resilient city initiatives have been more and more frequently employed to better link mitigation and adaptation strategies, by embedding them into the broader goal of sustainable development.

Robert Hopkins has started the transition initiative process in Totnes, England, with training and debate events focused on peak oil and climate change issues and addressed to solicit local community awareness (Hopkins, 2008, 2010, 2012). The transition towns movement has put large emphasis on low carbon development, focusing on renewable energy sources, energy consumption reduction, local cycles of production, and circular economy. Hence, this movement has been so far more oriented to climate change mitigation, although community building, positive envisioning, and participative approaches may also drive toward social resilience building in an adaptation perspective.

Hence, despite the significant commonalities linking the three urban metaphors, so far they have developed separately, moving as independent labels in addressing climate issues. In contrast, a strengthening of the potential synergies among them could lead toward more integrated approaches to climate issues as well to an improvement of current conceptual frameworks and operational tools.

In order to deal with climate issues in a sustainable perspective, indeed, urban complex systems have to enhance all the properties related to the concepts of persistence (ability to absorb change), adaptability (ability to adjust in changing condition), and transformability (ability to move toward new conditions). Dealing with climate challenges require to better focus on the different temporal perspectives: from current and short-term urgencies to long-term changes. Persistence, adaptability, and transformability are equally relevant to counterbalance climate change, since each of them gains relevance in different temporal spans: in the short term, persistence-based strategies may allow to improve cities' capacities to withstand current or expected climate-related impacts; in the medium term, adaptability may allow to adjust cities' responses in the face of changing climate conditions and uncertain impacts; in the long term, transformability may drive urban transition toward novel urban development patterns, capable of reducing, for example, the energy footprint of cities in order to prevent future climate-related impacts (Papa et al., 2015). Moreover, all three urban metaphors stress the key role of learning capacity in enhancing cities' capacities to adapt in the face of constantly changing conditions as well as the need for adaptive governance processes, engaging a wide range of actors along the different phases of the process—from the problems setting, to the decision making and the implementation phase.

In the following chapters of this book, the comparison among current practices and initiatives addressed to counterbalance climate change and often framed by one of these urban metaphors will allow us to better understand commonalities and peculiarities of the three considered urban metaphors, and to confirm or refuse, based on current experiences, the ones arising from scientific literature that have been discussed here in-depth.

References

Beck, U., 2008. Conditio humana. Il rischio nell'età globale. Laterza, Roma-Bari.

Carayannis, E.G., Barth, T.D., Campbell, D.F.J., 2012. The Quintuple Helix innovation model: global warming as a challenge and driver for innovation. Journal of Innovation and Entrepreneurship 1 (2). https://doi.org/10.1186/2192-5372-1-2.

Concilio, G., Rizzo, F., 2016. Preface: experimenting between design and planning. In: Concilio, G., Rizzo, F. (Eds.), Human Smart Cities, Rethinking the Interplay Between Design and Planning. Springer International Publishing, Switzerland, pp. V–X.

Cutter, S.L., Barnes, L., Berry, M., Burton, C., Evans, E., Tate, E., Webb, J., 2008. A place-based model for understanding community resilience to natural disasters. Global Environmental Change 18, 598–606. https://doi.org/10.1016/j.gloenvcha.2008.07.013.

Davoudi, S., Shaw, K., Haider, J.L., Quinlan, A.E., Peterson, G.D., Wilkinson, C., Fünfgeld, H., McEvoy, D., Porter, L., Davoudi, S., 2012. Resilience: a bridging concept or a dead End? "Reframing" resilience: challenges for planning theory and practice interacting traps: resilience assessment of a pasture management system in Northern Afghanistan urban resilience: what does it mean in planning practice? Resilience as a useful concept for climate change adaptation? The politics of resilience for planning: a cautionary note. Planning Theory & Practice 13 (2), 299–333. https://doi.org/10.1080/14649357.2012.677124.

Davoudi, S., Brooks, E., Mehmood, A., 2013. Evolutionary resilience and strategies for climate adaptation. Planning Practice & Research 28 (3), 307–322. https://doi.org/10.1080/02697459.2013.787695.

Ellström, P.E., 2010. Practice-based innovation: a learning perspective. Journal of Workplace Learning 22 (1/2), 27–40.

Feola, G., Nunes, R., 2013. Failure and Success of Transition Initiatives: A Study of the International Replication of the Transition Movement. Research Note 4. Walker Institute for Climate System Research, University of Reading. Available at: http://www.transitionresearchnetwork.org/uploads/1/2/7/3/12737251/walkerinresnote4.pdf.

Geels, F.W., 2002. Technological transitions as evolutionary reconfiguration processes: a multi-level perspective and a case-study. Research Policy 31 (8), 1257–1274.

Hopkins, R., 2008. The Transition Handbook. From Oil Dependency to Local Resilience. Green Books, Totnes (Devon).

Hopkins, R.J., 2010. Localisation and Resilience at the Local Level: The Case of Transition Town Totnes. PhD Thesis Dissertation. University of Plymounth. Available at: https://pearl.plymouth.ac.uk/bitstream/handle/10026.1/299/Hopkins%20R%20J_2010.pdf?sequence=4.

Hopkins, R., 2012. Peak oil and transition Towns. Architectural Design 82 (4), 72–77. https://doi.org/10.1002/ad.1432.

Komninos, N., 2011. Intelligent cities: variable geometries of spatial intelligence. Intelligent Buildings International 3 (3), 172–188. https://doi.org/10.1080/17508975.2011.579339.

Kramers, A., Höjer, M., Lövehagen, N., Wangel, J., 2014. Smart Sustainable Cities. Exploring ICT solutions for reduced energy use in cities. Environmental Modeling & Software 56, 52–62. https://doi.org/10.1016/j.envsoft.2013.12.019.

Loorbach, D., 2010. Transition management for sustainable development: a prescriptive, complexity-based governance framework. Governance 23 (1), 161–183.

Mehmood, A., Franklin, A., 2013. Sustainable urban land use and the question of eco-places. In: Zaidi, S.N. (Ed.), Proceedings of the Eighth Seminar on Urban and Regional Planning, pp. 172–194.

Meijer, A., Bolivar, M.P.R., 2016. Governing the smart city: a review of the literature on smart urban governance. International Review of Administrative Sciences 82 (2), 392–408. https://doi.org/10.1177/0020852314564308.

Papa, R., Galderisi, A., Vigo Majello, M.C., Saretta, E., 2015. Smart and resilient cities. A systemic approach for developing cross-sectoral strategies in the face of climate change. Tema. Journal of Land Use, Mobility and Environment 8 (1), 19–49. https://doi.org/10.6092/1970-9870/2883.

Rotmans, J., Loorbach, D., 2009. Complexity and transition management. Journal of Industrial Ecology 13 (2), 184–196. https://doi.org/10.1111/j.1530-9290.2009.00116.x.

Secchi, B., 2014. A new urban question: when, why and how some fundamental metaphors were used. In: Gerber, A., Patterson, B. (Eds.), Metaphors in Architecture and Urbanism. An Introduction. Transcript Verlag, ISBN 978-3-8376-2372-7.

Transition Network, 2016. The Essential Guide to Doing Transition. Your Guide to Starting Transition in Your Street, Community, Town or Organisation. Transition Network, Totnes. Transition Network Web site https://transitionnetwork.org.

UN, 2012. The Future We Want. Outcome Document of the United Nations Conference on Sustainable Development. Available at: https://sustainabledevelopment.un.org/content/documents/733FutureWeWant.pdf.

UN, 2017. New Urban Agenda — Habitat III. Available at: http://habitat3.org/wp-content/uploads/NUA-English.pdf.

SECTION II

LARGE-SCALE STRATEGIES TO COUNTERBALANCE CLIMATE CHANGE

CHAPTER 5

European Strategies and Initiatives to Tackle Climate Change: Toward an Integrated Approach

Adriana Galderisi[1], Erica Treccozzi[2]

[1]University of Campania Luigi Vanvitelli, Aversa, Italy; [2]Building Engineer, Naples, Italy

5.1 EUROPE AND CLIMATE CHANGE ISSUES

The main challenges that Europe is bound to encounter by not taking action against climate change are clearly described in the Intergovernmental Panel on Climate Change (IPCC) assessment reports. The latest evaluations, reported in the Fifth Assessment Report, identify extreme temperatures, extreme precipitation, and sea level rise as the three main climate impacts that Europe is exposed to (IPCC, 2014). These impacts entail the increase of annual heat waves and of coastal, fluvial, and pluvial floods affecting economic activities as well as the population, tourism, agriculture, and forestry (IPCC, 2014).

The alteration of climate patterns may affect both natural ecosystems—by reducing their capacity to deliver provisioning and regulating services with significant consequences on economy and general well-being of European countries—and cities, generally considered the main target of climate impacts. Cities indeed exacerbate the severity of climate-related events due, for example, to the high levels of soil sealing that impair rainwater drainage as well as to the prevalence of built areas that strengthen the urban heat island effect. Meanwhile, they represent the primary targets of climate impacts, due to the concentration of people and assets. Moreover, cities are largely dependent on their surrounding areas (e.g., for food and water supply), also showing an increasing vulnerability to the impacts of climate change on natural and rural ecosystems (Galderisi, 2014).

Due to the relevance of the expected physical and economic impacts of climate change (Ciscar and Iglesias, 2010), Europe has developed a growing commitment toward the development of effective policies, aimed both to prevent future climate-related hazards and to increase ecosystems' resilience to their unavoidable and already-occurring impacts. In detail, Europe has played a pivotal role, acting as one of the world leaders in global mitigation policies, by promoting more and more ambitious energy and climate targets and sustaining members states in achieving them. Also in terms of adaptation, despite a delayed start, in the last decade European countries have made great efforts to enhance and coordinate local initiatives.

Nevertheless, Europe's role is far from being easy, since in order to be successful, mitigation measures should homogeneously progress in each member state and all over the world; whereas adaptation strategies should be adequately tailored to the heterogeneous climate impacts that each geographical area has to deal with. In fact, both the type and severity of climate-related hazards strongly differ from region to region, according to the different climate features of European territories, and the factors affecting local exposure and vulnerability (e.g., urbanization patterns, population density, etc.).

Thus, despite the significant European commitment in counterbalancing climate change, the success of the European strategies is deeply affected on the global side by the international agreements on climate issues that have often been very slow and hampered by the variable local political wills; on the internal side they are affected by the effective involvement of and the actual cooperation among member states and cities that play a crucial role in addressing both mitigation and adaptation issues.

5.2 EUROPEAN STRATEGIES TO COUNTERBALANCE CLIMATE CHANGE

European response to climate change can be summarized in two main types of policies aiming, on the one hand, to prevent climate-related events from happening, and on the other hand, to increase ecosystems' ability to withstand the impacts of climate change. The key words describing these strategies are *mitigation* and *adaptation* that, respectively, stand for a precautionary approach, aimed to reduce the causes inducing changes in climatic conditions, and a pragmatic approach, aimed to contain the unavoidable and already perceivable impacts of climate change.

Even though both of these strategies would be required to counterbalance climate change, in 2000 Europe started its long path toward an effective climate strategy by exclusively focusing on mitigation issues. In fact, Europe's first decisions regarding climate change were based on the international guidelines (see Box 5.1), whose main goal was to contain the core cause of climate change—greenhouse gas (GHG) emissions. After this first phase, intended to reduce the alteration of the climatic natural patterns, Europe started transitioning toward a more pragmatic approach, aimed to cope with the impacts of climate change, while trying to prevent them from happening. This second phase, focused on adaptation strategies, started in 2007 and involved all member states and European cities in the development, respectively, of national adaptation strategies as well as of local adaptation plans.

These two phases, which carried on separately for some years, started to move toward integration in 2015, with the launch of a new program aimed at developing integrated strategies in the face of climate change, by focusing on the potential synergies between mitigation and adaptation measures. These strategies are still at an early stage, although they have triggered the third European phase toward a climate-proof strategy, paving the way to an integrated approach that might offer significant improvements both in the short and in the long term.

In the following paragraphs we will focus individually on the three phases of the European path to address climate change in order to understand how each phase evolved over time and how knowledge and experience guided Europe to shift from a sectoral toward an integrated approach to climate issues.

BOX 5.1 MAIN INTERNATIONAL STEPS TO COUNTERBALANCE CLIMATE CHANGE

1972: The environmental challenges were addressed for the first time during the "United Nations Conference on the Human Environment" of Stockholm. The biggest accomplishment of this event was the institution of the United Nations Environment Programme (UNEP), an environmental authority that helps European nations to pursue shared goals and encourages the development of coordinated programs (UN, 1972).

1988: In order to guide the climate-related decisions, the UNEP, together with the World Meteorological Organization, founded the Intergovernmental Panel on Climate Change (IPCC), which was responsible for providing scientific information on impacts and risks connected to climate change.

1994: The United Nations Framework Convention on Climate Change (UNFCCC) enters into force. This convention aims to reduce the concentration of GHGs, by controlling the level of emissions produced by human activities (UN, 1992). The strategies to be implemented, in order to achieve this goal, were discussed by the involved parties during the annual UNFCCC Conference of Parties (COP).

1995: The first COP, COP 1, took place in Berlin, and has since then met every year. During these sessions representatives of the countries involved discussed mitigation strategies based on the scientific support provided by the IPCC.

1997: During COP 3, quantified and binding goals were defined and reported in the Kyoto Protocol, wherein industrialized countries agreed to reduce "their overall emissions of such gases by at least 5% below 1990 levels in the commitment period 2008 to 2012" (UN, 1998).

2015: The Paris COP 21 sanctioned the end of a long decision-making process that lasted several years. This conference ended with the adoption of the Paris Agreement, that binds the 197 involved parties to "hold the increase in the global average temperature to well below 2°C above pre-industrial levels and pursuing efforts to limit the temperature increase to 1.5°C above pre-industrial levels" (Art. 2 UN, 2015). Unlike the Kyoto Protocol, involved parties are encouraged not only to reduce GHG emissions, but also to increase the environment's ability to absorb CO_2 by enhancing and safeguarding natural sinks (Art. 5 UN, 2015).

5.2.1 European Mitigation Goals and Strategies

European mitigation strategy was triggered by the international guidelines defined by the United Nations Framework Convention on Climate Change (UNFCCC) and by the decisions taken during the annual Conference of the Parties (COP). Indeed, the first COP focused on the causes of climate change as well as on the actions to engage in order to control those factors that, if left undisciplined, could lead to a temperature increase of 1.4–5.8°C in respect to the 1990 temperatures by the year 2100 (CEC, 2005). Therefore, the reduction of GHGs and the control of the activities responsible for these emissions were recognized as priority goals.

The first binding decisions were made with the Kyoto Protocol, adopted during the COP3 in 1997, which blamed developed countries and their 150 years of uncontrolled industrial activities, for the excessive GHG emissions. These countries were required to "reduce their emissions of such gases by at least 5 per cent below 1990 levels in the commitment period 2008 to 2012" (UN, 1998).

The European Union (EU) was requested to reduce its emissions by 8% between 2008 and 2014 (UN, 1998). The Kyoto Protocol, signed in New York on April 1998, was approved in Europe with the 2002/358/CE Council Decision on April 2002, which committed to define in December 2006 the tons of carbon dioxide each member state was responsible for. These quantities were later reported in the 2006/944/CE Commission Decision of December 2006.

Even though the official approval of the protocol was delayed, Europe's response to the Kyoto commitments started to develop in 2000 with the first European Climate Change Programme (ECCP).

The ECCP provided coordinated and compatible strategies for reducing emissions in different sectors. The decision-making process that led to these strategies was guided and coordinated by a steering committee that gathered the main issues in different "problem areas" (such as energy consumption, transport, industry, etc.). Each of these areas was thereafter assigned to a working group, composed mainly of specialized stakeholders, assembled to identify the feasible and economically compatible policies (CEC, 2000). Later on, these policies were extended by the Second European Climate Change Programme (ECCP II), started in 2005 and aimed to strengthen previous emission-control policies. The latter introduced new working groups and improved the actions foreseen by the first ECCP.

Nonetheless, the strongest set of actions versus climate change were promoted and achieved through the 10-year Programme Europe 2020. This Programme was outlined in 2010 as a response to the economic crisis that set back all development goals and endangered social and economic growth (EC, 2010). Besides improving and increasing the level of employment, the Programme encouraged a sustainable and green-oriented growth capable of achieving the following targets (EC, 2014a):

- 20% reduction of the GHG emissions (compared to 1990 levels);
- 20% of energy from renewable sources;
- 20% increase in energy efficiency.

Progress toward the first target showed significant results in terms of reduction of GHG emissions, approximately 18% by 2014 (EC, 2014b). Moreover, projections on future trends highlight that a reduction of about 24% could be achieved before 2020 (Fig. 5.1). However, even though Europe's overall progress has been remarkable, not all member states have equally contributed to achieve this goal. In fact, according to the available data, some states did not accomplish their national targets, and, in some cases, they have even increased their emissions (e.g., Ireland, +3.7%; Portugal, +6.5%; Spain, +15%) (EEA, 2016). Nevertheless, the poor performances of some states have been largely compensated by the overachievements of other member states, such as Germany and UK that "accounted for about 45% of the total GHG emission reduction at EU in 2014" (EEA, 2016).

The 2020 renewable energy target is being accomplished as well, especially thanks to the spread of solar and wind energy. In 2012 the share of renewable energy amounted to 14.4%, and a percentage higher than 20% was expected before 2020 (EC, 2014b).

In contrast, the third target is still far from being achieved. The 20% increase in energy efficiency implies a reduction of primary energy consumption to 1483 Mtoe. But, even though the crisis caused an abrupt reduction to 1583.5 Mtoe in 2007, the primary energy consumption started increasing again a few years later. Projections demonstrate that stronger measures and significant efforts are required in order to achieve the established goals (EC, 2014b).

However, based on the success of the 2020 Strategy, in 2011 the "Energy Roadmap 2050" introduced a new long-term goal: 80%–95% reduction of GHG emissions by 2050. This program aimed at guiding Europe through a long-standing decarbonization process, encouraging the development of innovative energy systems that involve a larger use of renewable resources, cheaper solutions, and more efficient systems (EC, 2011a).

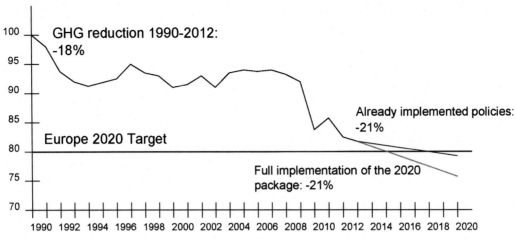

FIGURE 5.1 GHG reduction from 1990 to 2012 and future projections. *From EC, 2014b. Communication from the Commission to the European Parliament, the Council, the European Economic and Social Committee, and the Committee of the Regions. Taking Stock of the Europe 2020 Strategy for Smart, Sustainable and Inclusive Growth. Brussels, COM(2014) 130 Final/2, Annex 1–3. Available at: https://ec.europa.eu/info/sites/info/files/europe2020stocktaking_annex_en_0.pdf.*

In 2014, the 2030 Energy Roadmap outlined shorter-term goals. The Roadmap proposed to improve the 2020 goals and to draw closer to the 2050 decarbonization target, by promoting (EC, 2014c):

- 40% reduction of GHG emissions (compared to 1990 levels);
- 27% share of renewable energy consumption;
- 27% increase in energy efficiency.

However, the path toward the reduction of GHG emissions so far described and the longer-term decarbonization were based from the beginning on the clear recognition of the key role of local authorities in achieving effective mitigation goals. Thus, in 2008 Europe launched the Covenant of Mayors, aimed at promoting and coordinating local programs to address the ambitious European goals. This initiative, which counts today more than 5900 signatures,[1] represents a commitment of the local authorities to implement the European mitigation strategies and succeed in building up carbon-free cities.

In order to push local authorities to start the transition toward low-carbon cities, in 2011 the Smart Cities and Communities Initiative was also launched. The Initiative earmarked funds to encourage local actions capable of guaranteeing the spread of clean energy technologies, ensuring the CO_2 reduction in various sectors such as transport, building, and industry. The actions, carried out in pilot cities, would prove the real benefits of a low-carbon economy and promote the spread of these technologies (EC, 2011b).

In short, the followed path toward the establishment of ambitious mitigation goals and effective mitigation strategies (Fig. 5.2) proves a strong European commitment toward the reduction of GHG emissions and the widespread use of renewable energy. However, the promoted strategies focus on long-term goals, with benefits not immediately perceivable. Therefore, in order to face the unavoidable and already perceivable impacts of climate change, Europe started concentrating on adaptation measures, which could help member states to better face increasing climate-related impacts.

The process toward adaptation, represented in Fig. 5.2, stretches over a shorter time span and it is still on-going.

5.2.2 European Adaptation Strategies

The first step toward a European adaptation strategy occurred in 2007 when the green paper "Adapting to climate change in Europe — options for EU action" was published. The document encouraged European countries to take action against the risks induced by climate change; it emphasized, indeed, that the European average temperature "has risen by almost 1°C in the last century" and this condition contributed to the increase of droughts and floods in

---

[1] http://www.covenantofmayors.eu/IMG/pdf/CovenantLeaflet_web.pdf. It has to be noticed that the number of signatures refers to December 2016.

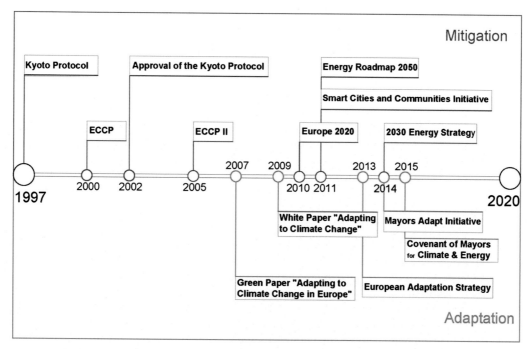

FIGURE 5.2 Europe mitigation and adaptation policies time line. *Authors' elaboration.*

different European regions (CEC, 2007). These criticalities, firstly highlighted and discussed in the green paper, led in 2009 to a white paper, "Adapting to climate change: Toward a European framework for action" (CEC, 2009). The latter outlined an intervention plan explaining how adaptation should be integrated in European development policies and which financial tools should be taken into account.

Since the release of the white paper, Europe started to work on a more detailed and operational document that took shape as the European Adaptation Strategy, adopted on April 2013. This document clearly underscored that "although climate change mitigation must remain a priority for the global community", we "have no choice but to take adaptation measures to deal with the unavoidable climate impacts and their economic, environmental and social costs" (EC, 2013a). This strategy aimed to increase European resilience to climate impacts by addressing three priority goals:

- To promote and support member states' actions; each member state should develop an adaptation strategy, taking into account local peculiarities. Europe, on the other hand, should coordinate these strategies and provide guidelines for developing adaptation plans on local scale;
- To support better-informed decision-making; Europe should improve climate-related knowledge, encouraging research and innovation and supporting decision-making at different geographical scales;
- To favor adaptation in key vulnerable sectors; adaptation strategies should be integrated in legislation regarding vulnerable sectors such as forestry, agriculture, biodiversity, transport, and disaster risk prevention and mitigation. For instance, all policies involving the energy, transport, and building sectors should be climate-proofed and aligned with the adaptation criterion.

Following the European Adaptation Strategy and according to its first priority goal, numerous member states have issued their national adaptation strategies. Moreover, the launch in 2014 of the Mayors Adapt initiative has significantly increased the number of European cities that have focused on adaptation issues, by developing local adaptation plans or by mainstreaming adaptation strategies into existing local development plans. The latter approach seems the most responsive to the principles established by the latest document on sustainable development, the Rio+20 "The Future We Want," issued in 2012, which clearly stated the need for mainstreaming both disaster risk reduction and climate adaptation in urban planning (UN, 2012). However, just like the Covenant of Mayors, the Mayors Adapt initiative has provided important support for spreading adaptation strategies all over Europe.

The second goal has led to significant promotion of research and innovation in the field of climate adaptation. For example, the Research and Innovation Programme Horizon 2020 (EC, 2011c), launched in 2011, has devoted great effort to research and innovation projects focused on climate adaptation (e.g., Ramses, BASE, Helix, etc.) and more recently on nature-based solutions, largely interpreted in current literature as a novel idea to reframe "policy debates on biodiversity conservation, climate change adaptation and mitigation strategies, and the sustainable use of natural resources" (Potschin et al., 2015). Moreover, in order to support better-informed decision-making and to share methods, tools, and best practices, the European platform Climate-Adapt was established in 2012. The platform provides data and information regarding future climatic trends, vulnerable regions and sectors, as well as opportunities and best practices for developing adaptation policies and plans (EEA, 2013).

Finally, the third goal has pushed toward the development of plans addressed to integrate adaptation strategies in different sectors and programs. An example of this is the Common Agriculture Policy, whose reform in 2013 pushed toward a new coordinated program, addressed not only to enhance agricultural production but also to increase the resilience of rural areas in the face of climate change (EC, 2013b).

5.2.3 Europe Toward an Integrated Climate Strategy

Starting from 2015, Europe seems to mark a significant change of pace compared to the previous phases. The latest phase of the European path to counterbalance climate change, indeed, pursues an integrated approach to mitigation and adaptation issues, seeking strategies capable of maximizing synergies and trade-offs between them.

Therefore, this phase seems to mark a transition from complementarity toward a synergy-based perspective (Duguma et al., 2014). During the previous phase, in fact, mitigation and adaptation were clearly recognized as complementary, since they both address climate change (Yohe and Strzepek, 2007), but have been processed with separate strategies and tools, considering at most the cobenefits they produce for each other.

The synergy-based perspective should emphasize, in contrast, policies and strategies capable of promoting integrated mitigation and adaptation measures. Such perspective has to be interpreted as the result of a long scientific debate that has clearly highlighted not only the benefits arising from an integrated approach to climate issues (Klein et al., 2005; Duguma et al., 2014; Galderisi et al., 2016) but also its potential limits and constraints due, for example, to the fact that mitigation and adaptation measures can be in some cases mutually conflicting (Laukkonen et al., 2009) or to the institutional complexity arising from the heterogeneity of actors in charge of mitigation and adaptation measures on different geographical scales (Klein et al., 2005).

This latest phase has led to the launch, in 2015, of the new Covenant of Mayors for Climate & Energy, which aims to achieve the following interrelated goals[2]:

- mitigation, by accelerating the decarbonization process;
- adaptation, by strengthening the ability to adapt to climate impacts;
- secure, sustainable, and affordable energy, by encouraging the use of renewable sources and improving energy efficiency.

Although the new approach is still at an early stage and the debate on how to move from complementarity toward a synergy-based perspective is still under investigation, the already-undertaken actions have witnessed a progression of approaches, aimed at promoting integrated strategies in the face of climate change to be mainly developed on a local scale, although coordinated and supported at national and EU levels.

5.3 CONCLUDING REMARKS: WHICH FUTURE FOR CLIMATE POLICIES IN EUROPE?

According to the previous discussion, the European approach to climate issues has significantly evolved over time and still continues to. The starting point of this long path was the launch, in 2000, of the ECCP that marked the start of a mitigation-driven phase, mainly devoted to the reduction of GHG emissions.

A turning point in the European approach to climate change occurred only in 2007 with the publication of a green paper, which marked the beginning of a new phase, characterized by a growing awareness of the urgent need for adaptation. The green paper raised awareness on the numerous impacts due to climate change and clearly stated the need for effective adaptation strategies and measures to reduce climate-induced damage. Briefly, starting in 2007, Europe has adopted a complementarity perspective, by coping simultaneously, although separately, with both mitigation and adaptation issues.

---

[2] http://www.covenantofmayors.eu/.

In the last few years Europe has started a third phase along its path to counterbalance climate change, aimed to emphasize synergies between mitigation and adaptation strategies and measures; nevertheless, this phase is at an early stage and its potential results are still uncertain.

Summing up, the briefly described path proves Europe's strong commitment in counterbalancing climate change, but highlights also its numerous limits and constraints. With regard to mitigation, despite Europe being considered one of the world leaders in global mitigation policies, it has to deal with both internal and external constraints. European efforts toward emissions reduction have to be supported by a shared and large commitment on a global scale in order to be successful. The inactivity of numerous industrialized parties, such as the United States, which backed out of the Kyoto Protocol in 2013,[3] is likely to compromise the achievement of the established and shared targets on a global scale, despite the efforts of individual countries. Likewise, the poor commitment to reduction goals of some member states contrasts with the overachievements of others, undermining the overall European performance.

As for adaptation policies, it is worth noting that, despite the key role played in terms of mitigation, Europe has started its path toward adaptation with a significant delay, since the first European Adaptation Strategy was adopted only in 2013. Hence, despite the increasing costs of climate-relate hazards, the local response to the Mayors Adapt initiative didn't reach the numbers and the results achieved by the Covenant of Mayors.

However, starting in 2015, Europe launched the new Covenant of Mayors for Climate & Energy that marked a significant change of pace compared to the previous phases, by pursuing an integrated approach to mitigation and adaptation issues. Unfortunately, it is still at an early stage and the debate on how to shift from complementarity toward a synergy-based approach and how to deal with the numerous obstacles and barriers is still open.

Hence, despite its strong commitment on mitigation issues and, more recently, on adaptation, Europe seems to find itself constrained in between global decisions—that do not always converge on the need to pursue ambitious goals to counterbalance climate change—and national and local tardiness, which have partially undermined European achievements in terms of adaptation issues. However, the last phase could bring interesting developments allowing for minimizing costs and maximizing benefits of climate policy thanks to the shift toward an integrated approach to mitigation and adaptation issues.

References

CEC (Commission of the European Communities), 2000. Communication From the Commission to the Council and the European Parliament. EU Policies and Measures to Reduce Greenhouse Gas Emission: Toward a European Climate Change Programme (ECCP). Brussels, COM(2000) 88 Final. Available at: http://eur-lex.europa.eu/legal-content/EN/TXT/PDF/?uri=CELEX:52000DC0088&from=EN.

CEC, 2005. Communication From the Commission to the Council, the European Parliament, the European Economic and Social Committee and the Committee of the Region. Winning the Battle Against Global Climate Change. Brussels, COM(2005) 35 Final. Available at: http://eur-lex.europa.eu/legal-content/EN/TXT/PDF/?uri=CELEX:52005DC0035&from=EN.

CEC, 2007. Green Paper From the Commission to the Council, the European Parliament, the European Economic and Social Committee and the Committee of the Regions. Adapting to Climate Change in Europe- Options for EU Action. Brussels, COM(2007) 354 Final. Available at: http://eur-lex.europa.eu/legal-content/EN/TXT/PDF/?uri=CELEX:52007DC0354&from=EN.

CEC, 2009. White Paper. Adapting to Climate Change: Toward a European Framework for Action. Brussels, 3. Available at: https://ec.europa.eu/health/archive/ph_threats/climate/docs/com_2009_147_en.pdf.

Ciscar, J.C., Iglesias, A., 2010. Physical and economic consequences of climate change in Europe. PNAS 2678–2683. https://doi.org/10.1073/pnas.1011612108.

Duguma, L.A., Minang, P.A., van Noordwijk, M., 2014. Climate change mitigation and adaptation in the land use sector: from complementarity to synergy. Environmental Management 54, 420–432. https://doi.org/10.1007/s00267-014-0331-x.

EC (European Commission), 2010. Communication From the Commission. Europe 2020. A Strategy for Smart, Sustainable and Inclusive Growth. Brussels, COM(2010) 2020 Final. Available at: http://eur-lex.europa.eu/legal-content/EN/TXT/PDF/?uri=CELEX:52010DC2020&from=EN.

EC, 2011a. Communication From the Commission to the European Parliament, the Council, the European Economic and Social Committee and the Committee of the Regions. Energy Roadmap 2050. Brussels, COM(2011) 885 Final. Available at: http://eur-lex.europa.eu/legal-content/EN/TXT/PDF/?uri=CELEX:52011DC0885&from=EN.

EC, 2011b. Public Consultation Report. Report of the Public Consultation on the Smart Cities and Communities. Brussels. Available at: https://ec.europa.eu/energy/sites/ener/files/documents/public_consultation_report.pdf.

EC, 2011c. Communication From the Commission to the European Parliament, the Council, the European Economic and Social Committee and the Committee of the Regions. Horizon 2020-The Framework Programme for Research and Innovation. Brussels, COM(2011) 808 Final. Available at: http://ec.europa.eu/research/horizon2020/pdf/proposals/communication_from_the_commission_-_horizon_2020__the_framework_programme_for_research_and_innovation.pdf.

EC, 2013a. Communication From the Commission to the European Parliament, the Council, the European Economic and Social Committee and the Committee of the Regions. An EU Strategy on Adaptation to Climate Change. Brussels, COM(2013) 216 Final. Available at: http://eur-lex.europa.eu/legal-content/EN/TXT/PDF/?uri=CELEX:52013DC0216&from=EN.

---

[3] http://edition.cnn.com/=.

EC, 2013b. Overview of CAP Reform 2014-2020, Agricultural Policy Perspectives Brief, 2-3. Available at: https://ec.europa.eu/agriculture/sites/agriculture/files/policy-perspectives/policy-briefs/05_en.pdf.

EC, 2014a. Communication From the Commission to the European Parliament, the Council, the European Economic and Social Committee and the Committee of the Regions. Taking Stock of the Europe 2020 Strategy for Smart, Sustainable and Inclusive Growth. Brussels, COM(2014) 130 Final/2. Available at: https://ec.europa.eu/info/sites/info/files/europe2020stocktaking_annex_en.pdf.

EC, 2014b. Communication From the Commission to the European Parliament, the Council, the European Economic and Social Committee and the Committee of the Regions. Taking Stock of the Europe 2020 Strategy for Smart, Sustainable and Inclusive Growth. Brussels, COM(2014) 130 Final/2, Annex 1–3. Available at: https://ec.europa.eu/info/sites/info/files/europe2020stocktaking_annex_en_0.pdf.

EC, 2014c. Communication From the Commission to the European Parliament, the Council, the European Economic and Social Committee and the Committee of the Regions. A Policy Framework for Climate and Energy in the Period From 2020 to 2030. Brussels, COM(2014) 15 Final. Available at: https://eur-lex.europa.eu/legal-content/EN/TXT/PDF/?uri=CELEX:52014DC0015&from=EN.

EEA (European Environment Agency), 2013. Adaptation in Europe. Addressing Risks and Opportunities From Climate Change in the Context of Socio-economic Developments. EEA Report, 3/2013. Available at: https://www.scribd.com/document/164062974/European-Environmental-Agency-Report-No-3-2013-Adaptation-in-Europe-Addressing-risks-and-opportunities-from-climate-change-in-the-context-of-socio#.

EEA (European Environment Agency), 2016. Annual European Union Greenhouse Gas Inventory 1990–2014 and Inventory Report 2016. EEA Report, 15/2016. Available at: https://www.eea.europa.eu/publications/european-union-greenhouse-gas-inventory-2016.

Galderisi, A., Mazzeo, G., Pinto, F., 2016. Cities dealing with energy issues and climate-related impacts: approaches, strategies and tools for a sustainable urban development. In: Papa, R., Fistola, R. (Eds.), Smart Energy in the Smart City. Urban Planning for a Sustainable Future. Springer International Publishing, pp. 199–218. https://doi.org/10.1007/978-3-319-31157-9_11.

Galderisi, A., 2014. Climate change adaptation. Challenges and opportunities for Smart urban growth. TeMA. Journal of Land Use, Mobility and Environment 7 (1), 43–67. https://doi.org/10.6092/1970-9870/2265.

IPCC, 2014. Summary for Policymakers. Climate Change 2014: Impacts, Adaptation and Vulnerability. Part a: Global and Sectoral Aspects. Contribution of Working Group II to the Fifth Assessment Report of the Intergovernmental Panel on Climate Change, vol. 22, pp. 11–12. Available at: http://www.ipcc.ch/pdf/assessment-report/ar5/wg2/ar5_wgII_spm_en.pdf.

Klein, R., Schipper, L., Dessai, S., 2005. Integrating mitigation and adaptation into climate and development policy: three research questions. Environmental Science & Policy 8, 582–583.

Laukkonen, J., Blanc, P., Lenhart, J., Keiner, M., Cavric, B., Kinuthia-Njenga, C., 2009. Combining climate change adaptation and mitigation measures at a local level. Habitat International 33, 287–292.

Potschin, M., Kretsch, C., Haines-Young, R., Furman, E., Berry, P., Baro, F., 2015. Nature-based solutions. OpenNESS. In: Potschin, M., Jax, K. (Eds.), Ecosystem Service Reference Book, p. 1. Available at: www.openness-project.eu/library/reference-book.

UN (United Nations), 1972. Report of the United Nations Conference on the Human Environment. Stockholm, 5–16 June. Available at: http://www.un-documents.net/aconf48-14r1.pdf.

UN (United Nations), 1992. United Nations Framework Convention on Climate Change. FCCC/INFORMAL/84 GE.05-62220 (E) 200705, Available at: https://unfccc.int/resource/docs/convkp/conveng.pdf.

UN, 1998. Kyoto Protocol to the United Nations Framework Convention on Climate Change. Available at: https://unfccc.int/resource/docs/convkp/kpeng.pdf.

UN, 2012. The Future We Want (Art. 135). Rio +20 United Nations Conference on Sustainable Development. Available at: http://www.un.org/disabilities/documents/rio20_outcome_document_complete.pdf.

UN, 2015. Paris Agreement. Available at: http://unfccc.int/paris_agreement/items/9485.php.

Yohe, G., Strzepek, K., 2007. Adaptation and mitigation as complementary tools for reducing the risk of climate impacts. Mitigation and Adaptation Strategies for Global Change 12, 727–739. https://doi.org/10.1007/s11027-007-9096-3.

CHAPTER 6

The American Approach to Climate Change: A General Overview and a Focus on Northern and Arctic Regions

Catherine Dezio
Politecnico of Milan, Milan, Italy

6.1 STRATEGIES AND INITIATIVES IN THE FACE OF CLIMATE CHANGE: THE AMERICAN APPROACH

On September 3, 2016, the United States and China, the two countries that produce the most carbon dioxide emissions (the world's largest greenhouse gas producers), announced the ratification of the Paris Climate Agreement. The agreement has been signed by 195 countries and it stipulates that all these countries undertake to reduce polluting emissions, in order to: keep the temperature rise below 2°C, stop increasing greenhouse gas emissions, and finance the poorest countries to help them develop fewer polluting sources of energy. Then-President Barak Obama's adherence to the agreement confirmed his intentions to tackle incisive environmental policies. In fact, on August 3, 2015, he presented the "Clean Power Plan"—a series of new regulations formally introduced by the US Environmental Protection Agency, which have as their primary goal the reduction of CO_2 emissions produced in the United States. The purpose of this plan was to speed up the transition to renewable energy production and set ambitious targets for cutting emissions; the demand has been a 32% reduction by 2030 and the reward of the states and of energy companies that would move quickly to increase their investments in solar and wind energy. If it remained standing, it could change the United States' energy system deeply. Obama described it as "the biggest and most important step we can take to combat climate change" (2015).

With the election of Donald Trump to the US presidency (he took office January 20, 2017), this framework has a complete reversal of its route, with direct repercussions on the rest of the planet as well.

Already during the election campaign, Trump had announced that he would take the United States out of the Paris Agreement, defining it as dangerous for the US economy. At the end of the 43rd G7 summit held in Taormina (Sicily, Italy), on May 26 and 27, 2017, these intentions were confirmed.

In spring 2017, Trump also signed the "Energy Independence" document with which he started the process of reformulating the "Clean Power Plan" (EPA, 2017), stating that the measure is necessary to promote US energy independence and revitalize the local coal industry, with the restoration of thousands of jobs that have disappeared in recent years. However, several economists have argued that Trump's policies may not be adequate to achieve that goal—partly because the United States is already heavily dependent on coal and natural gas, partly because the coal energy system will, in the future, use a growing number of machines and fewer workers (Godby et al., 2015).

The consequences of Trump's antienvironmental orientation will have a global reverberation: from a political point of view, on maintaining commitments from other States (i.e., China and India may decide not to abide the commitments made with the Paris Agreement); from the climatic point of view because the global thermometer will quickly exceed the 2°C threshold and extreme climatic events will change the geographies and dynamics of many countries; many ecosystems will be destroyed and access to food and water will be increasingly difficult for many developing countries; finally, according to simulations carried out by "climate interactive" experts, by

2030 there could be 3 billion tons more carbon, and the USA would only increase average temperatures by 0.3°C by the end of the 21st century (2017).

However, the complete dismantling of the Obama "Clean Power Plan" may take some time by the Trump administration. Many companies are still determined to maintain their plans to produce solar panels, wind turbines, and other systems to exploit renewable sources. They are companies that during the 8 years of the Obama administration have obtained funds and facilities and have seen great opportunities in pursuing clean energy. So, the federal tax incentives for solar and wind power decided by the Obama administration will continue for many years and they will even have the support of Republican politicians who are Congressional representatives of wind power producers, such as Texas and Iowa. According to US law experts, not only could it take years to change internal policies, but the Trump administration could encounter various obstacles, including the opposition of individual states.

The states of New York and of California have already said that they will oppose the new environmental policy; their two governors, as well as the governor of Washington State, have signed an accord that forces them to remain bound to the Paris Agreement. This is the "US Climate Alliance" (2017), which already has members among other American states and cities, and it is creating a real revolt against Trump's decision. Other governors are joining to this protest, from Colorado to Massachusetts to Connecticut, and the mayors of 61 cities, including New York, Chicago, Los Angeles, Philadelphia, New Orleans, Seattle, and Boston, who do not want to betray the nearly 200 countries of the Paris Agreement.

These climate-friendly policies are not a surprise in the US state policy framework. Already in 2009, in the northeastern United States, eight states (Maine, New Hampshire, Vermont, Connecticut, New York, New Jersey, Delaware, and Massachusetts) had set up a regional initiative to limit greenhouse gas emissions—the "Regional Greenhouse Gas Initiative." This initiative is a system that limits carbon dioxide emissions at state and interstate levels and pressures the federal government to meet these limits.

In 2007, five Western states (California, Arizona, New Mexico, Oregon, Washington) also launched the "Western Regional Climate Action Initiative" to develop an emission limitation system, emphasizing the necessity of alternative energy sources.

Even at the municipal level, orientation toward environmental policies is incisive. Five cities of the San Francisco Bay (Albany, Bates, Emeryville, Union City, and Piedmont) have studied London City's methods of reducing pollution, providing the first example of transatlantic collaboration on the subject.

Despite efforts at both municipal and individual levels to initiate environmental policies, according to Joann Carmin, a professor in the Department of Urban Studies and Planning at MIT, US cities are lagging behind countries that have been heavily affected by the effects of climate change. Among them, there are the countries of South America. In an article in the *Journal of Planning Education and Research*, Carmin and coauthors have analyzed local climate policy planning at Quito and Durban (South America), stating that, following severe storms, floods, or fatal heat waves, certain states of South America are facing the issue of climate change as a real urgent and priority political problem (2012).

Some countries, such as Colombia and Brazil, are increasingly affected by heavy floods that result in deaths and displaced citizens; in Mexico, drought and fires hit crops and farms, causing huge economic losses. That framework has been recently addressed in the "Migration and Climate Change" report (2015) by the Italian World Wide Fund for Nature (WWF), highlighting how climate change has already exposed hundreds of millions of people to their impacts—"from 2008 to 2014, more than 157 million people have moved as result of environmental disasters."

6.2 FOCUSING ON NORTHERN AND ARCTIC REGIONS

The same situation concerns the indigenous communities of Alaska, who have inhabited the Arctic forests for millennia. Shishmaref, Kivalina, Shaktoolik, and Newtok are facing a very critical situation because of their geographical position on the west coast of Alaska. Here heating is faster than in the rest of the planet, the Arctic ice is rapidly retreating, and the coastal communities are in serious and growing danger. Newtok is a Yupik Eskimo village located on the Ninglick River, next to the Bering Sea; Shishmaref and Kivalina are Inupiat Eskimo villages and are further north on the Chukchi Sea; Shaktoolik is a Malemiut Eskimo village located in Norton Sound. These four villages have existed on the Alaska coast for thousands of years, but environmental studies indicate that a catastrophic climate event could overwhelm them within the next 15 years (Hermann and Keene, 2017). Government agencies have responded to an increase in coastal erosion through their traditional methods of controlling erosion and flooding. However, because of the severity of erosion, these adaptation strategies have proven to be ineffective and a permanent migration remains the only appropriate solution, although no public funding has yet been specifically devised and there is no state strategy planning. This uncertainty toward the future has prompted the Inuit to give international resonance to the serious problems that endanger their own survival. In 2005, the Inuit Circumpolar Conference (ICC), a nongovernmental organization (NGO) representing about 155,000 people from the Arctic regions of Canada, the United States, Greenland, and Russia, prepared a petition to the Inter American Commission on Human Rights (IACHR) to denounce the violation of human rights, caused by the emissions of the United States. According to a report drawn up by the Arctic Council (intergovernmental forum formed by Canada, Russia, Norway, Denmark, United States, Sweden, and Finland), Arctic annual average temperatures are rising at a rate more than twice speed of the temperatures recorded in the rest of the world, causing devastating consequences on the Arctic region ecosystem.

Based on these data, strongly confirmed by the scientific community, the ICC considered that there was enough evidence to declare that some states are responsible for climate change problems, violating Inuit Human Rights; so it asked to IACHR to make explicit recommendations to the United States to: (1) take measures to limit greenhouse gas emissions; (2) cooperate internationally to mitigate the effects of climate change; (3) take into account the effects of its policies on Arctic populations; (4) consider a plan to protect the Inuit culture and resources; and (5) help the Inuit to adapt to the inevitable impacts of climate change.

On November 16, 2006, IACHR announced their decision, which refused the petition submitted by the ICC. However, in the following year, the Commission called on ICC, Center for International Environment Law (CIEL), and Earth Justice representatives to provide their testimony about the link between climate change and human rights, and the president of the ICC, Sheila Watt-Cloutier, emphasized that the Inuit economic and cultural system is going to succumb to the effects of climate change. If actual remedies will not be conceived and immediately realized, the Inuit will be forced to abandon the lands that have lived on for centuries.

Since then, representatives of indigenous peoples continue to attend international events on climate change to officially and continuously reaffirm the need to recognize respect for their human rights.

At the same time, the devastating consequences for marine ecosystems and for local communities are dangerous also for the rest of the world.

The Arctic permafrost contains almost double the amount of carbon that is in the atmosphere, so the thawing of these soils increases the release of carbon dioxide and methane and, consequently, the overall heating of the planet. The weather conditions and the ocean circulation in all parts of the globe are influenced by the events at the poles. Therefore, the changes in this region are a global problem that needs in every case a solution, even with the migration of the local community.

Managing these changes involves attention to the socioecological system as a whole, including both environment and community (both institutional and not); we could say even that every ecological problem is directly connected with the social systems, with their policies as well as with their resilience. If the community were to adapt to climate warming by reducing the emissions of fossil fuels, it would cause the reduction of the global warming rate; similarly, if the Arctic populations were to adapt to the difficulties of today's hunt, they would find new sources of food and this could stabilize their food supply. In reference to these examples, we can summarize in two issues the range of possibilities to find strategies and actions in this regard: the first one is the issue of maritime traffic (fossil fuel reduction and the respect for the marine ecosystem), and the second one is food security (the resources procurement and the accessibility to drinking water).

About the first point, a high-risk activity for the area is its own sea transport, even more for the low infrastructure of the region. Local economic development is based on the extraction of mineral resources (35% of Alaskan jobs are related to the energy sector) and on tourism by ship (1000 visitors to the North Pole every year), pushing maritime transport to an extreme (United States Coast Guard's Arctic Strategy, 2013). The US Coast Guard has set up a secure and protected maritime governance strategy for the Arctic region. That strategy has been designed to be a reference point and it establishes objectives and instruments with decades of perspective. In particular, the document argues three strategic objectives: (1) improving knowledge and awareness, through a system of monitoring and of information sharing; (2) the modernization of governance, i.e., the actors and skills system; (3) partnership extension, both public and private, both national and international. In addition to these three strategic objectives, there are many other additional factors, for example, national awareness, the improvement of relations between public and private groups, the prevention of terrorism, and risk management. The experience of the Coast Guard as an active subject of the strategy is essential to balance the commercial and economic activities with the marine environment point of view; moreover, the Coast Guard's relations with all stakeholders (companies, academia, environmental groups, local and national governments, indigenous communities) make it an entity that can ensure long-term success of choral governance.

About that, the Bering Strait is the most famous example; it is the most northern part of the Pacific Ocean, a water passage of 85 km that separates Alaska from Russia, whose coast is characterized by a rich biodiversity and it hosts subsistence communities that are dependent on marine life for their cultural and nutritional survival. The International Union for Conservation of Nature (IUCN) has identified 13 fragile areas in the Arctic region, from ecological and biological points of view, three of which are here in Bering.

The oil industrial activities and the reduction of the ice soil threaten native communities, wildlife, and the entire local ecosystem. For the near future, it is expected that there will be a significant increase in temperature, as well as an increase of industrial activities and of maritime traffic; in the North Sea, in 2012, 47 ships transported 1.3 million tons of cargo; in 2007, there were only two ships (United States Coast Guard's Arctic Strategy, 2013). We've gone from what was called "experimental navigation activities" to a more routine use of the Northern Sea Route. That increase of navigation in the Strait has caused the death of whales and the strike of ships, and these strikes could also have a negative impact on food security and on political systems. International responses regarding that problem have come from the International Convention for the Safety of Life at Sea and from the International Convention for the Prevention of Pollution from Ships. Moreover, the International Maritime Organization (IMO) tends to aim for a balance between the principle of "freedom of the seas" and the need to regulate people's safety, ships, and the environment; therefore, it requires to the coastal states of the Strait to agree to protective measures before going to the IMO. At the same time, the Arctic Marine Shipping Assessment (The Arctic Council, 2009) has provided a complete image of Arctic marine activity, representing a strategic guidance and a political framework. The assessment provides 17 recommendations on three themes: improving Arctic maritime security, protecting Arctic communities and environment, and building infrastructure.

Compared to the second theme, we know that food security exists "when all people, at all times, have physical and economic access to sufficient, safe and nutritious food that meets their dietary needs and food preferences for an active and healthy life".

Food security in the Arctic region, as elsewhere, also includes political aspects such as "food sovereignty," which highlights the right of peoples to define their own policies and strategies about production, distribution, and consumption of food. However, a definition of food security for the Arctic environment should also take into account cultural and social aspects of food for indigenous peoples, such as the ability to use and teach the traditional techniques and the food-sharing practices and the role that food plays in fostering the links of small communities like these.

However, to date there is no quantitative evaluation that specifically focuses on food security in the Arctic region, although some efforts in this direction are being taken.

During the Swedish Presidency of the Arctic Council (2011–13), for example, food security was a priority issue; in this period, working groups (such as the Arctic Monitoring and Assessment Programme and the Sustainable Development Working Group of the Arctic Council) have conducted a study that aims to define indicators for food security in the Arctic area. The Inuit Circumpolar Council Alaska has started to develop a framework for how to assess food security from an Inuit perspective, which includes interlinkages between cultural and environmental systems, a study based on literature reviews, community meetings, and interviews; the goal is to arrive at an instrument for assessing, measuring, and monitoring food security.

Besides all that, to make a framework that goes toward resilience on climate change issues would be required: (1) identification of mechanisms behind disruptions of food supply; (2) knowledge of alternative sources of supply and of water sharing; (3) social systems that help the poorest with state financial support; (4) mapping of factors that can have an impact on food security; and (5) the identification of food security projects that can present examples consistent with Arctic environments, to be applied beforehand. The spontaneous adaptation strategies of indigenous communities are interesting case studies on that issue.

New environmental conditions have suggested for some time the need to question the local lifestyles, reformulating the Arctic as a dynamic adaptive system. For example, as noted in the report of Chatham House Lloyd (Lahn and Emmerson, 2012), rural populations that historically lived by hunting have discovered new practices for food supply such as fishing, due to the reduction of the ice soil as a platform of hunting. At the same time, these new weather patterns threaten to slowly destroy the traditional heritage of indigenous communities, and the only possibility for keeping it is delegated to the institutions. Based on 26 cases studies of communities in the polar region, the "Caviar" project (Smit et al., 2010) has proved what had been said: the ability to adapt to changes in socioecological systems is closely related to the governance of communities. The same studies have shown that where transverse governance networks (such as comanagement) were developed, new and effective strategies have been applied.

Therefore, strengthening resilience of those systems can be achieved thanks to adaptive governance, which is capable of challenging the traditional systems of land management. Adaptive governance can be achieved when: we work with different scales; we implement both mitigation strategies and adaptation strategies, in a complementary and integrated manner; we apply decision-making mechanisms with horizontal dialogue between different governments, citizen groups, government agencies, NGOs, and companies. Adaptive governance also has a good system of monitoring and collecting data, from formal and informal sources, but above all it is based on openness toward learning, flexibility, and reorganization of all community actors.

Introducing this type of governance—but, first of all, an assumption of responsibility by the most powerful states through appropriate global and state policies—is an urgent and pressing need for many fragile areas of the world.

References

Carmin, J., Anguelovski, I., Roberts, D., 2012. Urban climate adaptation in the global South. Journal of Planning Education and Research 32 (1), 18–32. http://journals.sagepub.com/doi/abs/10.1177/0739456X11430951.

Remarks by the President in Announcing the Clean Power Plan, 2015. Available at: https://obamawhitehouse.archives.gov/the-press-office/2015/08/03/remarks-president-announcing-clean-power-plan.

EPA, 2017. Clean Power Plan and the Energy Independence. Available at: https://www.epa.gov/energy-independence.

Godby, R., Coupal, R., Taylor, D., Considine, T., 2015. Potential impacts on Wyoming coal production of EPA's greenhouse gas proposals. The Electricity Journal 28 (5), 68–79.

Hermann, V., Keene, E., 2017. A Continual State of Emergency: Climate Change and Native Lands in Northwest Alaska. Available at: https://www.thearcticinstitute.org/continual-state-emergency-climate-change-native-lands-northwest-alaska/.

Johnston, E., Jones, A., Siegel, L., Sterman, J., 2017. Analysis: U.S. Role in the Paris Agreement. Available at: https://www.climateinteractive.org/analysis/us-role-in-paris/.

Lahn, G., Emmerson, C., 2012. Arctic Opening: Opportunity and Risk in the High North, Chatam House, Lloyd's Risk Insight Report. Available at: https://www.chathamhouse.org/sites/files/chathamhouse/publications/0412arctic.pdf.

Paris Agreement, 2016. Available at: https://unfccc.int/files/essential_background/convention/application/pdf/english_paris_agreement.pdf.

Smit, B., Hovelsrud, G., Wandel, J., Andrachuk, M., 2010. Introduction to the CAVIAR Project and Framework, Community Adaptation and Vulnerability in Arctic Regions, pp. 1–22.

The Arctic Marine Shipping Assessment, 2009. https://oaarchive.arctic-council.org/handle/11374/54.

The Regional Greenhouse Gas Initiative (RGGI), 2009. Available at: https://www.rggi.org.

The, 2017 The U.S. Climate Alliance, 2017. Available at: https://www.usclimatealliance.org.

The Western Climate Initiative, 2007. Available at: http://www.westernclimateinitiative.org.

United States Coast Guard. Arctic Strategy, 2013. The U.S. Coast Guard's Vision for Operating in the Arctic Region: Ensure Safe, Secure, and Environmentally Responsible Maritime Activity in the Arctic. Available at: https://www.iho.int/mtg_docs/rhc/ArHC/ArHC4/ARHC4-3.1.3_INF_USCG_Arctic_Strategy_May_2013.pdf.

WWF Italia, "Migrazioni e cambiamento climatico", 2015. Available at: http://www.focsiv.it/wp-content/uploads/2015/10/WWF-Report.pdf.

Further Reading

Chapin III, F.S., Trainor, S.F., Cochran, P., Huntington, H., Markon, C., McCammon, M., McGuire, A.D., Serreze, M., 2014. Alaska. Climate change impacts in the United States: the third National climate assessment. In: Melillo, J.M., Richmond, T.C., Yohe, G.W. (Eds.), U.S. Global Change Research Program, pp. 514–536. https://doi.org/10.7930/J00Z7150.

FAO, 2006. Food Security. http://www.fao.org/forestry/13128-0e6f36f27e0091055bec28ebe830f46b3.pdf.

Inuit Petition to the Inter-American Commission on Human Rights for Dangerous Impacts of Climate Change. Available at: http://www.inuitcircumpolar.com.

Stockholm Environment Institute and the Stockholm Resilience Centre, 2013. Arctic Resilience Interim Report. https://www.sei-international.org/mediamanager/documents/Publications/ArcticResilienceInterimReport2013-LowRes.pdf.

CHAPTER 7

Climate Change Mitigation and Adaptation Initiatives in Africa: The Case of the Climate and Development Knowledge Network "Working With Informality to Build Resilience in African Cities" Project

Kwame Ntiri Owusu-Daaku[1], Stephen Kofi Diko[2]

[1]University of West Florida, Pensacola, FL, United States; [2]University of Cincinnati, Cincinnati, OH, United States

7.1 INTRODUCTION

Some of the main impacts of climate change on the African continent as they affect human settlements include sea-level rise, increased temperature, and reduced and irregular precipitation resulting in effects such as erosion and inundation, drought and depletion of groundwater sources, and food insecurity (Magadza, 2000). Many of these climatic conditions combine with nonclimatic factors such as rapid urbanization to produce impacts such as the urban heat island effect (Patz et al., 2005) and increased flooding (Douglas et al., 2008). We begin this chapter with a background to resilience as a convening concept to enable the discussion of climate change adaptation in tandem with mitigation strategies, as cities work toward promoting sustainable urban development. We then discuss the "Working With Informality to Build Resilience in African Cities" project to highlight one such initiative of integrating climate change adaptation and mitigation within the context of urban development. Though the selected project is not the only urban development initiative on the continent that combines both mitigation and adaptation strategies, we selected this project because of its focus on informality, which we view as a common characteristic of many African cities. As such, lessons from this project can provide more cross-cutting ideas for many other African contexts. We summarize the discussion of the project with some lessons learned and conclude with some reflections on the project's approach for building the resilience of cities to climate change impacts in Africa.

7.2 BACKGROUND: RESILIENCE AS A CONVENING CONCEPT FOR INTEGRATING CLIMATE CHANGE MITIGATION AND ADAPTATION[1]

Mitigation appears conspicuously hard to find when reading about resilience. Climate change resilience is often used in tandem and sometimes interchangeably with the term *adaptation*. Such climate change discussions usually highlight that the roots of resilience to climate change stem from disaster risk reduction, ecology, and sociology and

---

[1] This section is reproduced with some modification from a blog post written by the first author, available at http://www.hurdl.org/is-mitigation-beginning-to-enter-the-resilience-conversation/.

suggest a concept of "bouncing back" or "recovery" from some shock or stress (Bahadur et al., 2010; Green, 2009; Tyler and Moench, 2012). We would like to review some definitions of the terms *resilience, mitigation, adaptation,* and *sustainable development* as they relate to climate change by the Intergovernmental Panel on Climate Change (IPCC). We focus on IPCC definitions because the IPCC constitutes an international body of experts who assess the science related to climate change. The *Glossary of the Working Group (WG) II Contribution* to the IPCC's Fifth Assessment Report (AR5) defines resilience as "the capacity of social, economic, and environmental systems to cope with a hazardous event or trend or disturbance, responding or reorganizing in ways that maintain their essential function, identity, and structure, while also maintaining the capacity for adaptation, learning, and transformation" (IPCC, 2014, p. 1772). The glossary also defines climate change mitigation as "a human intervention to reduce the sources or enhance the sinks of greenhouse gases" (IPCC, 2014, p. 1769); adaptation as "the process of adjustment to actual or expected climate and its effects" (IPCC, 2014, p. 1758); and sustainable development (citing the 1987 World Commission on Environment and Development definition) as "development that meets the needs of the present without compromising the ability of future generations to meet their own needs" (IPCC, 2014, p. 1774). We argue that definitions of mitigation, adaptation, and sustainable development can all be placed under the broad definition of resilience that the IPCC puts forth. Mitigation, in reducing greenhouse gases, relates to the aspect of the definition of resilience in which a "system responds or reorganizes in ways that maintain their essential function, identity and structure" (IPCC, 2014, p. 1772). Adaptation, in adjusting to actual or expected climate impacts, relates to the aspect of resilience that discusses coping "with a hazardous event or trend or disturbance" while "maintaining the capacity for adaptation…" (IPCC, 2014, p. 1772). Lastly, sustainable development, in meeting present needs without compromising the capabilities of future generations, denotes a process of "learning, and transformation" (IPCC, 2014, p. 1772).

Based on this interpretation, we wonder why is a strategy (such as mitigation) that seeks to reduce the "stress" of greenhouse gases on the earth's atmosphere not considered a form of resilience? Could the reason that mitigation has not emerged as a popular strategy for building climate change resilience be because the earth has not been able to retain its basic structure and function in absorbing greenhouse gases and so can basically never "bounce back" or "return to a former state"? Or perhaps mitigation's absence from the discussion of resilience is because resilience has been a more popular term in the disaster risk reduction sector (which is characterized by more violent and extreme events). In other words, resilience has come to be framed more narrowly in relation to the immediate and extreme impacts of climate change such as floods or hurricanes (which are commonly viewed as disasters) rather than slower onset or more long-term changes such as droughts or sea level rise (that would be more quickly attributed to climate change). Furthermore, the IPCC's distinction between mitigation of climate change and mitigation of disaster risk and disaster (IPCC, 2014) further raises more questions. Is this framing, the reason resilience is commonly applied in light of adaptation alone—as adaptation is a more immediate actionable strategy than mitigation? If so, within climate change issues that already suffer from the back burner effect (the tendency for issues that do not have an immediate impact to be tabled to be addressed later) (Abrahams, 2014), mitigation then becomes the action that gets relegated to the background. However, in different geographical contexts, like in Europe, mitigation is front and center in the discussion on climate change resilience, and this prioritization is as a result of the embraced perspective of climate change resilience as including both mitigation and adaptation strategies.

Our problem, with mitigation not being integral to the definition of climate change resilience, is that language matters when it comes to climate change interventions (Owusu-Daaku and Diko, 2017). Since the concept of mitigation is not apparent in a number of definitions and conceptualizations of climate change resilience, it seems that so far, mitigation actions are being entirely left out of some of the efforts to build resilience to climate change. Nevertheless, the IPCC in the Summary for Policy Makers (SPM) of its special report on "Managing the Risks of Extreme Events and Disasters to Advance Climate Change Adaptation" (SREX) stated that "adaptation and mitigation *can* (emphasis ours) complement each other and together can significantly reduce the risks of climate change" (IPCC, 2012, p. 2). We are of the opinion, in tandem with much of the research in this chapter, that adaptation and mitigation should complement each other as extensive scholarship already shows (Nyong et al., 2007; Laukkonen et al., 2009; Kane and Shogren, 2000; Hamin and Gurran, 2009). Granted, mitigation was not the focus of the SREX report (as the authors of the report point out) but the statement on complementarity advances the notion that mitigation has not necessarily been viewed widely enough by scholars and practitioners as an ultimate complement to adaptation efforts in building resilience to climate change.

There appears to be hope, as this chapter and other chapters in this book point toward. In the IPCC AR5, the Working Group II (WG II) Contribution in its SPM discusses the concept of climate-resilient pathways, which the WG II defines as "sustainable-development trajectories that combine adaptation and mitigation to reduce climate change and its impacts." (IPCC, 2014, p. 28). Could a concept described as resilient that combines both adaptation and mitigation, in the context of sustainable development, be an indication that mitigation is beginning to enter the conversation about resilience? We most certainly would like to think so.

7.3 CASE STUDY: THE CLIMATE AND DEVELOPMENT KNOWLEDGE NETWORK "WORKING WITH INFORMALITY TO BUILD RESILIENCE IN AFRICAN CITIES" PROJECT

Now we turn to the "Working With Informality to Build Resilience in African Cities" project of the Climate and Development Knowledge Network (CDKN) as an illustrative example of how cities can play a vital role in promoting resilience as an integrative concept for combining both mitigation and adaptation strategies to climate change. The CDKN program is a group of research and implementation organizations that seek to make development and climate change mitigation and adaptation goals more compatible ("climate compatible development"). It does this through the promotion of "triple wins" and "cobenefits" between adaptation, mitigation, and development (CDKN, n.d.). Fig. 7.1 presents a graphical representation of the concept of climate-compatible development.

According to Fig. 7.1, the stated focus of this chapter would be on primarily "cobenefits"—the integration of both mitigation and adaptation strategies. However, as we discussed earlier, a more comprehensive term such as resilience could encompass more than merely adaptation and mitigation and would include strategies to reduce poverty and promote sustainability. There is a consensus in the urban climate change resilience literature that poverty reduction and promoting urban sustainability must be part of efforts to increase urban resilience to climate change in order to make cities truly resilient to climate and other shocks and stresses (Leichenko, 2011). Hence climate compatible development, and this project in particular, will demonstrate how cities are a key player in building resilience to climate change by embracing the presence of informality existing in many cities across Africa.

The proceeding analysis centers on the "Working with Informality to Build Resilience in African Cities" project—the need for the project, the project's conception and implementation, and findings. We achieve the aforementioned by assessing project briefs, workshop minutes, blog posts, a video, and a summary report of the project. Time and accessibility constraints did not permit engagement with project participants—an effort that would have otherwise enriched our analysis. We will conclude the analysis and this section in general with lessons we can learn about this particular approach of embracing contextually relevant characteristics (in this case, such as informality) to build the resilience of cities to climate change.

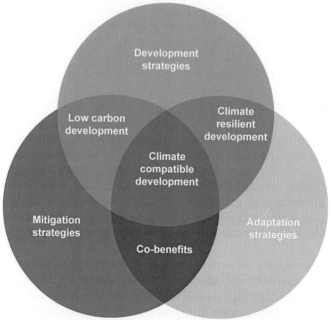

FIGURE 7.1 An illustration of climate compatible development. *From Mitchell, T., Maxwell, S., 2010. Defining Climate Compatible Development. CDKN Policy Brief.*

African cities have been marked by informality (Myers and Murray, 2006; Hansen and Vaa, 2004; Lindell, 2010; Heintz and Valodia, 2008). One definition of informality provided by Roy (2009) is "a state of deregulation, one where the ownership, use, and purpose of land cannot be fixed and mapped according to any prescribed set of regulations or the law". CDKN, in realizing this characteristic of African cities, conceptualized the "Working With Informality to Build Climate Resilience in African Cities" project. Much of this informality, according to the project's justification, is characterized by slums and shanty towns, unsurveilled economies, unemployment, youthful populations, low industrialization, high growth, and poverty rates (CDKN, n.d.). The project recognizes that engaging with informality is one avenue through which to respond to the challenge of climate change vulnerability and impacts in cities in an African context. This focus on informality highlights a pro-poor and people-focused approach to the combined climate change mitigation and adaptation, along with sustainable development, strategy of building resilience. This focus on people, and poor people for that matter, contrasts with a more institutional-centric approach in building resilience and reducing vulnerability to climate change (Adger et al., 2013). A more people-centric and pro-poor approach is necessary within the African urban context due to (as has already been outlined) the lack of formalized systems and structures to manage urbanization—in other words, the presence of informality.

This project ran from January 2013 to February 2015, with an allocated budget from CDKN of £360,000 and was implemented in partnership with the African Centre for Cities (ACC) at the University of Cape Town (UCT), South Africa. The case study cities for the project were Kampala (Uganda), Accra (Ghana), and Addis Ababa (Ethiopia). The initial title of the project was "Building Climate Resilience Through Tackling Informality and Promoting Integrated Urban Development and Management in African Cities" (CDKN et al., 2013, p. 1). However, this title changed after a scoping workshop organized from July 9 to 11, 2013 in Cape Town, South Africa to the latter title of "Working With Informality to Build Resilience in African Cities" (CDKN et al., 2013). This change from "tackling informality" to "working with informality" denotes a shift in the view of the project initiators and implementers to informality not simply as a problem that needs to be addressed but as a phenomenon that should be worked with (CDKN et al., 2013). To quote popular computer programming parlance, "it (informality) is not a bug, it's a feature."

During the conception stages of the project, the project viewed informality as a problem that needed to be addressed. This is evident in the use of the word *tackle* in the project title and further articulations that postulated that climate change interacts with existing pressures on the urban poor to make these poor people more vulnerable to climate change impacts and effects (CDKN et al., 2013). As such, the initial goal of the project was to determine "what systematic-level interventions" could "deal with informality and slum urbanization in African cities" (CDKN et al., 2013, p. 17). Again, note the use of the word *deal* used in relation to informality and slum urbanization. We do not highlight the use of such language to argue that informality or slum urbanization are necessarily good things that should remain. However, we do argue that they are contextual characteristics of African cities (many of which have origins in the more formalized and less-slummed cities of the "more developed" world) that need to be, as this project now rightly recognizes, "worked with" to achieve goals of poverty reduction, sustainable livelihoods, and just and equitable urban development. The project at the conception stages reflected this pro-poor focus, as one outcome of the initial goal was to ensure sustainable basic services in the households of the urban poor in Africa (CDKN et al., 2013).

In order to achieve the aforementioned goal, the project had the specific objectives of satisfying the needs of urban development and management through a crosscutting focus (such as informality), and integrating bottom-up programs and project interventions, top-down government initiatives, civil society objectives, and private sector interests (CDKN et al., 2013).

The conveners of the scoping workshop in July 2013 sent out these initial thoughts on the project to the workshop participants in a preworkshop concept note, before the name was changed (CDKN et al., 2013). The participants were from representatives of the three project cities—Accra, Addis Ababa, and Kampala—and the objectives of the workshop were to: (1) learn about the urban climate change–related vulnerabilities in each city particularly as expressed in informal settlements, (2) share ongoing projects and programs that addressed these climate-related challenges, (3) outline possible financing models to implement the project, (4) discuss the institutional reality of scaling the project up and out, and (5) identify areas in which to design inclusive and systemic-level interventions (CDKN et al., 2013, p. 19).

Key concepts from the workshop included possible future scenarios for urban development in Africa within the context of climate change and the fact that one-size-fits-all solutions were definitely not the way forward. The future scenarios challenged preconceived notions about what an ideal African city should look like, i.e., not necessarily a city high on technology and operating a green economy (which only tended to fail to address deeper structural issues contributing to marginality and informality) but a city that was adaptive, connected, accessible, low tech, and localized in terms of its renewal (CDKN Africa, Roux, 2013). Such an ideal city would also be consistently working toward low-carbon emissions with an eye toward equity in access and the ability to pay for low-carbon technologies and/or development (CDKN et al., 2013).

The workshop participants also visited Langrung, an informal settlement with in-situ upgrading, and the Hout Bay Recycling Cooperative, to experience examples of the kind of work the "Working with Informality" project would like to achieve (CDKN et al., 2013). Some workshop participant lessons from the site visits included the integral role of vibrant community-level leadership, making populations in informal settlements visible and legible through mapping technologies, and the need to consider social and ethical issues before and during demolition in the name of slum upgrading (CDKN Africa, 2014). The next steps after the July 2013 workshop were to implement the workshop outcomes in the project cities. After the implementation, a framework of guidelines and best practices would be published from the implementation undertaken in the three project cities (CDKN et al., 2013).

CDKN and ACC published this framework manual in May 2014; the manual intended to provide those responsible for city planning, management, research institutions, and donor agencies the knowledge base and practical action steps to work with informality to build the resilience of African cities to current and expected changes in the prevailing climate (Taylor and Peter, 2014). In working with informality, the framework outlined a stepwise approach to operationalize in situ development in slums and similar settlements and consistently advocated for a participatory approach to planning and reflexive stance throughout the document (Taylor and Peter, 2014). The manual was particularly vocal in the relationship between adaptation and mitigation in addressing informality in African cities. While the document declared that adaptation should be the priority for many African cities, these climate change adaptation initiatives should be selected with an eye toward simultaneously meeting the goals of mitigation (Taylor and Peter, 2014). For example, a waste management strategy of composting instead of the use of landfills reduces methane emissions (mitigation) while providing compost to improve soil fertility affected by higher temperatures or reduced rainfall (adaptation) (Taylor and Peter, 2014). To this end, the manual has "assessing mitigation cobenefits" as one of the pillars in implementing climate-compatible development in African cities (Taylor and Peter, 2014, p. 10). The framework concludes with a call for reviews and evaluations of applications of the approach in other African cities in order to refine and strengthen the framework (Taylor and Peter, 2014). Unfortunately, we were unable to access any final or midterm evaluations of this project, so we are unable to provide an assessment of how the ideals of this project played out on the ground in the case study cities. However, an Internet search of the manual shows its publication on many other platforms other than CDKN's. This prevalence of the manual indicates that the document gained some purchase among the city planning and climate change adaptation and resilience-building circle of practitioners and researchers.

7.4 LESSONS LEARNED

The "Working with Informality" project provides a number of lessons not only for African cities but also for many other cities all around the world who are seeking to integrate climate change mitigation and adaptation strategies. The lessons learned are from the case study itself and not so much from the discussions of the ambiguity surrounding the term *resilience* and the possibility of resilience as a convening concept to combine mitigation and adaptation. However, the project in building resilience by working with informality and promoting adaptation, mitigation, and sustainable pro-poor urban development illustrates the resolution of this ambiguity.

The first lesson this case study provides is the need to identify the characteristics of the context within which climate change resilience is sought to be achieved, and work with such essential characteristics. It is imperative that city/urban planners do not impose their own worldviews or opinions on what a resilient city looks like but rather work with the current state of the city and see how all stakeholders can make that particular context resilient.

The second lesson of this project is the need for a crosscutting involvement of stakeholders. Not only should there be active engagement of the local, subnational, and national governments with local residents but the government should also involve the private sector and civil society in these engagements. Governments here play a key mediating and facilitative role, and the outcomes of just resilience can most likely be achieved if the focus of all stakeholders continually comes to rest on the poorest and most marginal in the urban context.

The third lesson, which relates more directly to integrating mitigation and development strategies, is to go for the low-hanging fruit first while making strides toward longer-term or harder-to-reach goals. This project began with existing development approaches of providing basic services and infrastructure to the urban poor, assessed how these services were being or could be impacted by climate change, and then designed interventions that could tackle the service and climate change needs concurrently. The project also kept the focus of low-carbon emissions and development in view and sought out technologies such as solar energy or low-combustion cooking stoves that would also simultaneously advance the goals of climate change mitigation and adaptation.

7.5 CONCLUSION

In this chapter, we have demonstrated the example of one approach to integrate climate change adaptation and mitigation strategies in the context of urban development, by operationalizing resilience as a convening concept to discuss not only adaptation and mitigation but also poverty reduction and sustainable development. We hope that the approach of CDKN in the "Working with Informality" project and the lessons we can extrapolate from this project will serve as fodder not just for cities in other areas of the world commonly considered as the global South but for other cities around the world that seek to complement mitigation and adaptation strategies in the name of building resilience to climate change.

References

Abrahams, D., 2014. The Back Burner Effect. HURDL Blog. Available at: http://www.hurdl.org/the-back-burner-effect/12/07/2016.

Adger, W.N., Barnett, J., Brown, K., Marshall, N., O'Brien, K., 2013. Cultural dimensions of climate change impacts and adaptation. Nature Climate Change. 3, 112–117.

Bahadur, A., Ibrahim, M., Tanner, T., 2010. The Resilience Renaissance? Unpacking of Resilience for Tackling Climate Change and Disasters. IDS Working Paper.

CDKN (n.d). PROJECT: Working with informality to build resilience in African cities. Project Reference: AAAF-0009. Available at: http://cdkn.org/project/working-with-informality-to-build-climate-resilience-in-african-cities/?loclang=en_gb.

CDKN, South North, African Centre for Cities, 2013. Workshop Report July 2013. Available at: http://cdkn.org/wp-content/uploads/2013/08/Workshop-report_climate-resilience-and-informality-in-African-cities_final.pdf.

CDKN Africa, Roux, J.P., 2013. Opinion: Opportunities in Urban Informality, Development and Climate Resilience in African Cities. Available at: http://cdkn.org/2013/08/opinion-opportunities-in-urban-informality-development-and-climate-resilience-in-african-cities/?loclang=en_gb.

CDKN Africa, 2014. Film: Working with Informality to Build Climate Resilience in African Cities. Available at: http://cdkn.org/2014/06/film-working-with-informality-to-build-climate-resilience-in-african-cities/?loclang=en_gb.

Douglas, I., Alam, K., Maghenda, M., McDonnell, Y., McLean, L., Campbell, J., 2008. Unjust waters: climate change, flooding and the urban poor in Africa. Environment and Urbanization 20 (1), 187–205.

Green, K.P., 2009. Climate Change: The Resilience Option. Energy and Environment Outlook 4 (October). American Enterprise Institute for Public Policy Research.

Hamin, E.M., Gurran, N., 2009. Urban form and climate change: balancing adaptation and mitigation in the US and Australia. Habitat International 33, 238–245.

Hansen, K.T., Vaa, M. (Eds.), 2004. Reconsidering Informality: Perspectives from Urban Africa. Nordiska Afrikainstitutet.

Heintz, J., Valodia, I., 2008. Informality in Africa: A Review. WIEGO Working Paper, 3.

Intergovernmental Panel on Climate Change (IPCC), 2014. Climate change 2014: impacts, adaptation, and vulnerability. Part A: global and sectoral aspects. In: Field, C.B., Barros, V.R., Dokken, D.J., Mach, K.J., Mastrandrea, M.D., Bilir, T.E., Chatterjee, M., Ebi, K.L., Estrada, Y.O., Genova, R.C., Girma, B., Kissel, E.S., Levy, A.N., MacCracken, S., Mastrandrea, P.R., White, L.L. (Eds.), Contribution of Working Group II to the Fifth Assessment Report of the Intergovernmental Panel on Climate Change. Cambridge University Press, Cambridge, United Kingdom and New York, NY, USA.

Intergovernmental Panel on Climate Change (IPCC), 2012. Summary for policymakers. In: Field, C.B., Barros, V., Stocker, T.F., Qin, D., Dokken, D.J., Ebi, K.L., Mastrandrea, M.D., Mach, K.J., Plattner, G.K., Allen, S.K., Tignor, M., Midgley, P.M. (Eds.), Managing the Risks of Extreme Events and Disasters to Advance Climate Change Adaptation. A Special Report of Working Groups I and II of the Intergovernmental Panel on Climate Change. Cambridge University Press, Cambridge, UK, and New York, NY, USA, pp. 1–19.

Kane, S., Shogren, J.F., 2000. Linking adaptation and mitigation in climate change policy. Climatic Change 45, 75–102.

Laukkonen, J., Blanco, P.K., Lenhart, J., Keiner, M., Cavric, B., Kinuthia-Njenga, C., 2009. Combining climate change adaptation and mitigation measures at the local level. Habitat International 33, 287–292.

Leichenko, R., 2011. Climate change and urban resilience. Current Opinion in Environmental Sustainability 3 (3), 164–168.

Lindell, I., 2010. Between exit and voice: informality and the spaces of popular agency. African Studies Quarterly 11 (2&3).

Magadza, C.H.D., 2000. Climate change impacts and human settlements in Africa: prospects for adaptation. Environmental Monitoring and Assessment 61, 193–205.

Mitchell, T., Maxwell, S., 2010. Defining Climate Compatible Development. CDKN Policy Brief.

Myers, G.A., Murray, M.J., 2006. Introduction: situating contemporary cities in Africa. In: Murray, M., Myers, G. (Eds.), Cities in Contemporary Africa. Palgrave Macmillan, US, pp. 1–25.

Nyong, A., Adesina, F., Elasha, B.O., 2007. The value of indigenous knowledge in climate change mitigation and adaptation strategies in the African Sahel. Mitigation and Adaptation Strategies for Global Change 12, 787–797.

REFERENCES

Owusu-Daaku, K.N., Diko, S.K., 2017. The Sea Defense Project in the Ada East District and its Implications for Climate Change Policy Implementation in Ghana's Peri-urban Areas. Annual Reducing Urb Pov Grad Pap Comp, Urban Sustainability Laboratory, Washington DC. Woodrow Wilson International Center for Scholars.

Patz, J.A., Campbell-Lendrum, D., Holloway, T., Foley, J.A., 2005. Impact of regional climate change on human health. Nature 438, 310–317.

Roy, A., 2009. Why India Cannot Plan Its Cities: Informality, Insurgence and the Idiom of Urbanization. Planning Theory. 8 (1), 76–87.

Taylor, A., Peter, C., 2014. Strengthening Climate Resilience in African Cities: A Framework for Working with Informality. African Centre for Cities and Climate and Development Knowledge Network.

Tyler, S., Moench, M., 2012. A framework for urban climate resilience. Climate and Development 4, 311–326.

CHAPTER 8

Addressing Climate Change in China: Policies and Governance

Gørild Heggelund

Fridtjof Nansen Institute, Lysaker, Norway

8.1 INTRODUCTION: CLIMATE CHANGE AND ENERGY IN CHINA

China has in the past decades experienced rapid economic growth that has lifted hundreds of millions of Chinese out of poverty (Heggelund and Nadin, 2017; World Bank, 2016). This economic growth has relied heavily on coal as its main energy source. In 2014, the Chinese energy mix constituted coal (66%), oil (17%), hydroelectricity (8%), natural gas (6%), renewables (2%), and nuclear energy (1%) (BP, 2015). It was reported in 2007 that China had surpassed the United States as the world's largest greenhouse gas (GHG) emitter in absolute terms (PBL, 2007). China's CO_2 emissions accounted for 28% of the global emissions in 2016 (Le Quéré et al., 2018). China's climate policy and efforts are of global relevance. In the past decades addressing climate change has evolved from viewing it as a largely scientific concern and a foreign policy issue primarily discussed in international negotiations to one that is now regarded as fundamental to the nation's socioeconomic development (Heggelund and Nadin, 2017). Moreover, due to the reliance on coal, climate change has traditionally been seen through the prism of energy and economic development (Heggelund et al., 2010; Heggelund et al., 2018b). Energy and climate change eventually became targets in China's Five Year Plans (FYPs). With growing awareness of climate risks, mitigation and adaptation are now equally important as illustrated in the 13th FYP (2016–20) (13th Five Year Plan, 2016). Besides, air pollution in cities and elsewhere has in recent years become one of the key drivers for climate action in China and one of the strongest incentives for reform of the country's energy sector.

The purpose of this chapter is to give a brief overview of China's climate policies, starting with China's climate governance, followed by climate strategies and policies, cities and climate change, and concluding remarks.

8.2 CLIMATE GOVERNANCE

In the past decade, China's climate policy and governance have evolved rapidly; central policy documents that include climate objectives are the Five-Year Plans[1] and other climate strategies, such as the National Climate Programme (2007). The National Development and Reform Commission (NDRC), responsible for economic and social development policy, coordinates climate change work in China and heads the country's delegations to the international climate conferences[2]. Developing and implementing policy requires cooperation and understanding

---

[1] The Five-Year-Planning (FYP) process is crucial in China's national economic planning, as it lays out the framework and even legislation for socioeconomic development over 5 years. Given the political significance of FYP, the mainstreaming of climate change mitigation and adaptation into the national economic development framework has demonstrated strong political will, as well as integration at the highest possible level. NDRC is the commission in charge of coordinating, compiling, and soliciting input from the line ministries for the Five Year Plans.

[2] The 13th National Peoples' Congress (NPC) in March 2018 announced significant structural changes to the State Council ministries, including moving the climate change portfolio from NDRC to the Ministry of Ecology and Environment (MEE).

between the many ministries. Consequently, to address these challenges and to coordinate national climate work, the National Climate Change Leading Group, chaired by the premier, was established in 2007 during the 11th FYP (2006–10). It illustrated the need for deeper understanding among policy makers for climate change, and demonstrated that the climate issue was on the agenda of the highest levels of the Chinese government (He, 2014). Members are representatives from ministries and commissions with relevant responsibilities to carry out climate policies. Also, in 2007, China decided to elevate the Climate Office within the National Development Reform Commission into a Climate Change Department, thereby giving more clout to the climate change officials in the government.

China's international engagement through the United Nations Framework Convention on Climate Change (UNFCCC) as well as multilateral and bilateral collaboration have been important for domestic climate policy making. China submitted its Intended National Determined Contribution (INDC) to the UNFCCC before COP21 in Paris 2015 (NDRC, 2015). China's pledges to curb growth in its CO_2 emissions by around 2030 or earlier[3] as well as enhance risk and resilience work, such as strengthening assessment and risk management of climate change, improve national monitoring, early warning, and communication systems on climate change (NDRC, 2015). President Xi Jinping during his state visit to the United States in 2015 committed to the climate conference in Paris in 2015 (The White House, 2015a). China ratified the Paris Agreement in 2016, which indicated growing ambitions and leadership role in addressing global climate change.

8.3 STRATEGIES AND INITIATIVES IN CLIMATE CHANGE

China has mainstreamed climate change into the central economic development framework such as the FYPs in response to a number of core drivers such as energy consumption, energy security, and air pollution. Increasing awareness of the need for mitigation, climate change was introduced in the 10th FYP in 2001 (2001–05). Energy savings and energy consumption of the country were key targets of the 10th FYP (Andrews-Speed, 2012, p. 16, Li et al., 2011 p. 294). The targets set by the plan were not reached, and consequently the 11th FYP (2006–10) continued to focus on energy efficiency and energy intensity (Heggelund et al., 2010, 2018a). China's coal consumption nearly trebled in the period 2000–13, an energy intensive growth period that links closely with China's growing greenhouse gas emissions (Green and Stern, 2016). Large volumes of energy were required in a period characterized by high levels of investment in heavy industry sectors such as steel and cement production (Green and Stern, 2016). Seeing the impact of energy consumption on China's environment, China's 12th FYP in 2011 (2011–15) for the first time integrated climate change action as one of the key principles for the country's national sustainable development. The National People's Congress approved the 13th FYP (2016–20) in March 2016 that sets binding goals for a 15% reduction in energy intensity and an 18% decline in carbon intensity by 2020. Hence, efforts continue on energy efficiency, energy consumption reduction, and promotion of renewable energy. For instance, the specific FYP for addressing climate change and GHG emissions control issued by the State Council (2016) established a cap on coal consumption of 4.2 billion tons in 2020, and the share of coal in energy consumption to 58% in 2020 as opposed to the current 64% (BP, 2016).

China's efforts to address and control increasing emissions from fossil fuels and to address air pollution have been to set energy efficiency/intensity goals, and carbon intensity goals in its last two FYPs, yet these were hard to reach by command-and-control methods. In line with deepening of market reform and economic restructuring as decided at the 18th Communist Party Congress in November 2012 (China.org., 2012), China has therefore turned to more market-based approaches such as the carbon market to curb emissions. The carbon market was launched in December 2017 and is set to play an important role in addressing the country's emissions (NDRC 2017). Moreover, energy efficiency remains one of China's top energy and climate change policy agendas (He, 2014; State Council, 2016) and increasing energy efficiency reduces GHG emissions and air pollution, reducing overall energy consumption, which also improves China's energy security (Green and Stern, 2015).

From largely focusing on mitigation and energy-related issues, climate policy has evolved in the past decade to a growing awareness of resilience and vulnerability to climate change. For instance, adaptation was included alongside mitigation and energy efficiency in the National Action Plan on Climate Change in 2007 (Heggelund and Nadin, 2017; NDRC, 2007). In particular, the plan emphasized the importance of addressing both adaptation

---

[3] The INDC goals are to lower CO_2 emissions per unit of GDP by 60%–65% from the 2005 level; to increase share of nonfossil energy up to 20% by 2030; to increase its forest stock volume by around 4.5 billion cubic meters from the 2005 level; and to control coal consumption by setting a cap on coal use.

and climate change due to its impact on social and economic sectors (NDRC, 2007). China launched the National Adaptation Strategy in November 2013, which demonstrates a more targeted and strategic approach to managing climate risk, and emphasizes priority work areas outlined in the 2007 National Program (Heggelund and Nadin, 2017). A follow-up to the strategy is the Action Plan for Climate Change Adaptation in Cities (NDRC & MOHURD, 2016). Adaptation and resilience receive growing attention reflected in research and reports such as the Third National Assessment Report launched in 2015. The report illustrates the severity of the climate impact for China, compiles the latest science and policy options from state-appointed experts exemplified by rising sea levels, as well as shifting rainfall and snow patterns (Buckley, 2015; The Third Assessment, 2015). This is also reflected in policy implementation; for instance, China's ministries and provinces are required to make National Sector Adaptation Plans, Urban Adaptation Plans, and Provincial Adaptation Plans (Heggelund and Nadin, 2017).

8.4 MITIGATION AND ADAPTATION IN CITIES

More than half the Chinese population now lives in cities, and the government aims to increase the urbanization rate to 60% by 2020; by 2030 more than 1 billion Chinese are expected to live in cities (Pan et al., 2013). Cities will play a key role in addressing climate change. Rapid urbanization and industrial development have created environment pressures and pollution, as well as vulnerabilities and increased exposure to risks of natural disasters. Recent reports show that China's coastal cities are among the most vulnerable in terms of socioeconomic developments and assets (Doig and Ware, 2016). China's challenges are numerous as it needs to address both the risks and opportunities presented by climate change while at the same time addressing long-term development needs. Especially, it needs to address in more sustainable ways the disparity between rural and urban vulnerable populations (Nadin et al., 2015).

Low-carbon development is one policy approach in urban areas. China has followed a pattern to first test a policy in "pilots" before introducing it as a national strategy. A first batch of low-carbon pilots were introduced in cities and provinces in 2010, and a second batch in 2012, and include 36 cities and 6 provinces that were introduced (Fig. 8.1; Hu et al., 2015). The pilot cities' low-carbon plans have set targets as well as key achievements and specific measures for reducing CO_2 emissions, industrial structure adjustment, energy structure optimization (i.e., structural change), energy efficiency improvement, and the increase of carbon sinks (Heggelund et al., 2018b). After 5 years of experience, among these pilots, nine cities formed the Alliance of Peaking Pioneering Cities in 2015 by committing carbon emissions peak year goals, such as Beijing in 2020 (The White House, 2015b). In 2015 and 2016, more cities joined and signed the China-U.S. Climate Leaders Declaration in which they committed to establishing ambitious targets, reporting GHG inventories, establishing climate action plans, and enhancing bilateral partnership and cooperation (LBNL, 2016). Moreover, China decided in 2011 to gradually establish a carbon Emissions Trading Scheme (ETS) in its 12th FYP period (2011–15) (NDRC, 2011). Five cities (Beijing, Chongqing, Shanghai, Shenzhen, and Tianjin), as well as two provinces (Guangdong and Hubei), would initiate carbon-trading pilots (NDRC, 2011; Heggelund et al., 2018a). The decision was a move to establish a GHG emissions control target for 2020 through market mechanisms with lower costs and accelerated transformation of economic development patterns and industrial upgrading. The national carbon market is scheduled to begin operation in 2017 as announced by President Xi Jinping during his 2015 US visit (The White House, 2015a).

8.5 CONCLUDING REMARKS

Climate change has moved to the top of China's political agenda. The country's climate policies have gradually strengthened in the FYPs, in particular since 2007, and strengthened nationally binding targets for energy consumption per unit gross domestic product (GDP) (energy intensity), carbon emissions per unit GDP (carbon intensity), as well as increased nonfossil energy share. China also has demonstrated growing ambition and leadership in addressing global climate change through global negotiations and international collaboration. The country's vulnerability to climatic changes and subsequent social impacts and economic losses have brought attention to resilience, and the 13th FYP (2016–20) places adaptation alongside mitigation in importance. Cities are central in addressing climate change, and the low-carbon pilots have played an important role in promoting low-carbon strategies. The year 2017 was a milestone year for China as the national ETS began operation. The highest political leadership has stated support for the ETS and expectations for its success are high (Fig. 8.2).

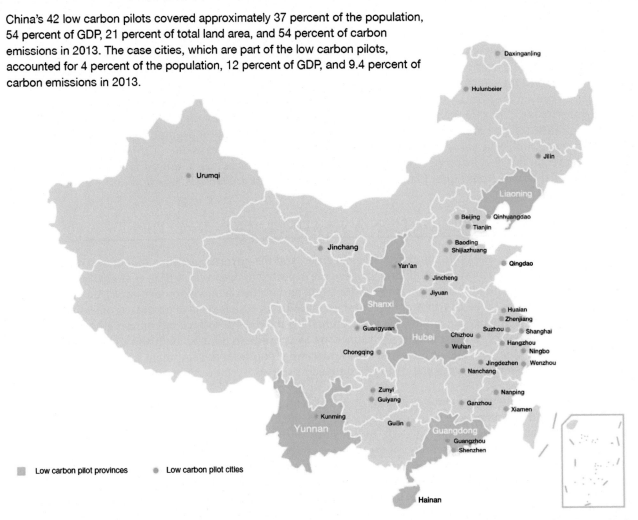

FIGURE 8.1 China's low carbon pilots. *Courtesy: Hu, M., Li, Y., Ang, L., Shuang, L., Lingyan, C., 2015. Energy Foundation China, Innovative Green Development Program, Low Carbon Cities in China: National Policies and City Action Factsheets. Available at: http://www.efchina.org/Attachments/Report/report-cemp20151020/iGDP_CityPolicyFactsheet_EN.pdf.*

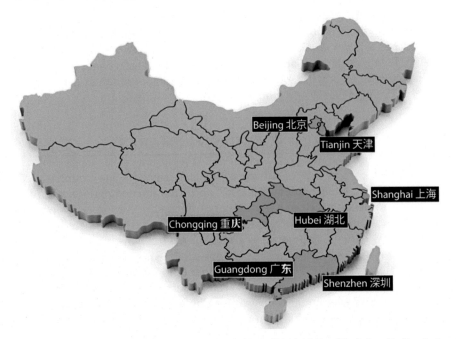

FIGURE 8.2 China's seven Emissions Trading Scheme pilots. *Swartz, J., 2016. China's National Emissions Trading System: Implications for Carbon Markets and Trade; ICTSD Global Platform on Climate Change, Trade and Sustainable Energy; Climate Change Architecture Series, Issue Paper, 6; International Centre for Trade and Sustainable Development, Geneva, Switzerland. Available at: http://www.ieta.org/resources/China/Chinas_National_ETS_Implications_for_Carbon_Markets_and_Trade_ICTSD_March2016_Jeff_Swartz.pdf.*

References

13th Five-year plan, 2016. The People's Republic of China's 13th Five-year Plan of the National Economic and Social Development). Available at: http://www.sdpc.gov.cn/fzggzz/fzgh/ghwb/gjjh/201605/P020160516532684519514.pdf.

Andrews-Speed, P., 2012. The Governance of Energy in China. Transition to a Low-carbon Economy. Palgrave Macmillan, London.

BP, 2015. BP Statistical Review of World Energy. British Petroleum Co, London.

BP Statistical Review, 2016. China's Energy Market in 2015. Available at: http://www.bp.com/content/dam/bp/pdf/energy-economics/statistical-review-2016/bp-statistical-review-of-world-energy-2016-china-insights.pdf.

Buckley, C., November 29, 2015. Chinese report on climate change depicts somber scenarios. The New York Times.

China.org.cn, 2012. Amendments Reflect CPC's Resolve. Available at: http://china.org.cn/china/18th_cpc_congress/2012-11/15/content_27118842.htm.

Doig, A., Ware, J., 2016. Act Now or Pay Later: Protecting a Billion People in Climate-threatened Coastal Cities, Christian Aid, May. Available at: http://www.christianaid.org.uk/Images/act-now-pay-later-climate-report-may-2016.pdf.

Green, F., Stern, N., 2015. China's "new Normal": Structural Change, Better Growth, and Peak Emissions. Policy Report, Centre for Climate Change Economics and Policy (CCCEP). University of Leeds.

Green, F., Stern, N., 2016. China's changing economy: implications for its carbon dioxide emissions. Climate Policy 1–15.

He, G., 2014. Engaging emerging countries: implications of China's major shifts in climate policy. In: Putra, A., Han, E. (Eds.), Governments' Responses to Climate Change: Selected Examples From Asia Pacific. Springer-Verlag Singapur, pp. 11–24.

Heggelund, G., Nadin, R., 2017. Climate change policy and governance. In: Sternfeld, E. (Ed.), Routledge Handbook of Environmental Policy in China. Routledge, London and New York, pp. 97–112.

Heggelund, G., Andresen, S., Fritzen, B.I., 2010. Chinese climate policy: domestic priorities, foreign policy and emerging implementation. In: Harrison, K., McIntosh Sundstrom, L. (Eds.), Global Commons, Domestic Decisions. MIT Press, Cambridge (MA), USA/London, UK, pp. 239–261.

Heggelund, G., Stensdal, I., Maosheng, D., 2018a. China's development of ETS as a GHG mitigating policy tool: a case of diffusion or domestic drivers? Review of Policy Research forthcoming.

Heggelund, G., Price, L., Zhou, N., Min, H., 2018b. Addressing China's Emissions: Low Carbon Cities and the Alliance of Peaking Pioneer Cities forthcoming.

Pan, J., Gomez-Echeverri, L., Heggelund, G., 2013. In: Sustainable and Liveable Cities: Toward Ecological Civilization, UNDP, China National Human Development Report. China Translation & Publishing Corporation, Beijing. Available at: http://hdr.undp.org/sites/default/files/china_nhdr_2013_en_final.pdf.

LBNL (Lawrence Berkeley National Laboratory), 2016. US-China Climate Smart Cities Initiative. Declaration. Available at: https://ccwgsmartcities.lbl.gov/declaration.

Le Quéré, C., et al., 2018. Global Carbon Budget 2017. Earth System Science Data 10, 405–448. https://doi.org/10.5194/essd-10-405-2018.

Li, Y., Yang, X., Zhu, X., Mulvihill, P.R., Mathews, D.H., Sun, X., 2011. Integrating climate change factors into China's development policy: adaptation strategies and mitigation to environmental change. Ecological Complexity 8, 294–298.

Hu, M., Li, Y., Ang, L., Shuang, L., Lingyan, C., 2015. Energy Foundation China, Innovative Green Development Program, Low Carbon Cities in China: National Policies and City Action Factsheets. Available at: http://www.efchina.org/Attachments/Report/report-cemp20151020/iGDP_CityPolicyFactsheet_EN.pdf.

Nadin, R., Opitz-Stapleton, S., Yinlong, X., 2015. Climate Risk and Resilience in China. Routledge, Taylor and Francis Group, London.

NDRC (National Development and Reform Commission), 2007. China's National Climate Change Programme. Beijing. Available at: http://www.ccchina.gov.cn/WebSite/CCChina/UpFile/File188.pdf.

NDRC (National Development and Reform Commission), 2011. Notice on Initiating Carbon Trading Pilots in China. Available at: http://www.sdpc.gov.cn/zcfb/zcfbtz/201201/t20120113_456506.html (in Chinese).

NDRC (National Development and Reform Commission), 2015. China's Intended Nationally Determined Contributions Enhanced Actions on Climate Change. Available at: http://www4.unfccc.int/submissions/INDC/Published%20Documents/China/1/China's%20INDC%20-%20on%2030%20June%202015.pdf.

NDRC (National Development and Reform Commission), MOHURD (Ministry of Housing and Urban-Rural Development), 2016. Action Plan for Climate Change Adaptation in Cities. FGQH, 245 (in Chinese). Available at: http://bgt.ndrc.gov.cn/zcfb/201602/t20160216_774739.html.

NDRC, 20 December 2017. Initiation of the national trading system. Chinese national carbon trading scheme. http://www.ccchina.org.cn/Detail.aspx?newsId=70154&TId=93. Last accessed 9 February 2018, in Chinese.

PBL, Netherlands Environmental Assessment Agency, 2007. China Now No. 1 in CO_2 Emissions: USA in Second Position. Available at: http://www.pbl.nl/node/47363.

State Council, 2016. State Council Notice on the Work Plan for Controlling GHG Emissions in the 13th FYP. Available at: http://www.gov.cn/zhengce/content/2016-11/04/content_5128619.htm.

Swartz, J., 2016. China's National Emissions Trading System: Implications for Carbon Markets and Trade; ICTSD Global Platform on Climate Change, Trade and Sustainable Energy; Climate Change Architecture Series, Issue Paper, 6; International Centre for Trade and Sustainable Development, Geneva, Switzerland. Available at: http://www.ieta.org/resources/China/Chinas_National_ETS_Implications_for_Carbon_Markets_and_Trade_ICTSD_March2016_Jeff_Swartz.pdf.

The Third National Assessment of Climate Change Writing Committee, 2015. The Third National Assessment of Climate Change. Science Press (in Chinese).

The White House: Office of the Press Secretary, 2015a. The United States and China Issue Joint Presidential Statement on Climate Change with New Domestic Policy Commitments and a Common Vision for an Ambitious Global Climate Agreement in Paris. Available at: https://www.whitehouse.gov/the-press-office/2015/09/25/fact-sheet-united-states-and-china-issue-joint-presidential-statement.

The White House: Office of the Press Secretary, 2015b. U.S.-China Joint Presidential Statement on Climate Change. Available at: https://www.whitehouse.gov/the-press-office/2015/09/25/us-china-joint-presidential-statement-climate-change.

World Bank, 2016. World Bank Indicators. Available at: http://data.worldbank.org.

CHAPTER 9

Mitigation and Adaptation Strategies in the Face of Climate Change: The Australian Approach

Giuseppe Forino, Jason von Meding, Graham Brewer
University of Newcastle, Callaghan, NSW, Australia

9.1 CLIMATE CHANGE ISSUES IN AUSTRALIA

Several climate-related hazards such as tropical cyclones, floods, bushfires, droughts, and hailstorms occur every year in Australia. According to the Commonwealth Scientific and Industrial Research Organisation (Cleugh et al., 2011) and the Fifth Assessment by the Intergovernmental Panel on Climate Change (Reisinger et al., 2014), climate change has the potential to contribute to increasing the frequency and duration for some of these extreme events in Australia, making it one of the developed countries most vulnerable to climate change—related hazards (Cleugh et al., 2011; Garnaut, 2011; Hobday and McDonald, 2014; Forino et al., 2017). The climatic diversity of the country ranges from tropical monsoonal to arid, temperate, and alpine conditions, with regional differences in relation to impacts by climate change—related hazards on economy, environment, and society (Cleugh et al., 2011; Reisinger et al., 2014). Also, most of the major population centers and about 90% of the population are along the coast and are likely to experience more frequent and intense sea level rise, inundation, floods, coastal erosion, and heatwaves (Cleugh et al., 2011; Garnaut, 2011; Reisinger et al., 2014).

Due to these issues, the federal government of Australia undertook a long commitment in order to tackle climate change, balancing the necessity of reconciling environmental benefits with the achievement of economic goals (Bulkeley, 2001). The Australian government system is a liberal democracy including three levels, that are the federal government, the State/Territory government, and the local governments (Nalau et al., 2015). The federal government allocates responsibilities to the State/Territory governments, and therefore delegates legal mandates and schemes of local governments to the legislation of related State/Territory governments (Measham et al., 2011; Nalau et al., 2015). States and territories, however, also have their own legislatures able to set broad or sector-specific mitigation policies, as well as retain varying degrees of ownership and regulatory oversight (e.g., over the electricity supply sector) (Kember et al., 2013).

Within such a scenario, the Council of Australian Governments (COAG) is the primary intergovernmental platform for the negotiation between state/territory governments and the federal government. For example, the COAG has instituted a process for streamlining mitigation and adaptation strategies based on broad principles of complementarity; however, each government is free to interpret COAG's recommendations, suggesting the process will be difficulty harmonized (Kember et al., 2013). Therefore, dispersion and duplication of responsibilities, jurisdictional disputes, and lack of trust usually occur within and among government levels (Mukheibir et al., 2013; Howes et al., 2015). This leads to the creation of isolated areas for intervention, and inhibits an integrated response to climate change across levels and sectors. In the Australian government system, furthermore, short-term electoral mandates do not coincide with the necessity for long-term decisions about climate change.

Therefore, politics is often reluctant in proactively undertaking initiatives for responding to climate change (Mukheibir et al., 2013; Forino et al., 2017).

Notwithstanding this, different aspects of the federal portfolio including environmental issues, natural resource management, and foreign affairs were mainstreamed into the goal of addressing climate change—related issues. Meanwhile, state/territory governments and local governments, together with social, economic, and environmental stakeholders, started to be interested in these issues (Bulkeley, 2001), and since the late 1980s mitigation strategies were promoted to reduce greenhouse gas (GHG) emissions by reducing the carbon dependence of the national economy. Additionally, in the last decade adaptation strategies started to be considered, particularly for coping with hazards such as sea level rise and floods. Against this background, this chapter provides a brief overview of the main mitigation and adaptation strategies by the Australia federal government.

9.2 SIGNIFICANT MITIGATION AND ADAPTATION STRATEGIES BY THE AUSTRALIAN FEDERAL GOVERNMENT

One of the first strategies[1] enacted by the federal government to cope with climate change issues was the "Interim Planning Target," adopted in 1990 and aimed at reducing GHG emissions to 1988 levels by 1990 and at cutting emissions by a further 20% by the year 2005. These targets were accompanied by the intention of not triggering adverse effects on the Australian economy, and particularly upon trade competitiveness, in the absence of similar actions by other countries (Bulkeley, 2001). In 1992, the launch of the National Greenhouse Response Strategy followed (Commonwealth of Australia, 1992), was a milestone occasion for promoting initial research and assessment of climate change-related vulnerability into regions and economic sector of Australia, as well as for applying related findings into planning practice and environmental management (Howes et al., 2012). In 1998, the National Greenhouse Strategy (Commonwealth of Australia, 1998) proposed a national framework combining mitigation and adaptation to be applied in and assessed for different regions and economic sectors; however, it targeted just some key strategic sectors, missing the opportunity for providing for the first time an integrated perspective to the national climate change response (Howes et al., 2012). In 2005, a National Climate Change Adaptation Programme made available AUD14 million over 4 years, to be linked to the strategy National Climate Change Risk and Vulnerability: Promoting an Efficient Adaptation Response in Australia (Australian Government, 2005). This program identified vulnerable sectors/systems/regions, which required the assessment of climate change—related impacts and the support of their adaptation priorities to increase coping capacities (Forino et al., 2017). In 2007, the milestone National Climate Change Adaptation Framework (COAG, 2007) identified key sectors/regions with potential areas for promoting adaptation. This framework aimed at identifying and implementing adaptive capacities and providing specific strategic direction to address adaptation and reducing vulnerability (Forino et al., 2017). In 2008, the Australian Government created the National Climate Change Adaptation Research Facility (NCCARF) 2008—13 (2014-2017 at the time of writing) to prepare for and manage the risks posed by climate change—related hazards. The NCCARF addresses the adaptation needs of decision makers and practitioners in order to deal with projected impacts such as more frequent and more intense heatwaves, increasing risk of sea level rise and flooding from rivers and the sea, and increasing coastal erosion.[2] Selected themes include terrestrial, marine, and freshwater biodiversity, human health, settlements and infrastructure, emergency management, primary industries, indigenous communities, and social, economic, and institutional dimensions. In 2009, the Australian Government assessed main climate change-related risks in coastal areas (Australian Government, 2009) and identified main areas threatened by climate change and main challenges and priorities for promoting an effective response also through CCA (Howes et al., 2012; Forino et al., 2017). In 2011, the National Strategy for Disaster Resilience (COAG, 2011) recognized climate change as one of the drivers of disaster risk, and proposed the implementation of mitigation measures and the support to adaptive capacities by local communities. On December 2, 2015, the Australian Government released a National Climate Resilience and Adaptation Strategy to address climate-related risks for the benefit of the community, economy, and environment. This strategy combines mitigation (to avoid risks of a changing climate by reducing the emission of GHGs) and adaptation (to manage risks caused by climate change already locked in and from the potential for more severe changes), and looks at strategies across key sectors, including: coasts; cities and the built environment;

---

[1] For a more comprehensive list of climate and climate change strategies in Australia, see Talberg et al. (2015).

[2] https://www.nccarf.edu.au/.

agriculture, forestry, and fisheries; water resources; natural ecosystems; health and well-being; disaster risk management; and, resilient and secure regions (Australian Government, 2015a).

9.3 BEYOND MITIGATION AND ADAPTATION: CLIMATE CHANGE AS A CONTESTED POLITICAL ISSUE

The mitigation and adaptation strategies briefly presented are just some among the most significant promoted by the federal government. The reduction of GHG emissions, the identification of key vulnerable social, economic, and environmental sectors, as well as the recent consideration of climate change within a resilience framework, represent promising strategies that ensure the long-term commitment of Australia in tackling climate change-related issues. Nevertheless, reflections are necessary about the contradictions into the Australian society, as on the one side it urges mitigation and adaptation, while on the other side it still promotes an economic growth driven by fossil fuel extraction and GHG emissions production, therefore largely contributing to global climate change. For example, in 2005, Australia ranked 15th in the global production of GHG emissions, as well as in 2010–11 its per capita emissions were the highest among countries of the Organisation for Economic Co-operation and Development (OECD) (Head et al., 2014). In 2013–14, oil represented the largest share of national energy consumption (38%), while coal remained the second largest primary consumed fuel (black and brown coal accounted for 32% of total energy consumption) (Department of Industry and Science, 2015). Conversely, renewable energy sources accounted just for 6% of total energy consumption in 2013–14 (Department of Industry and Science, 2015).

Australia is also one of the major net exporters of energy by fossil fuels, and the global largest exporter of coal. In 2007–08, Australia exported more than two-thirds of the geologically stored energy that it extracted (Hobday and McDonald, 2014; Forino et al., 2017). In 2013–14, black coal exports increased by 12% (Department of Industry and Science, 2015) with global demand stimulating investment capacity. The Australian government is also approving further development for coal and coal seam gas mining activities, particularly in areas devoted to agriculture or presenting high environmental values, therefore exacerbating conflicts between different forms of local economies and leading to further stress to ecosystems and biodiversity (Hobday and McDonald, 2014).

Therefore, this contradiction makes mitigation and adaptation in Australia as highly volatile and polarized issues that have undergone several reversals (Forino et al., 2014, 2017), based on the positions about climate change expressed by the two major political parties, the Australian Labor Party and the Liberal Party of Australia (Head et al., 2014). Across electoral mandates of these major political parties, indeed, significant and tangible advancements in mitigation and adaptation strategies alternated with regressions or mere political statements by both parties' governments (Talberg et al., 2015). An enlightening example is that of the "Carbon Tax," a mitigation strategy under the form of a carbon-pricing scheme which was effective from July 1, 2012. This scheme was part of the Clean Energy Plan reform, which aimed at reducing GHG emissions by 5% below 2000 levels by 2020, and 80% below 2000 levels by 2050. Such a scheme required Australia's largest polluters to buy permits for CO_2 emissions. The price for such permits was initially fixed for 3 years, after which it would have been regulated by the market (Head et al., 2014; Forino et al., 2017). The scheme was strongly opposed by the Liberals supported by oil and energy corporations. With the shift in the federal government from Labor to Liberal in 2013, the Liberal Prime Minister Tony Abbott prioritized the repeal of the Carbon Tax into the government agenda. Abbott initially dismantled the Department of Climate Change, and incorporated climate change issues and related mitigation and adaptation strategies within the Environment Department. Eventually, on July 17, 2014, the Carbon Tax was repealed (Forino et al., 2017). Likewise, in 2016 the federal government (under the ad interim Liberal Prime Minister Malcolm Turnbull) did pressures on the UNESCO to remove emblematic Australian cultural heritage sites (such as the Great Barrier Reef and the Kakadu and Tasmanian forests) from those considered as severely threatened by climate change within the report *World Heritage and Tourism in a Changing Climate* (Markham et al., 2016), claiming that these mentions could compromise the national tourism sector (Forino et al., 2016; Slezak, 2016).

In July 2016, Malcolm Turnbull was elected prime minister. One of the main climate change–related challenges for his electoral mandate and the whole of Australia in the near future will be that of meeting the mitigation economy-wide targets defined by the Paris Agreement following the Conference of Parties by the United Nations Framework Convention on Climate Change (UNFCCC). These targets aim at reducing GHG emissions by 26%–28% below 2005 levels by 2030, in order to contribute to the global UNFCCC's objective of limiting global average temperature rise to below 2°C (Australian Government, 2015b).

9.4 CONCLUSIONS: WHICH APPROACH FOR MITIGATION AND ADAPTATION IN AUSTRALIA?

In Australia, the combination of mitigation and adaptation strategies by the federal government moves toward the achievement of ambitious targets such as those claimed under the Paris agreement and the ultimate provision of benefits for the well-being of local communities. Such combination, therefore, has been and will continue to be pivotal in strengthening coping capacities of the Australian society at all the government levels and for all stakeholders. Mitigation strategies such as reducing GHG emissions by e.g., the transition toward a low-carbon economy, the increasing use of renewable energy, and the reactivation of a carbon-pricing scheme may provide effective results in the short term. Similarly, strategies for adapting to, e.g., floods and sea level rise risks through structural measures, building design, or land-use planning able to minimize risks, become necessary particularly for medium-large urban areas. Such strategies would also require strengthening the collaboration and the mutual exchange among different levels of government (federal, states and territories, and local governments) as well as the involvement of local communities and of the broad range of economic actors, particularly those who mostly contribute to GHG emissions.

Nevertheless, it is worthwhile mentioning that a leading role by the federal government (Measham et al., 2011; Head et al., 2014; Nalau et al., 2015) in driving and supporting the contributions by multi-level governments, local communities, and market actors is imperative to achieve mitigation and adaptation goals and targets. Such leadership is also required for promoting an approach able to understand the contribution of the current Australian economic paradigm and its related productive system to climate change (Forino et al., 2017). This should ultimately lead to reflect about new (and certainly unpopular) socioeconomic trajectories of production and consumption, able to move immediately and quickly away from the use of fossil fuels and to represent environmentally viable and sustainable long-term solutions for ensuring well-being to individuals and communities, particularly to the most disadvantaged ones (e.g., the Aboriginal and Torres Strait Islanders, see Reisinger et al., 2014). If such leadership persists to be missed, mitigation and adaptation strategies such as those described herein will just provide limited and impromptu results, which will not address the long-term challenges posed by the current production system on places and communities across Australia.

Acknowledgments

Giuseppe Forino is supported by a Ph.D. scholarship from the University of Newcastle.

References

Australian Government, 2005. Climate Change Risk and Vulnerability. Promoting an Efficient Adaptation Response in Australia. Canberra.
Australian Government, 2009. Climate Change Risks to Australia's Coast. A First Pass National Assessment. Department of Climate Change, Canberra.
Australian Government, 2015a. National Climate Resilience and Adaptation Strategy 2015. Canberra.
Australian Government, August 2015b. Australia's Intended Nationally Determined Contribution to a New Climate Change Agreement.
Bulkeley, H., 2001. No regrets?: economy and environment in Australia's domestic climate change policy process. Global Environmental Change 11 (2), 155–169.
Cleugh, H., Smith, M.S., Battaglia, M., Graham, P., 2011. Climate Change: Science and Solutions for Australia. CSIRO.
COAG, 2007. National Climate Change Adaptation Framework. Canberra.
COAG, 2011. National Strategy for Disaster Resilience. Building the Resilience of Our Nation to Disasters. Canberra.
Commonwealth of Australia, 1992. National Greenhouse Response Strategy. Department of the Arts, Sport, Environment, Tourism and Territories, Canberra.
Commonwealth of Australia, 1998. National Greenhouse Strategy. Canberra.
Department of Industry and Science, 2015. Australian Energy Update. Canberra.
Forino, G., von Meding, J., Brewer, G., Gajendran, T., 2014. Disaster risk reduction and climate change adaptation policy in Australia. Procedia Economics and Finance 18, 473–482.
Forino, G., MacKee, J., von Meding, J., 2016. A proposed assessment index for climate change-related risk for cultural heritage protection in Newcastle (Australia). International Journal of Disaster Risk Reduction 19, 235–248.
Forino, G., von Meding, J., Brewer, G., 2017. Climate change adaptation and disaster risk reduction integration. In: Madu, C.N., Kuei, C. (Eds.), Australia: Challenges and Opportunities. Handbook of Disaster Risk Reduction & Management. World Scientific Press & Imperial College Press, London (Chapter 29).
Garnaut, R., 2011. The Garnaut Review 2011: Australia in the Global Response to Climate Change. Cambridge University Press.
Head, L., Adams, M., McGregor, H.V., Toole, S., 2014. Climate change and Australia. Wiley Interdisciplinary Reviews: Climate Change 5 (2), 175–197.
Hobday, A.J., McDonald, J., 2014. Environmental issues in Australia. Annual Review of Environment and Resources 39, 1–28.

Howes, M., Grant-Smith, D., Reis, K., Bosomworth, K., Tangney, P., Heazle, M., McEvoy, D., Burton, P., 2012. The Challenge of Integrating Climate Change Adaptation and Disaster Risk Management: Lessons From Bushfire and Flood Inquiries in an Australian Context.

Howes, M., Tangney, P., Reis, K., Grant-Smith, D., Heazle, M., Bosomworth, K., Burton, P., 2015. Towards networked governance: improving interagency communication and collaboration for disaster risk management and climate change adaptation in Australia. Journal of Environmental Planning and Management 58 (5), 757–776.

Kember, O., Jackson, E., Chandra, M., 2013. GHG Mitigation in Australia: An Overview of the Current Policy Landscape. Working Paper. World Resources Institute, Washington, DC. Available at: http://pdf.wri.org/ghg_mitigation_in_australia_overview_of_current_policy_landscape.pdf.

Markham, A., Osipova, E., Lafrenz, S.K., Caldas, A., 2016. World Heritage and Tourism in a Changing Climate. UNESCO Publishing.

Measham, T.G., Preston, B.L., Smith, T.F., Brooke, C., Gorddard, R., Withycombe, G., Morrison, C., 2011. Adapting to climate change through local municipal planning: barriers and challenges. Mitigation and Adaptation Strategies for Global Change 16 (8), 889–909.

Mukheibir, P., Kuruppu, N., Gero, A., Herriman, J., 2013. Overcoming cross-scale challenges to climate change adaptation for local government: a focus on Australia. Climatic Change 121 (2), 271–283.

Nalau, J., Preston, B.L., Maloney, M., 2015. Is adaptation a local responsibility? Environmental Science & Policy 48, 89–98.

Reisinger, A., Kitching, R., Chiew, F., Hughes, L., Newton, P., Schuster, S., Tait, A., Whetton, P., 2014. Chapter 25: Australasia. Intergovernmental Panel on Climate Change Fifth Assessment Report, Working Group II, Impacts, Adaptation & Vulnerability. IPCC, Geneva.

Slezak, M., 2016. Australia Scrubbed From UN Climate Change Report After Government Intervention. Available at: https://www.theguardian.com/environment/2016/may/27/australia-scrubbed-from-un-climate-change-report-after-government-intervention.

Talberg, A., Hui, S., Loynes, K., 2015. Australian Climate Change Policy: A Chronology. R. P. S. Parliament Library. Department of Parliamentary Services, Canberra.

SECTION III

CITIES DEALING WITH CLIMATE CHANGE: INSTITUTIONAL PRACTICES

CHAPTER 10

European Cities Addressing Climate Change

Giada Limongi
Engineer, Naples, Italy

10.1 THE PIVOTAL ROLE OF CITIES IN COUNTERBALANCING CLIMATE CHANGE

Measures to counterbalance climate change can be brought back to two different lines of action, which complement each other: mitigation measures that address the causes of climate change and include actions to reduce greenhouse gas (GHG) emissions; and adaptation measures that aim to reduce its negative impacts and comprise actions for strengthening systems' capacities to face heterogeneous climate-related hazards.

Cities are widely considered as key players in the challenge against climate change; they are mainly responsible for GHG emissions and meanwhile a vulnerable target of climate impacts, hosting the majority of population, strategic activities, and infrastructures.

Hence, local authorities have been more and more widely recognized as crucial actors for carrying out both mitigation and adaptation strategies. In fact, they may take actions in key sectors to counterbalance climate change, such as urban planning, civil protection, risk management, energy, transport, and construction (Luise, 2014). Moreover, even though the issue of climate change has to be addressed on different geographical scales, local authorities are required to translate global objectives and strategies into specific measures, tailored to the peculiarities of local contexts.

On a global scale, the first step toward an effective engagement of cities in preventing climate change was the establishment, in 2005, of the C40 initiative, a global network of large cities, promoted by the mayor of London and mainly addressed to reduce GHG emissions worldwide. From the very beginning, this initiative has had great success, expanding the number of participating cities from 18 to 40 in 1 year and progressively enlarging its objectives. Currently, about 90 cities support this initiative, including 19 European cities. Even though the C40 network has involved a limited number of large cities, it has contributed to create a wider community, a global city network sharing information, best practices, and innovative tools, enabling in so doing the transfer of innovative models and practices from participant cities to other cities all over the world. Moreover, the success of the C40 network has significantly fueled the awareness that cities have the responsibility, as well as the capacity, to create solutions to climate change, by "acting both locally and collaboratively" (C40 Cities, 2015).

Therefore, in 2008, the European Union (EU) started to bet on cities to meet the challenge of climate change. Hence, after the adoption of Climate and Energy Package 2020, it launched the Covenant of Mayors aimed to involve cities in developing action plans to reduce GHG emissions, also based on the best practices successfully carried out by large cities, both within and outside Europe.

However, it is worth underlining that, while the C40 initiative was started by cities themselves, the Covenant of Mayors, despite having significantly contributed to widen the number of cities actively engaged in mitigation policies, including numerous small and medium-sized European cities, has been marked by a strong institutional leadership. The EU has in fact significantly guided cities to meet their commitments, by setting the specific targets to be achieved, supporting the exchange of information and good practices, and providing detailed guidelines to carry out effective sustainable energy action plans (SEAPs) (Davide et al., 2013). As we detail in the following discussion, the Covenant of Mayors responded enthusiastically, and Europe is now considered as one of the world leaders in tackling climate change and limiting global warming.

Nevertheless, as clearly noted by numerous scholars, addressing climate change involves a twofold challenge, mitigation and then adaptation, which means acting on the effects by reducing climate impacts. As highlighted by Kress (2007), indeed, "even if we were to stop all greenhouse gas emissions today, we would still feel the impacts of climate change for decades to come."

Also in respect to adaptation issues, Europe has followed, both in timing and in content, previous initiatives implemented on a global scale.

It is worth emphasizing that, although cities play a crucial role in the field of adaptation, requiring the latter tailored to the site measures to reduce vulnerabilities to the heterogeneous climate impacts, they have begun to tackle the challenge of adaptation thanks to the strong commitment of large international organizations. In May 2010, the Office of the United Nations for Disaster Risk Reduction (UNISDR) announced the Making Cities Resilient campaign to support sustainable urban development by promoting cities' resilience and increasing the understanding of disaster risk, including climate-related risks, on a local level (UNISDR, 2015). However, by recognizing that adaptation strategies have to be developed according to the peculiarities of local contexts, rather than providing cities with specific targets to be achieved and detailed guidelines, UNISDR pursued the objective to encourage the exchange of information and best practices among the involved local authorities.

In 2013, the Rockefeller Foundation launched another global initiative, the 100 Resilient Cities (100RC), aimed at financially supporting cities to adopt development strategies capable of increasing urban resilience in the face of heterogeneous challenges, including climate impacts. From December 2013 to May 2016, 100RC expanded from 32 to 100 involved cities.

Following these global initiatives, and mainly due to the increasing impacts of climate-related hazards on European countries, in April 2013, the EU adopted the European Strategy on Adaptation to Climate Change and, in 2014, promoted a new initiative, the Mayors Adapt, similar in its structure to the Covenant of Mayors and aimed to support local authorities in developing comprehensive local adaptation strategies or integrating these into relevant existing plans.

However, again in 2014, the UN Secretariat and the largest world's city networks (C40, ICLEI, etc.) launched a new global initiative, the Compact of Mayors, which marked a further step forward in the fight against climate change (Fig. 10.1).

The new global coalition of mayors and city officials was committed, indeed, to reduce local greenhouse gas emissions, enhancing meanwhile cities' resilience to climate change.

The change of pace started on a global scale and directed to address both mitigation and adaptation, and was quickly taken up by the EU such that, on October 2015, it was decided to merge the two main European initiatives, the Covenant of Mayors and the Mayors Adapt, into the new Covenant of Mayors for Climate and Energy, which adopted the EU 2030 objectives and an integrated approach to mitigation and adaptation issues.

Once more, following the path traced by the world's large city networks, the EU provides a consistent framework capable of involving a large number of European cities, and above all the medium and small cities that represent the majority of the European cities, in developing an integrated approach to mitigation and adaptation issues, by contributing in doing so to the implementation of the EU 40% GHG-reduction target by 2030.

FIGURE 10.1 The timeline of the global and European initiatives to counterbalance climate change. *Author's elaboration.*

Nevertheless, in the face of a global challenge, as climate change is, a new initiative has been started on a global level aimed to merge the two main world initiatives addressing climate change: on June 2016, the Compact of Mayors and the Covenant of Mayors announced the new Global Covenant of Mayors for Climate and Energy, which was officially started in January 2017. The Global Covenant represents the largest global coalition of cities committed to tackle, based on an integrated perspective, three key issues: climate change mitigation; adaptation to its adverse effects; and access to secure, clean, and affordable energy.

Such a global initiative, although at a very early stage, seems to be a significant turning point to strengthen the role of cities, driving their action beyond the boundaries of national or regional targets and providing local authorities with a stronger voice in international climate policy and action.

10.2 EUROPEAN CITIES ADDRESSING MITIGATION ISSUES: THE COVENANT OF MAYORS INITIATIVE

In 2008, after the adoption of the EU 2020 Climate and Energy Package, the European Commission launched the Covenant of Mayors, aimed to promote, support, and coordinate local authorities in addressing the ambitious European goals in terms of climate mitigation. The Covenant's signatories committed to implement the European mitigation strategies, addressing the goals established by the EU (20% reduction of the GHG emissions; 20% of energy from renewables; 20% increase in energy efficiency), through the implementation of the SEAPs.

At the end of 2017, the number of cities[1] that had joined the Covenant of Mayors was about 5900; the number of signatory cities largely varies among the different countries (Fig. 10.2), ranging from the very few participating cities

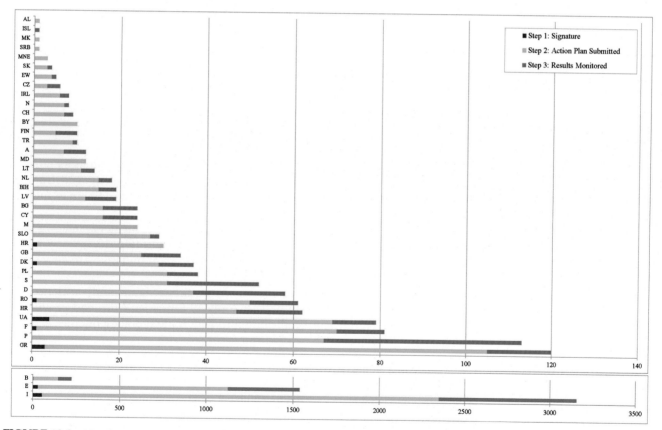

FIGURE 10.2 Number of Covenant signatories for each state. *Author's elaboration based on data provided by the Covenant of Mayors website (http://www.covenantofmayors.eu).*

---

[1] Regions, provinces, aggregation of municipalities and municipalities could join Covenant of Mayors; data and information reported in this chapter refer only to aggregations of municipalities and municipalities.

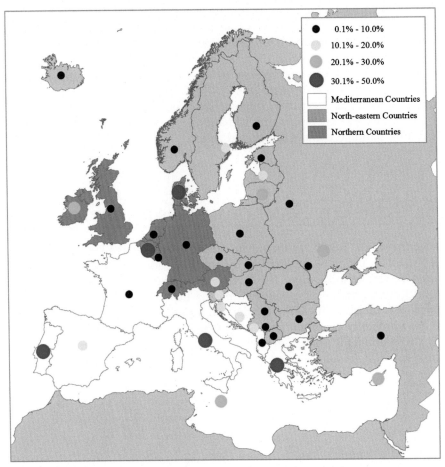

FIGURE 10.3 Covenant of Mayors: signatory cities for each member state (% on the total number of cities). *Author's elaboration based on data provided by the Covenant of Mayors website (http://www.covenantofmayors.eu).*

recorded, for example, in Albania, Island, Serbia, and Macedonia, up to the very high number of signatory cities in Italy (more than 3000). In detail, it has be noticed that there is a very high number of cities engaged from all Mediterranean countries (Italy, Spain, Portugal, Greece); the number of signatory cities in Italy and Spain is, indeed, significantly higher than the average of all the other European countries.

By weighting the number of signatory cities based on the overall number of cities in each member state, we have to remark that also Denmark and Belgium show a significant percentage of involved cities (Fig. 10.3).

It is also noticeable that over 50% of the European cities that have joined the Covenant initiative are very small towns, with populations less than 10,000 inhabitants (Fig. 10.4).

Thus, the Covenant initiative has achieved a twofold result: on the one hand, it has significantly widened the overall number of European cities actively involved in mitigation policies (only 19 European cities had previously joined the network C40); on the other hand, it has significantly contributed to widen the typology of cities actively

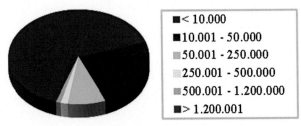

FIGURE 10.4 Classification of the cities that joined the Covenant of Mayors by population size. *Author's elaboration based on data provided by the Covenant of Mayors website (http://www.covenantofmayors.eu).*

involved in mitigation policies, by involving a large number of medium-sized and small cities, which represent the majority of the European cities.

Finally, it is also interesting to notice that, according to the available data, only 98 cities out of 5900 (slightly more than 1.5%), despite having joined the initiative, did not carry out the SEAP, whereas more than 1500 cities (about the 25%) have also started the monitoring phase, after the implementation of the plan (Fig. 10.2).

These data seem to confirm the effectiveness of this initiative that, as mentioned above, has allowed Europe to achieve and improve the 2020 goals, establishing itself as one of the world leaders in tackling climate mitigation.

10.3 EUROPEAN CITIES ADDRESSING ADAPTATION ISSUES: THE MAYORS ADAPT INITIATIVE

The Mayors Adapt initiative, similarly, to the Covenant of Mayors, was addressed to support local authorities in carrying out comprehensive local adaptation strategies or integrating adaptation issues into relevant existing plans, increasing in doing so cities' resilience in the face of climate impacts.

The initiative was launched in April 2014; it has been joined mostly by cities that had already joined the Covenant of Mayors (125) along with another 23 cities[2], and has involved, above all, coastal cities in Italy, Spain, Portugal, and Greece. Also in this case, as in the Covenant initiative, most of the signatory cities—approximately the 80%—are small- and medium-sized cities (Fig. 10.5), being very often large cities already engaged in global adaptation initiatives.

It is worth noting that the significantly lower involvement of European cities in the Mayors Adapt initiative could be attributed, not to lower attention of European cities to adaptation issues but to the relatively short life of this initiative (2 years) as well as to the higher difficulty in developing adaptation plans compared to the SEAPs.

Most of the signatory cities, after joining the initiative, are indeed still at an early stage of the adaptation process—the preparation phase (Fig. 10.6). In detail, the response of cities to the Mayors Adapt initiative can be described with reference to three wide geographical areas:

- Northeastern Europe, where the participation level is very low and the number of signatory cities that have prepared the adaptation plan is very limited;
- Mediterranean basin, where cities' engagement is relatively high, but the number of cities that have prepared and implemented an adaptation plan is quite limited;
- Northern Europe, where the number of participating cities is high and most of them have already presented an adaptation plan and are currently moving forward to the monitoring phase.

By taking into account the main difference between mitigation and adaptation strategies—that the former is addressed to achieve established targets in terms of GHG emissions as well as of energy saving, and the latter is aimed at reducing local vulnerabilities to climate-related hazards—it is important to emphasize the close dependency of adaptation strategies on the features of the urban context at stake, both in terms of type of climate impacts and related vulnerabilities that have to be considered and in terms of results that can be achieved. Every city, in

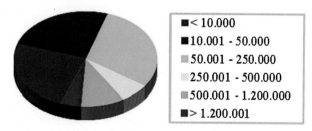

FIGURE 10.5 Classification of the cities that joined the Mayors Adapt by population size. *Author's elaboration based on data provided by the Mayors Adapt website (http://mayors-adapt.eu).*

---

[2] Data and information about Mayors Adapt initiative are updated to April 2017 (http://mayors-adapt.eu/). The website is no longer available.

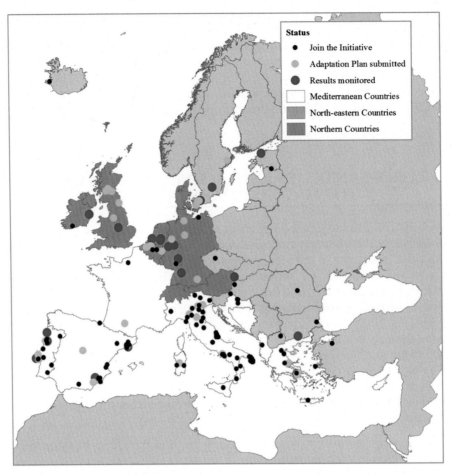

FIGURE 10.6 Mayors Adapt: signatory cities and current status. *Author's elaboration based on data provided by the Mayors Adapt website (http://mayors-adapt.eu).*

fact, is unique, since it is affected by different hazards, characterized by heterogeneous spatial, functional, and social characteristics that significantly determine its vulnerability, requiring being tailored to the site strategies.

This partly explains the greater difficulty to outline effective adaptation strategies as well as the different role played by the EU: to guide cities toward the development and implementation of effective adaptation policies and measures. Europe has not provided goals and targets to be achieved, but only platforms and guidelines to share information and best practices as well as to guide the adaptation process.

10.4 EUROPEAN CITIES EMBRACING AN INTEGRATED CLIMATE STRATEGY: THE NEW COVENANT OF MAYORS FOR CLIMATE AND ENERGY

The European initiatives discussed herein addressed mitigation and adaptation as separate, although complementary issues—both of them considered as crucial to counterbalance climate change (Yohe and Strzepek, 2007), but addressed through different strategies and tools. By the end of 2015, the Covenant of Mayors and the Mayors Adapt were merged into the new Covenant of Mayors for Climate and Energy, based on an integrated approach to climate issues.

Hence, starting from November 2015, the cities joining the new Covenant decided to commit to developing a Sustainable Energy and Climate Action Plan (SECAP) aimed at both cutting CO_2 emissions by at least 40% by 2030 and increasing urban systems' resilience to climate change.[3]

---

[3] http://www.covenantofmayors.eu/about/signatories_en.html.

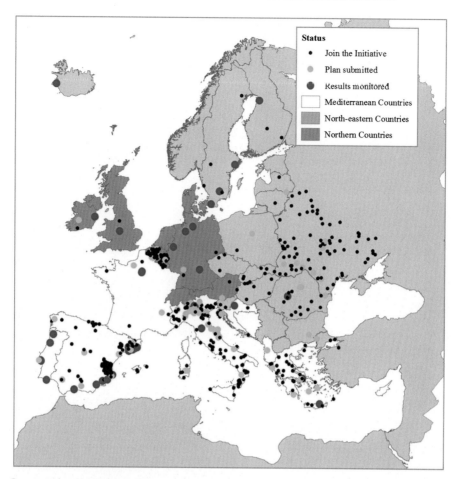

FIGURE 10.7 New Covenant for Climate and Energy: signatory cities and current status. *Author's elaboration based on data provided by the Covenant of Mayors website (http://www.covenantofmayors.eu).*

By the beginning of October 2017, about 750 cities had already joined this new European initiative. It is worth noting that among these cities, 100 had already joined the Covenant of Mayors, four had joined the Mayors Adapt, and only seven had previously joined both initiatives.

It has to be noted that some European cities, by taking part in all three initiatives launched by the EU, have progressively specified their path toward an effective climate strategy. For example, the city of Ghent, Belgium, which signed the Covenant of Mayors in 2009 and the Mayors Adapt in 2014, has progressively improved its strategies to counterbalance climate change, being the first Belgian city to embrace an integrated approach to climate issues, by joining, in November 2015,[4] the New Covenant of Mayors for Climate and Energy (Fig. 10.9).

Fig. 10.7 shows the distribution and the status of the European cities that have currently joined the New Covenant. It is worth noting that in northern Europe, the number of participating cities is quite low, both because major European cities (e.g., Amsterdam, Copenhagen, London, Berlin, Paris) had previously joined global initiatives, such as the C40 initiative and the 100RC, and because most of them had already joined previous European initiatives. In the Mediterranean area this new initiative is having a significant response in terms of membership, although most of the signatory cities have yet to submit the SECAP. In northeastern Europe, cities' engagement is significantly higher with respect to previous initiatives, although most of the cities are still at a very early stage. However, the increasing number of involved cities seems to testify to the growing attention paid by cities located in northeastern Europe to climate issues, which are increasingly considered as a priority issue in cities' agendas all over the Europe.

Similarly to previous initiatives, most of the signatory cities—approximately 80%—are represented by small- and medium-sized cities (Fig. 10.8).

---

[4] http://www.covenantofmayors.eu/news_en.html?id_news=691.

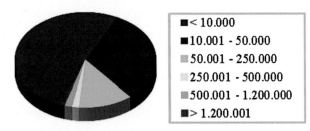

FIGURE 10.8 Classification of the cities participating in the New Covenant by population size. *Author's elaboration based on data provided by the Covenant of Mayors website (http://www.covenantofmayors.eu).*

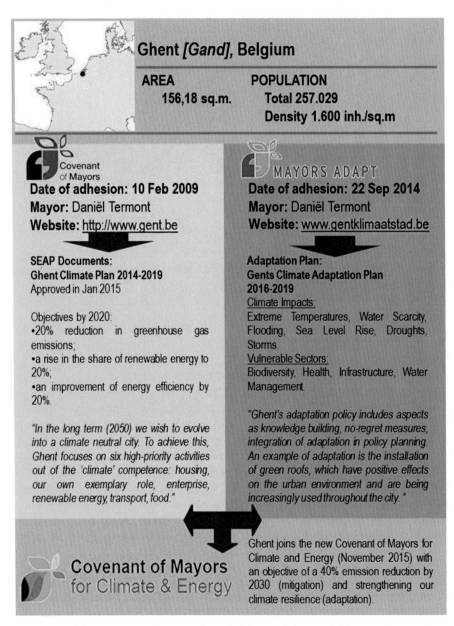

FIGURE 10.9 City of Ghent: the steps toward an integrated climate strategy. *Author's elaboration.*

10.5 CONCLUDING REMARKS

This chapter has shown the gradual transition of European initiatives aimed at counterbalancing climate issues from a complementary perspective, based on the idea that mitigation and adaptation are both crucial to counterbalance climate issues but they can be pursued separately, toward an integrated approach (Galderisi et al., 2016), based on comprehensive climate strategies capable to emphasize synergies and trade-offs between mitigation and adaptation measures.

Moreover, the heterogeneous responses of the European cities to the different initiatives that, starting from 2008, have been launched by the EU have been analyzed in depth.

By comparing the cities' response to these initiatives, it is clear that the first one, the Covenant of Mayors, had a response far greater than the others; while it involved, indeed, about 6000 cities, only 150 cities joined the Mayors Adapt, and 750 have so far joined the New Covenant. These differences are partly explained by the short life of the Mayors Adapt that, after only 1 year, was replaced by the New Covenant, but also by the fact that the latest initiative is still very young, having been launched only in 2015. Furthermore, as previously mentioned, the remarkable differences in cities' engagement could also be explained by the differences, in terms of contents, between SEAPs and adaptation plans.

The latter, indeed, cannot be based on well-established targets, and, so far, standardized methods and procedures to adapt cities to climate impacts are not available, since adaptation requires tailored-to-the-site measures as well as strong cooperation among different policy sectors.

One of the main results achieved by the three initiatives launched by the EU has to be recognized in the involvement of a large number of small- and medium-sized cities, which represent the majority of the European cities, recognizing in doing so their pivotal role in the fight against climate change. As mentioned earlier, in fact, most of the signatory cities are small- and medium-sized cities within all the three considered initiatives.

Therefore, it can be emphasized that, whereas large European cities have directly promoted or joined global initiatives aimed to carry out mitigation and/or adaptation plans, most European cities have started to address climate issues only thanks to the significant support ensured by the EU and based on the experience gained by large European cities, such as London, which acted as pioneers in promoting global initiatives to address climate change.

With respect to the effective outcomes arising from the substantial involvement of the European cities, it is worth noting that while only a few cities have joined the Covenant of Mayors without submitting and implementing a SEAP, the number of cities that have only joined the Mayors Adapt and the New Covenant, without submitting a plan, is significantly higher, although this number largely varies according to different geographical areas (Baffo et al., 2009) (Fig. 10.10).

The limited effectiveness of the three European initiatives, in terms of number of plans submitted and implemented, in the Mediterranean countries and in the Eastern countries can be explained, on the one hand, by the lower

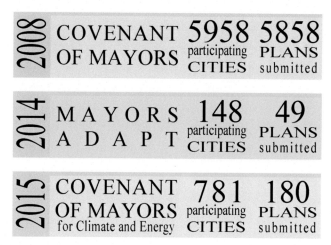

FIGURE 10.10 Number of cities involved and that submitted plans in the three European initiatives addressing climate change. *Author's elaboration.*

priority accorded in these areas to environmental issues; on the other hand, it could be explained by the lack of available resources to develop innovative strategies against climate change (European Union - Committee of the Regions, 2011).

However, the increasing response to the latest initiative, the New Covenant of Mayors for Climate and Energy, bodes well, highlighting both growing attention all over Europe to the adaptation issues and increasing awareness of the centrality of climate policies among Eastern European cities.

References

Baffo, F., Gaudioso, D., Giordano, F., 2009. L'Adattamento ai cambiamenti climatici: strategie e piani in Europa. ISPRA Rapporto 94/2009, 29–32. Available at: http://www.isprambiente.gov.it/contentfiles/00003500/3580-rapporto94-2009.pdf/.

C40 Cities, 2015. C40 Fact Sheet Why Cities? Available at: http://c40-production-images.s3.amazonaws.com/fact_sheets/images/5_Why_Cities_Dec_2015.original.pdf?1448476459.

Davide, M., Giannini, V., Venturini, S., Castellari, S., 2013. Questionario per la Strategia Nazionale di Adattamento ai Cambiamenti Climatici: elaborazione dei risultati. Rapporto per il Ministero dell'Ambiente e della Tutela del Territorio e del Mare, Rome, 12–17. Available at: http://www.minambiente.it/sites/default/files/archivio/allegati/clima/snac_rapporto_questionario.pdf.

European Union - Committee of the Regions, 2011. Adaptation to Climate Change. Policy Instruments for Adaptation to Climate Change in Big European Cities and Metropolitan Areas, European Union and the Committee of the Regions, pp. 25–31. https://doi.org/10.2863/30880.

Galderisi, A., Mazzeo, G., Pinto, F., 2016. Cities dealing with energy issues and climate-related impacts. Approaches, strategies and tools for a sustainable urban development. In: Papa, R., Fistola, R. (Eds.), Smart Energy in Smart City. Urban Planning for a Sustainable Future, pp. 215–216. https://doi.org/10.1007/978-3-319-31157-9.

Kress, A., 2007. Adaptation and Mitigation an Integrated Climate Policy Approach. Climate Alliance, 3. Available at: http://www.amica-climate.net/.

Luise, D., 2014. The challenge of Mayors Adapt: the answers expected from the Italian reality. Urbanistica 3 may-august. In: Filpa, A., Ombuen, S. (Eds.), Understanding Climate Change Planning for Adaptation, vol. 2014, pp. 15–16. Available at: http://www.urbanisticatre.uniroma3.it/dipsu/wp-content/uploads/2015/01/U3_quaderni_05.pdf.

UNISDR, 2015. What Is the "Making Cities Resilient Campaign"? About the Campaign. Available at: https://www.unisdr.org/campaign/resilientcities/home/about.

Yohe, G., Strzepek, K., 2007. Adaptation and mitigation as complementary tools for reducing the risk of climate impacts. Mitigation and Adaptation Strategies for Global Change 12, 727–739. https://doi.org/10.1007/s11027-007-9096-3.

CHAPTER 11

Adaptation and Spatial Planning Responses to Climate Change Impacts in the United Kingdom: The Case Study of Portsmouth

Donatella Cillo
Environment Agency, England, United Kingdom

11.1 CLIMATE CHANGE IMPACTS IN THE UNITED KINGDOM

The increase in global temperature and the effects of the warming of the climate system are already felt in the United Kingdom (UK). The United Kingdom has experienced a rise in its average temperature by about 1°C since the 1970s (BEIS, 2014; Jenkins et al., 2009). Under the UK Climate Change Act 2008, the country has committed to reduce its greenhouse gas (GHG) emissions by 80% below 1990 levels by 2050 (Davoudi, 2013). However, even in a carbon stabilization scenario achieved through mitigation measures, climate-related hazards would still be experienced throughout the country. Heavier rainfall events associated with an enhanced risk and intensity of flooding, higher sea levels, drought, and a higher frequency of heat and cold waves are some of the impacts that have already and will continue to affect the United Kingdom, making the country's natural and human systems highly vulnerable. Adaptation measures aimed at the adjustment of natural and human systems in response to climate stimuli so as to moderate potential harms or exploit beneficial opportunities (IPCC, 2012; Klein et al., 2003) are therefore necessary. In this adaptation effort, spatial planning is widely recognized in England as playing an important role (Davoudi et al., 2009). This chapter will describe the adaptation and spatial planning responses that the city of Portsmouth, England, (Fig. 11.1) is adopting to respond to climate change challenges and to increase its resilience.

11.1.1 Portsmouth

The city of Portsmouth represents a complex living system, which due to its singular features, ecosystems, and heavy concentration of population will be extremely exposed to climate change impacts affecting the United Kingdom. Portsmouth lies on Portsea Island on the south coast of England and is surrounded by Portsmouth Harbor to the east and Langstone Harbor to the west (Fig. 11.2).

Both harbors are ecologically significant and protected by international designations (PSAG, 2010a). The relationship with the sea, the geographical position, and the presence of the Royal Navy and the dockyard have influenced the city's character, distinguishing it with a vibrant, historic, and diverse waterfront, and acting as catalysts for its growth, economy, and tourism activities (PCC, 2010a; 2012). Indeed, Portsmouth is one of the major economic cities of the subregion of urban south Hampshire, along with Southampton (PCC, 2010b). On a land area of 40 km^2 and being home to approximately 203,500 inhabitants (PCC, 2010a), "Portsmouth is England's second most densely populated city (5000 people per km^2) outside of inner London" (PSAG, 2010a). The city is extremely flat and low lying with almost the entire land of the shoreline being 0 m above sea level. The entire coastline has been protected by man-made coastal defenses (PSAG, 2010a), whereas other parts have been artificially altered through reclamation land (PCC, 2011).

FIGURE 11.1 Portsmouth location in Europe and United Kingdom. *From: Google Maps-Engine.*

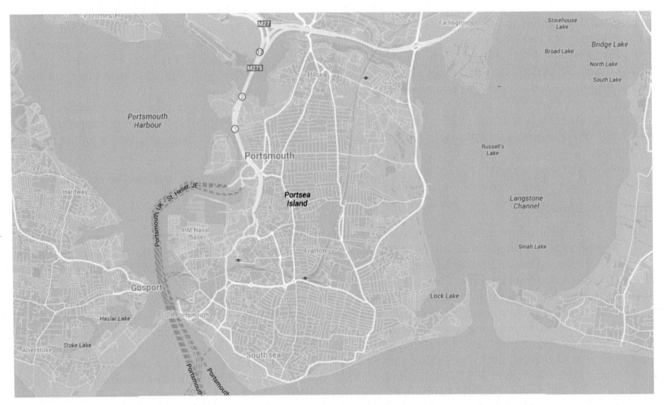

FIGURE 11.2 Portsmouth location within the harbors. *From: Google Maps-Engine.*

11.1.2 Climate Change Impacts on Portsmouth

Due to the city's territorial characteristics outlined herein and its geographical position, flooding from the sea, storm events, wave overtopping, and hot summer and heavy winter precipitation have always been major challenges to Portsmouth. However, climate change will only increase the severity and frequency of these issues.

Currently about 25,700 properties and major infrastructures are at risk of flooding from the sea (PCC and EA, 2011). However, it is predicted that in the next 70 years, sea level around Portsmouth will rise by 70 cm and that "by 2100 storm surge events can occur up to 20 times more frequently" (PCC, 2010c). This will result in an increased risk of coastal flooding with up to 60,000 residents living in flood risk areas, limited availability of land for new development (PCC, 2010c; PSAG, 2010a), and higher pressure on coastal flood defenses (Fig. 11.3). Summer mean temperature in the short term is likely to increase up to 2.9°C, whereas the likelihood of summer precipitation could decrease by 31% by 2030s and by 50% by 2090s (PCC, 2010c). The impacts of these changes will range from the increase in urban heat island effect to heat-related health impacts, increase in tourism, which in turn will lead to higher pressure on local infrastructure, and to an increase of summer droughts resulting in potential water shortages and loss of valuable ecological species and habitats (PCC, 2010c; PSAG, 2010a). Furthermore, it is predicted that by the 2080s mean winter temperature is also likely to increase by up to 4.8°C, with over a 50% increase in the amount of precipitation (PCC, 2010c). As a consequence, the frequency and intensity of flooding from surface water will be higher, impacting the capacity of the water infrastructures to cope with intensive rainfall (PCC, 2010c; PSAG, 2010a).

The set of slow- and quick-moving phenomena (Galderisi, 2014) impacting on Portsmouth as described before will make the city's built and natural environment, as well as its economy and tourism, extremely vulnerable.

Portsmouth will therefore need to adapt to the impacts of climate change "by resisting or changing in order to reach and maintain an acceptable level of functioning and structure" (UNISDR, 2004) and consequently in order to become a city that is resilient to climate change.

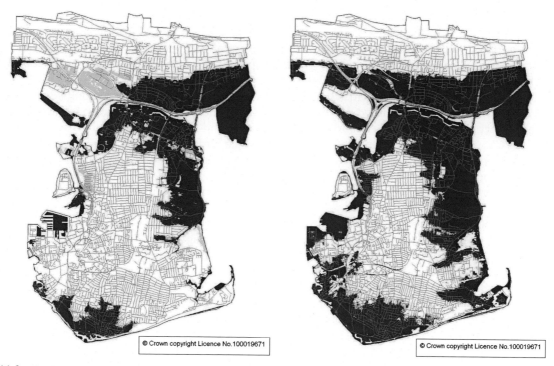

FIGURE 11.3 Environment Agency Flood zones 2009 (*left*) and predicted 2115 (*right*). *From: Portsmouth City Council (PCC) and Environment Agency (EA), 2011. Local Development Framework — Development and Tidal Flood Risk, Statement of Common Ground. Available at: https://www.portsmouth.gov.uk/ext/documents-external/pln-local-dev-climate-change-floodrisk-statement.pdf — Crown Copyright License No. 100019671.*

11.2 PORTSMOUTH'S CLIMATE CHANGE ADAPTATION AND PLANNING RESPONSES

There is a responsibility to show that Portsmouth is a city that takes climate change seriously. **Portsmouth Climate Change Strategy (2010)**

The legislative changes to the English planning system introduced by the coalition government and its Localism Bill[1] in 2011 have placed climate change adaptation more firmly at the center of the national spatial planning agenda. Indeed, in accordance with the National Planning Policy Framework 2012 (NPPF), which constitutes guidance in producing local plans and is a material consideration in planning decisions, it is a statutory requirement for local planning authorities to proactively plan, work with communities and adopt policies to adapt to climate change. Therefore, in responding to the city's climate change challenges, Portsmouth has produced a series of adaptation and planning responses where the ways in which the city is spatially configured and the land is used and sustainably developed have a pivotal role (Davoudi et al., 2009).

In 2010, the multiagency document "Portsmouth Climate Change Strategy" was published. It was produced by the Portsmouth Sustainability Action Group (PSAG), which put together community representatives and key agencies in order to deliver key sustainability objectives (PSAG, 2010b) and to produce a coordinated approach to tackle climate change. The strategy commits to adapt to climate change through the achievement of the following objectives:

- Enhanced cognitive dimension of the major problems affecting the city;
- Protected coastline and reduction of the impacts of flooding on developments;
- Minimized impacts of any emergency with a rapid and appropriate response;
- Implemented water efficiency measures (PSAG, 2010a).

Alongside this strategy, which is mainly setting up an adaptation framework, several planning documents and policies have been produced, and they will have a key role in delivering the Portsmouth Climate Change Strategy adaptation objectives (Fig. 11.4).

In 2012 Portsmouth adopted its "Portsmouth's Core Strategy" (PCC, 2012a), which in order to deliver sustainable development, sets out strategic policies that take full account of climate change impacts on the city's natural and human systems. The flood risk policy for example sets out how the city will cope with the increase in flood risks from the sea, surface, and foul water. Whereas, the strategic policy "A Greener Portsmouth" defines the council's commitment in protecting and enhancing green infrastructure assets, which will contribute to reduce the likelihood and severity of surface water flooding and the probability of urban heat island effect (PCC, 2012a). Moreover, water-efficiency measures have to be provided in new development proposals in accordance with the plan policy "Sustainable Design and Construction" (PCC, 2012a).

The council has also produced an Infrastructure Delivery Plan (IDP), which sitting alongside "Portsmouth's Core Strategy," identifies physical (i.e., flood defenses, waste water, and drainage) and green infrastructure needed to support the delivery of new development and to protect existing and future development from the challenges of climate change (PCC, 2011b). The Portsea Island Coastal Defence Strategy is at the heart of the IDP and it also "forms the backbone of the coastal flood risk management planning for the city" (ESCP, 2011). The implementation of the proposals recommended in the Coastal Defence Strategy will depend upon the ability of the council in securing the necessary funds and working in partnership with operating authorities and land owners (ESCP, 2011; PCC, 2011b).

In supporting the flood risk management strategy and the Core Strategy, there are also two further documents: the Strategic Flood Risk Assessment (SFRA) and the Surface Water Management Plan (SWMP). The former is an evidence-based document providing detailed assessment of the coastal flooding risks in the city and informing strategic decision-making in relation to spatial planning (PUSH, 2007). Whereas the latter details flooding from sewers, drains, and runoff from land occurring as a result of heavy rainfall, and its objective is to raise awareness, identify flood risks and assets, and agree on mitigation measures (PCC, 2012b). Finally, the city also has in place a Flood Response Plan and a Multi-agency Response Plan, whose aims are to set out the management and response arrangements "in preparation for and response to a flood event" (PCC, 2011c).

---

[1] The Localism Bill is the legislative foundation for the shift of power from the centralized state to local communities. The bill sets how local authorities benefit from decentralization and how they have a vital role in passing power to communities and individuals.

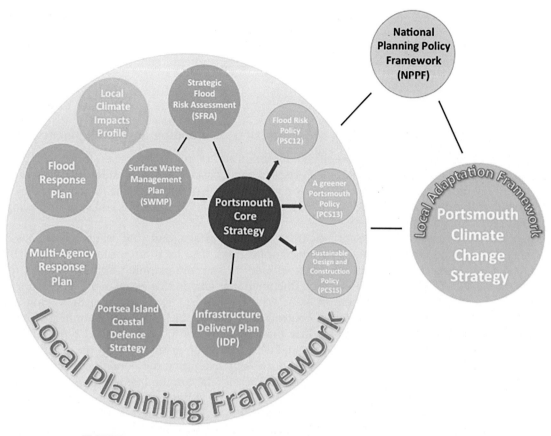

FIGURE 11.4 Adaptation and spatial planning frameworks. *Author's elaboration.*

11.3 TOWARD A RESILIENT PORTSMOUTH

The array of planning documents and policies produced by Portsmouth and described herein shows how the city has fully adhered to the national goal of adapting to climate change challenges. The following analysis of the Climate Change Strategy objectives shows how the city planning responses are the means through which Portsmouth will implement its climate change objectives, deliver sustainable development, minimize vulnerability, and become a city that is resilient to the impacts of climate change.

11.3.1 Enhanced Cognitive Dimension of the Major Problems Affecting the City

The climate change strategy objective of enhancing the knowledge and awareness of the climate change impacts affecting Portsmouth is embedded in several planning and climate change documents. In supporting its climate change strategy, Portsmouth has produced a Local Climate Impacts Profile document (PCC, 2010d). This provides an overview of how local communities, assets, infrastructure, and services have been impacted in the past by heavy rainfall, higher temperatures, and storms. The scope is to understand how and where the city can build and increase its resilience to climate change (PCC, 2010d). The "Guide for services 2012 — Building resilience in a changing climate, How resilient is your service?" and the series of interviews undertaken by the council to collect information and understand how past extreme events have affected service delivery (PCC, 2012c), constitutes another example of the city's commitment to have a better idea of how future events might impact the city services and what actions can be taken to prevent future disruptions.

The SFRA and SWMP then provide information on the frequency, impact, speed of onset, depth, and velocity of coastal flooding and surface water flooding from sewers and runoff from land for the present day and for the potential impacts of climate change over the next 100 years (PUSH, 2007). This robust knowledge enables the local planning authority to prepare appropriate flood risk policies, make informed decisions on the allocation of land for new development, and plan for emergency responses. Finally, the Portsea Island Coastal Defence Strategy includes a comprehensive study of the current and future coastal processes to enable the production of an informed coastal flood risk management strategy (ESCP, 2011). The accumulation of knowledge carried out as part of the city adaptation response to climate change impacts proves to be a solid basis for the preparation and implementation of planning and policies responses, strategies, and emergency plans that can all lead toward a resilient Portsmouth.

11.3.2 Protected Coastline and Reduction of the Impacts of Flooding on Developments

This objective appears to be at the heart of the city's adaptation and planning responses. Indeed the flood risk and green infrastructure policies, SFRA, SWMP, and the Portsea Island Coastal Defence Strategy all play a pivotal role in enhancing the ability of the city to persist and adapt to an increased risk of flooding. The flood risk policy PSC12 sets up how the council will reduce and manage flood risk issues impacting on existing and future developments by following a flood risk management hierarchy.

The first step will be to assess the level of risk from the sea, surface, and foul water when considering the allocation of sites for new development and planning applications. The local authority's SFRA and SWMP and site-specific Flood Risk Assessments produced by developers to support planning applications, as required by the NPPF (paragraphs 100–103), constitute essential documents to enable this assessment. The council will then avoid flood risk by locating new development in areas with the lowest risk of flooding, following a sequential approach. This approach is "the central element of controlling flood risk in UK spatial planning" (Coleman, 2009) and it is the way in which local authorities use their planning powers to guide certain types of development away from the areas at the highest flood risk (Coleman, 2009). However, given the built-up nature of the city, the existing flood risk, and that sea level rise will increase this risk by extending the flood zones (PCC and EA, 2011), Portsmouth is not able to follow a sequential approach and therefore accommodate the level of expected development (i.e., new houses and employment sites) in the areas at the lowest flood risk.

For this reason, controlling flood risk and mitigating any residual risks through flood resilient design measures and effective flood warning and evacuation plans (PCC, 2012a) are at the heart of the flood risk hierarchy of policy PSC12. Indeed, great emphasis has been given to the coastal strategy proposals of maintaining, replacing, or raising the existing man-made coastal defenses shaping the coastline (ESCP, 2011). These proposals, in addition to enabling the city's future development to withstand shocks and preserve their structure, will also protect the thousands of people, businesses, and property that are already at risk of flooding. To implement these proposals, the council will need, however, to secure public and private funding. These are set up in the city's IDP, which requires landowners and developers to pay directly for some defenses or to provide contributions (PCC, 2011b).

To protect a city like Portsmouth from coastal flooding, in addition to maintaining and raising hard infrastructures, consideration also should be given to controlling the risk of flooding from the sea through the use of more soft strategies. Although not included in the flood risk policy, but in policy PCS13 "A Greener Portsmouth," there is a commitment to protect and enhance the ecological integrity of the city's two main natural defenses: the Langstone and Portsmouth Harbors. Indeed, their saltmarshes and mudflats are ecosystems extremely important in providing contextual defenses to floods, storm surges, and high tides thanks to their ability to absorb changes and disturbances (11.5). However, the artificial defenses built along the shoreline have led to a continued loss of saltmarshes, coastal squeeze and erosion, and changes in tidal currents (NS, 2010). Therefore, it is extremely important that the implementation of controlling flood risk through man-made defenses goes together with the protection and enhancement of the two harbors. This will increase the capacity of the city to resist and therefore be resilient to climate change impacts if man-made coastal defenses fail.

In addition to coastal flooding, Portsmouth will also need to control an increased risk of flooding from sewers, drains, and runoff from land. To do so, the council will be permitting, in accordance with its flood risk policy, only new strategic development where "the necessary surface water drainage, foul drainage and sewage treatment capacity is available" (PCC, 2012a). Furthermore, the council will implement the mitigation measures identified in the SWMP, such as the provision of Sustainable Urban Drainage Systems or the separation of foul and surface water sewers, and it will improve the capacity problems of the existing sewer systems (PCC, 2012b).

The recognition that controlling flood risk through man-made and soft defenses is not the only way to deal with climate change and its effects is evident in the last step of the flood risk management hierarchy of policy PSC12.

This step sets out how to limit damage to new developments and deal with any residual risks when and if flood defenses are overtopped or fail, by implementing flood resilience and resistance measures in the design of buildings (i.e., raised floor levels) and having in place flood warning and safe evacuation procedures (PCC and EA, 2011).

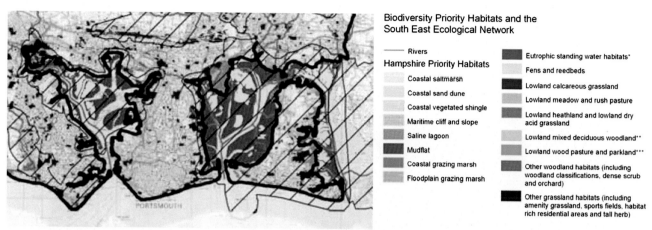

FIGURE 11.5 Langstone and Portsmouth Harbors' biodiversity. *Modified from: Urban south Hampshire (PUSH), 2008. Towards a Green Infrastructure strategy for South Hampshire: Advice to PUSH. Available at: http://www.push.gov.uk/towards_a_gi_strategy_-_advice_to_push_jul_08.pdf.*

11.3.3 Minimized Impacts of Any Emergency With a Rapid and Appropriate Response

The objective of minimizing impacts of any emergency through a rapid and appropriate response is embedded in the mitigation step of the flood risk policy PSC12 described earlier and implemented by the city's Flood Response and Multi-agency Response Plans. Indeed this step calls for mitigating residual flood risks in new and existing developments by having in place effective emergency response plans (PCC, 2012a). The council recognizes that in order to provide a rapid and therefore appropriate emergency response, some necessary and timely preventive actions are needed. These actions include: scheduled inspections, maintenance and clearance of the drainage system and of coastal defenses; closure of the flood gates in Old Portsmouth; closure of shoreline roads, especially where overtopping is a hazard to people and vehicles; and distribution of sandbags (PCC, 2011c). In addition to these actions, forecasting and early warnings are issued directly to the public, the media and responding services, local authorities, and other agencies though a 24-h warning system (PCC, 2011c). The measures included in the adopted emergency plans show how Portsmouth has planned for a timely capacity to avoid eventual losses and disturbances to the community and the city services in case of flood events.

11.3.4 Implemented Water Efficiency Measures

Finally, the objective of implementing water efficiency measures in response to a potential deficit in fresh water supply due to climate change is embedded in the Core Strategy policy "Sustainable Design and Construction" PCS15. It is indeed a requirement under the national planning policy that local authority should take full account of the water supply when adapting to climate change. For this reason, policy PCS15 requires new development to implement enhanced standards of water efficiency in accordance with national building regulations (PCC, 2012b). In addition to this, the local water company is seeking to increase the fresh water supply through a new reservoir and improvements to the city water treatment works (PSAG, 2010a). If fully implemented, these measures will ensure that Portsmouth is able to be resilient to potential water shortages due to an increase in summer mean temperature.

11.4 CONCLUSIONS

Portsmouth is one of several cities in the United Kingdom that will be extremely exposed to the major impacts of climate change. The city recognizes that climate change is now its greatest long-term threat. In line with the national spatial planning agenda, Portsmouth is committed to support a city resilient to the impacts of climate change, to deliver sustainable development, and minimize the vulnerability of its communities and assets. In order to do so, it has successfully adopted and implemented a series of local planning policies, strategies, and plans. Within this local planning framework of the city's Climate Change Strategy, four main objectives are embedded. The analysis carried out of the city climate change objectives and their implementation though spatial planning reveals that

FIGURE 11.6 Floating neighborhood of IJburg, Amsterdam. Credit: *Donatella Cillo.*

the integration between Portsmouth's adaptation and planning responses will support a city that is able to resist and manage climate change disturbances without crossing a particular threshold (Davoudi et al., 2013). Indeed, the city's planning responses of maintaining and strengthening man-made coastal defenses; protecting and enhancing the two natural defenses of the Langstone and Portsmouth Harbors; accumulating knowledge and awareness of the climate change impacts affecting the city; locating new development in areas with the lowest risk of flooding; and implementing flood resilient design measures, flood emergency responses, and water efficiency measures will ensure that existing and future residents and assets are safe and able to adapt to the most challenging climate change stresses that Portsmouth will need to cope with, namely sea level rise, the increase in coastal flooding, and summer droughts.

Nevertheless, in addition to proactively planning to increase its ability to persist and adapt in response to future stresses and in order to be fully resilient to climate change impacts, Portsmouth could have had also better planned to build its capacity of introducing innovative urban configurations and ways of living. Indeed, a resilient city is also a city that gives room to the development of new and more desirable urban forms. Examples of innovative scenarios can be found in the development of the floating neighborhood of IJburg, in the city of Amsterdam, the Netherlands, that has amphibious homes with jetties instead of paved footpaths (Fig. 11.6) or in the adaptation strategy of Rotterdam, which emphasizes the opportunity of creating new types of infrastructures that allow living with and on the water. The lack of innovation and capacity to create completely new urban landscapes can, however, be attributed to the statutory necessity of plans and policies to be in line with the English national planning policies. Indeed, the NPPF is a national spatial regulation that calls for controlling climate change risks rather than favoring more innovative and contextualized solutions (Markus and Savini, 2016). Therefore, planning in accordance with the NPPF has inhibited Portsmouth's capacity to find innovative solutions and create completely new urban landscapes. In a scenario of developable space shortness, Portsmouth has planned for the provision of new dwellings, jobs, and facilities by delivering more and higher flood defenses and implementing a series of mitigation measures. However, only by finding and embracing innovative urban configurations together with providing resistant infrastructures and adapting to changes and disturbances, it can be said that Portsmouth is a city that is resilient to climate change.

Disclaimer

The views expressed in this chapter are not necessarily the official position of the Environment Agency.

References

BEIS (The Department for Business, Energy, & Industrial Strategy), 2014. Climate Change Explained [online]. Available at: https://www.gov.uk/guidance/climate-change-explained#uk-government-action.

Coleman, A., 2009. Climate change and flood risk methodologies in the UK. In: Davoudi, S., Crawford, J., Mehmood, A. (Eds.), Planning for Climate Change: Strategies for Mitigation and Adaptation for Spatial Planners. Earthscan, London.

Davoudi, S., Crawford, J., Mehmood, A., 2009. Climate change and spatial planning response. In: Davoudi, S., Crawford, J., Mehmood, A. (Eds.), Planning for Climate Change: Strategies for Mitigation and Adaptation for Spatial Planners. Earthscan, London.

Davoudi, S., 2013. Climate change and the role of spatial planning. In: Knieling, J., Leal Filho, W. (Eds.), Climate Change Governance, Climate Change Management. Spinger-Verlag Berlin Heidelberg.

Davoudi, S., Brooks, E., Mehmood, A., 2013. Evolutionary resilience and strategies for climate adaptation. Planning, Practice and Research 28 (3), 307–322.

Department for Communities Local Government (DCLG), 2012. National Planning Policy Framework (NPPF).

Eastern Solent Coastal Partnership (ESCP), 2011. Portsea Island Coastal Strategy Study. Available at: https://www.portsmouth.gov.uk/ext/documents-external/pln-local-dev-portseaisland-coastal-strategy-study.pdf.

Galderisi, A., 2014. Urban Resilience: a framework for empowering cities in face of heterogeneous risk factors. ITU A|Z Journal 11 (1), 36–58.

IPCC, 2012. Managing the Risks of Extreme Events and Disasters to Advance Climate Change Adaptation, Special Report of Working Groups I and II of the Intergovernmental Panel on Climate Change. Cambridge University Press, Cambridge, United Kingdom and New York, NY, USA.

Jenkins, G.J., Murphy, J.M., Sexton, D.M.H., Lowe, J.A., Jones, P., Kilsby, C.G., 2009. UK Climate Projections: Briefing Report. Met Office Hadley Centre, Exeter, UK.

Klein, J.T., Nicholls, R.J.J., Thomalla, F., 2003. Resilience to natural hazards: how useful is this concept? Environmental Hazard 5, 35–45.

Markus, M., Savini, F., 2016. The implementation deficits of adaptation and mitigation: green buildings and water security in Amsterdam and Boston. Planning Theory and Practice 17 (4), 497–515.

North Solent SMP, 2010. North Solent Shoreline Management Plan. Available at: http://www.northsolentsmp.co.uk/index.cfm?articleid=10025&articleaction=nthslnt&CFID=14261252&CFTOKEN=dd82386ab9030d6c-C4F30D98-93CD-CF58-FC95BC134698162B.

Partnership for Urban South Hampshire (PUSH), 2007. Strategic Flood Risk Assessment. Available at: http://maps.hants.gov.uk/push/Reports/ReportList.htm.

Partnership for Urban South Hampshire (PUSH), 2008. Towards a Green Infrastructure strategy for South Hampshire: Advice to PUSH. Available at: http://www.push.gov.uk/towards_a_gi_strategy_-_advice_to_push_jul_08.pdf.

Portsmouth City Council (PCC), 2012a. The Portsmouth Plan. Portsmouth's Core Strategy. Available at: https://www.portsmouth.gov.uk/ext/documents-external/pln-portsmouth-plan-post-adoption.pdf.

Portsmouth City Council (PCC), 2012b. Surface Water Management Plan. Available at: https://www.portsmouth.gov.uk/ext/documents-external/cou-policies-flood-swmp.pdf.

Portsmouth City Council (PCC), 2012c. Guide for Services 2012 – Building Resilience in a Changing Climate, How Resilient Is You Service? Available at: https://www.portsmouth.gov.uk/ext/documents-external/cmu-climate-resilience-help.pdf.

Portsmouth City Council (PCC) and Environment Agency (EA), 2011. Local Development Framework – Development and Tidal Flood Risk, Statement of Common Ground. Available at: https://www.portsmouth.gov.uk/ext/documents-external/pln-local-dev-climate-change-floodrisk-statement.pdf.

Portsmouth City Council (PCC), 2011a. Strategy Unit's Research Function Briefing September 2011 – Sustainability, Climate Change and Carbon Management. Available at: https://www.portsmouth.gov.uk/ext/documents-external/cmu-response-to-national-sustainability-strategy.pdf.

Portsmouth City Council (PCC), 2011b. Infrastructure Delivery Plan. Available at: https://www.portsmouth.gov.uk/ext/documents-external/pln-cil-infrastructure-delivery-plan.pdf.

Portsmouth City Council (PCC), 2011c. Flood Response Plan. Available at: https://www.portsmouth.gov.uk/ext/documents-external/cou-flood-response-plan.pdf.

Portsmouth Sustainability Action Group (PSAG), 2010a. Portsmouth Climate Change Strategy. Available at: https://www.portsmouth.gov.uk/ext/documents-external/cmu-climate-strategy-full.pdf.

Portsmouth Sustainability Action Group (PSAG), 2010b. Partnership Agreement. Available at: https://www.portsmouth.gov.uk/ext/documents-external/cmu-sustainability-group-agreemeent.pdf.

Portsmouth City Council (PCC), 2010a. Shaping the Future of Portsmouth. A Strategy for Growth and Prosperity in Portsmouth. Available at: https://www.portsmouth.gov.uk/ext/documents-external/cou-regeneration-strategy.pdf.

Portsmouth City Council (PCC), 2010b. Portsmouth the Waterfront City – Southsea Seafront Strategy 2010-2026. Available at: https://www.portsmouth.gov.uk/ext/documents-external/dev-southseaseafrontstgy-2010-26.pdf.

Portsmouth City Council (PCC), 2010c. Planning to Adapt to Climate Change, A Headline Summary. Available at: https://www.portsmouth.gov.uk/ext/community-and-environment/green-living/how-climate-change-affects-portsmouth.aspx.

Portsmouth City Council (PCC), 2010d. Local Climate Impacts Profile. Available at: https://www.portsmouth.gov.uk/ext/documents-external/cmu-climate-impacts-profile.pdf.

UNISDR, 2004. Living with Risk: A Global Review of Disaster Reduction Initiatives. United Nations International Strategy for Disaster Reduction, New York, USA and Geneva, Switzerland. Available at: http://www.unisdr.org/files/657_lwr1.pdf.

CHAPTER 12

Land-Use Planning and Climate Change Impacts on Coastal Urban Regions: The Cases of Rostock and Riga

Sonja Deppisch
HafenCity University Hamburg, Hamburg, Germany

12.1 CLIMATE CHANGE IMPACTS AND LAND-USE PLANNING IN URBAN REGIONS

Many climate change impacts will have a spatial connotation and will affect land-use and related issues (Revi et al., 2014). Therefore, land-use planning can play a key role in dealing with the impacts of climate change in urban regions due to its task of balancing different interests and demands on land and of developing future-oriented as well as locally based strategies on its future physical structure and use. Land-use planning is attributed manifold further capacities for adapting to climate change (Hurlimann and March, 2012, p. 479f.), such as to provide formal legal instruments to organize future land use and therewith to provide for implementing adaptation measures. Also, land-use planning is known to have a major impact on the vulnerability of built and unbuilt structures and land uses (e.g., Bulkeley, 2013). It is not only the effects of climate change that alter urban regions; it is also their interaction with assets of urban regions and with other drivers of land-use development, such as demographic or economic change.

Within this problem setting, the following research questions are addressed: How can urban and regional planning meet the resilience challenges evoked by the specific climate change impacts on urban regions? What opportunities does land-use planning offer for tackling the impacts of climate change impacts in a resilient manner, and which barriers exist?

Presented are results of a 5-year research project on how urban and regional planning can tackle the impacts of climate change on coastal urban regions of the Baltic Sea. These results are discussed using a transformative resilience lens. This lens is based on socioecological resilience thinking applied to planning (Wilkinson, 2012). It is useful to refer to socioecological resilience thinking when addressing climate change because it focuses on practical problems and highlights complexity and socioecological interdependencies across time and space (Deppisch and Hasibovic, 2013). In addition, as planning is a deliberative act, it is claimed here that a pure socioecological resilience notion is not sufficient to be applied to the used research question, but that it is more insightful, especially with reference to land-use planning within urban regions, to refer also to the notion of transformative practices in planning (Albrechts, 2010). This is necessary, as urban regions are human-dominated systems, where a pure returning back to the original state might not be a socially or even socioecologically preferred aim, but the transformation of the given (e.g., eventually nonsustainable) development path would be desirable. The transformative resilience lens then allows putting the current state of the urban region in question, too.

In order to answer the research question, an abductive research design (Van de Ven, 2007, p. 101ff.) was adopted. The empirical material was obtained from two single case studies of urban regions on the Baltic Sea coast in Germany (in-depth single case study) and Latvia (smaller explorative case study). The case studies involved semistructured qualitative interviews that were conducted with key respondents from local and regional administrations, land-use planning and planning-related politics in Rostock between 2009 and 2012 and in Riga in 2012. Additionally, local and regional land-use planning documents were analyzed.

Depending on their location, coastal urban regions will probably be impacted by sea level rise and increases in intensity and frequency of storm surges. They will also be affected by temperature rise and altered precipitation patterns (as more inland urban regions, too). Although both case studies are located on the Baltic Sea coast, they exhibit differences in institutional and planning contexts.

What now follows is a generic outline of the cases.

12.1.1 Rostock Case

Rostock is a medium-sized city in the north of Germany, located on the Baltic Sea coast. The city, with around 200,000 inhabitants, covers an area of around 200 km². The urban region includes a further 50,000 or so inhabitants, who live in 22 neighboring smaller local communities and towns. In German planning terms, this urban region, covering more than 540 km², makes up an urban-suburban area characterized by functional interlinkages (SUR, 2011). According to subnational planning law, this area has to be considered together, and a common frame for land-use development must be devised. The current version of this common frame was adopted in 2011 (SUR, 2011). Demographic data suggests that the urban population decreased by around 20% between 1991 and 2008. At the same time, people moved to the suburbs, resulting in increased planning activities there. Planners are currently observing a return movement to the inner city. The urban population is ageing, with an increasing proportion of inhabitants over the age of 65. The core city dates back to the 13th century, and has been a hub for maritime trade ever since. Now, it is also a popular tourist destination.

Whilst the city is economically dependent on its coastal location, the same beneficial aspect harbors the risk of potential undesired climate change impacts following a rise in sea level and an increase in storm surges. Both the frequency and intensity of extreme events, such as storms, storm surges, and heavy precipitation events are expected to increase. Annual rainfall is expected to increase by approximately 8%, featuring increasingly dry summers and more humid winters (Richter et al., 2013). Average temperatures are expected to increase by 2.1–4.8°C by the end of the 21st century, with more hot days in summer and fewer frosty days in winter (Krämer et al., 2012). Already now, an explicit urban heat island effect can be observed in the city core (Richter et al., 2013). Several extreme weather events have hit this urban region in the past, including downtown flooding due to heavy rainfall and a severe storm that also hit the hinterland.

Land-use planning in Germany is organized around a general regional plan, which is binding for other, especially local planning authorities. However, municipalities are self-governing bodies that have the right to govern their own affairs, provided that they remain within the limits set by law. Developing local land-use plans is a mandatory task stipulated by law. Municipalities develop a general preparatory land-use plan for their whole territory and then further plans for specific smaller areas; the latter are legally binding for everyone. Since the national Land-Use-Planning Act was amended in 2009, climate change adaptation is a duty to be taken into account within new land-use plans, but no concrete goals were defined.

The city of Rostock is very active in promoting itself as a host for renewable energy, also aiming to achieve an energy turnaround. As early as 2005, a framework concept on climate change mitigation was developed by the urban administration, and an updated version was adopted by the City Council in 2010. This topic is very prominent in local politics and also appears in general land-use planning. It was therefore astonishing that the impacts of climate change and adaptation were of no specific local or regional interest in 2009, when the research commenced. The topics were mentioned only briefly in the regional plan (adopted 2011) and the urban preparatory land-use plan (2009). Due to the long-lasting procedures until a land-use plan is finally adopted, we must note here that the plans were already developed years before and also before and during the amendment of the Land-Use Planning Act and the rising political awareness toward adaptation. No general strategy for tackling potential climate change impacts was developed by land-use planning. Nevertheless, a number of measures could serve as unintended adaptation measures, such as keeping areas free from development (Albers et al., 2013, p. 17).

In contrast to the topic of the energy turnaround, the topics of climate change impacts and how to tackle them were neither perceived by the public nor debated at local and regional levels. The regional newspaper mainly reported about a subnational vulnerability assessment for the whole federal state of Mecklenburg-Western Pomerania, which dates back to 2007 and had no further implications for land-use planning.

Two research projects on climate change impacts and adaptation were conducted in the region, triggering further action. One project (the author was involved) initiated an intense collaborative science-practice dialogue on the topic of climate change impacts and future land-use development in the urban region of Rostock up to 2050 (for details, see Hagemeier-Klose et al., 2013). This process involved a discussion of future potential impacts of climate change on land-use development in the urban region; an assessment of further drivers of land-use change; the creation of different scenarios of future land-use development; and the development of strategies and measures for tackling these change processes. The main results were integrated in a framework concept of adaptation to climate change, drawn up by the urban administration's Department of the Environment in collaboration with scientists. This document, adopted by the City Parliament in October 2012, was intended to initiate a further strategy-building process for tackling the potential impacts of climate change.

12.1.2 Riga Case

Riga is the capital of Latvia with almost 700,000 inhabitants living in an area of 307 square kilometers (RPR, 2005, p. 8). Since the 1990s, Riga has been continually decreasing in terms of total population numbers due to low birth rates and migration. Riga has the main Latvian harbor and is the economic heart of Latvia, generating more than half of the gross national product (Albers et al., 2013, p. 35); Latvia was the first country in Europe to be hit by the economic crisis in 2008. The city is divided by the Daugava River, which flows into the Baltic Sea. Rising sea levels as well as groundwater levels and storm floods are expected to be the main severe potential impacts of climate change for the city. The whole planning region comprises five further communities and a total of 10,000 square kilometers with approximately 1.2 million inhabitants. For this region, an increase in temperature between 2.6 and 4°C is expected until end of the 21st century, as well as an increase in winter precipitation and a decrease in summer precipitation (Latvian Ministry of the Environment, 2009, p. 126f.). In 2005, Riga was hit by severe flooding with an estimated 60–70% of the urban territory being flooded.

After initial activities such as climate change impact assessments primarily at the national level, further explicit climate change—related activities were not pursued due to the economic crisis, but reference to climate change impacts and adaptation was made in other national documents. Land-use planning is not foreseen as a priority field of action. Due to a project cofunded by European funds (LIFE +), the city assessed hydrological risks and developed a strategy toward floods. Even if this 2-year project was hosted by an urban planning administration, it could not influence the static, already fixed and very detailed urban plan. The project ended with recommendations toward future land-use planning and floods.

The so-called "spatial plan" covers the whole urban territory and serves as a binding building plan, and providing details for implementation. Any small changes to the plan require formal procedures and can take up to years. The current plan does not address climate change impacts or adaptation, but floods and coastal erosion, which it addressed years before climate change became an issue and also a trigger for both of them. That way, it does already address climate change impacts, but not with an intended perspective on climate change. Also urban planning has thus far not worked on the topic of climate change, with the single exception of the mentioned LIFE - + project, but its findings were not yet integrated into planning.

12.2 CHALLENGES, OPPORTUNITIES, AND BARRIERS FOR RESILIENT LAND-USE PLANNING

12.2.1 Land-Use Planning Challenged by Specific Characteristics of Climate Change, Its Effects and Potential Impacts

What are the challenges facing those involved in land-use planning? And how can these challenges be judged, as seen through a resilience lens? First, climate change is perceived by practitioners as a geographically somehow distant (global) phenomenon that cannot be experienced directly. With more severe effects and impacts yet to come in the future, it is also temporally distant. This leads to the impression of many of the interviewees that climate change is not yet relevant for current land-use planning, which adopts only short-term planning horizons.

A second challenge is the inherent uncertainty of climate change scenarios that cannot be eliminated. This means that the potential impacts of climate change can only be described in ranges or eventualities. Here the challenge from a resilience perspective is that planning already now has to deal with this uncertainty regarding future developments and to provide for learning to live with change in the respective land-use plans. Planning practice is also faced with the challenge of capturing the potential impacts of climate change in their interdependencies with other changing social, technological, and ecological influencing factors on land-use development. So, the planning practitioners of Rostock expected to experience fewer future heat problems due to the coastal winds, but research showed, that already now, the city has an explicit urban heat island with higher temperatures than the peri-urban region (Richter et al., 2013). This illustrates the resilience challenge put to planning to consider socioecological interdependencies, nonlinear dynamics, and interdependent change processes.

Even if land-use planning takes an all-encompassing cross-sectoral approach, it is difficult to assess the aggregated impacts of climate change (Agard et al., 2014, p. 1758) and to find existing information on integrated or cross-sectoral impacts because few research results are available (see Kovats et al., 2014, p. 1304). In addition, it is virtually impossible to find specific localized information, or considerable research efforts are required to find such information. And as the cases show, real integrated all land-use relevant sector-spanning assessments are not usual. Instead, as for example, the Rostock case shows, every sector does its own assessments and these are then put together for generating the land-use plan.

Owing to these uncertainties, the ambiguities and potential states of not knowing, and the long-term horizon, as well as the manifold interdependencies across different scales, tackling impacts of climate change in land-use planning is considered to be weaker than meeting current land-use interests and their potential benefits (see also more general information in Underdal, 2010, p. 387).

12.2.2 Barriers and Opportunities Offered by Land-Use Planning to Address These Challenges

Considering the barriers and opportunities offered by land-use planning to address these challenges and to take on an orientation toward resilience and transformation, the planners within the cases have the added difficulty that the formal planning instruments available are inadequate. Both the German and Latvian cases revealed that the challenge of complexity and uncertainty cannot be tackled for the whole urban territory at the level of land-use planning. The plan is very static and therefore inflexible. Once adopted, it is binding and very difficult to change. The explorative reference case of Riga even revealed a more difficult situation—at least during the research period up to 2012. In this case, the land-use plan for the whole territory was simultaneously the development plan. This plan was detailed and fixed as a development building plan; it was legally binding, and provided no room for interpretation or further concretization, which the preparatory land-use plan in the Rostock case does to a certain extent.

The formal instruments available in the German planning system can be improved so as to make the situation less static. Improvements could include using spans instead of fixed margins or assigning only terminated land uses to specific areas in order to keep crucial areas free from fixed buildings (Othengrafen, 2014). In Latvia, there are now ongoing work to change the static and very detailed plan into a more flexible instrument.

While more severe impacts of climate change are predicted with a long-term horizon, the cases cross-cutting land-use planning only provided short-term planning horizons up to 20 years. However, the consequences of land-use planning can be long term in nature, such as affecting the infrastructure or buildings or altering the management of natural resources. It was mainly landscape and water planning as well as coastal hazard planning as part of sectoral planning that are viewed over a longer horizon. In Rostock, future vulnerabilities can be evoked through a planned new settlement area closed to the river front, in a potentially climate change–induced flood-prone area. In this case, practitioners perceive no pressure to act, because the potential impacts are detached from their own system and the current time frame of planning activities. And neither future generations nor nonhuman entities of an ethical nature are directly involved in planning processes (Underdal, 2010, p. 388; Hurrlimann and March, 2012, p. 484).

An additional barrier to taking up the challenges of climate change is the restricted framing of the problem due to specific practitioners' perceptions and attitudes. There was a dominant orientation toward risk and outcome in both cases. Concentrating on external and climate-induced risks only involves the danger of neglecting the socioecological context in which climate change impacts and processes of climate change adaptation manifest themselves. Uncertainty was not perceived as a systemic inherent characteristic of open socioecological systems. Shocks or disturbances are related to the external environment and are not linked to any (weak, nonintended, dysfunctional, etc.) internal system structures or functions. This reinforces the objective to preserve the system as it stands, and fails to provide transformative notions. It was nearly possible to detect an integrated view on the impacts of climate change in both cases, also within cross-cutting land-use planning. Mainly, however, there was a sectoral understanding of risks, vulnerabilities, and final impacts. Risks were mainly related to built infrastructure. There was little or no connection to the social sphere such as associated with demographic factors or social vulnerability.

In addition, most of the practitioners interviewed applied a merely linear perspective on their cities and the future development of drivers of land use, based on a scientific-technical ratio. The land-use planners interviewed stated that they depend on data from other sectors and that they have insufficient resources of their own to collect data themselves or that this is not provided for in the institutional planning procedure. Adaptation was also perceived as a task that mainly has to be brought forward at other levels, not at the local level.

12.2.3 Barriers and Opportunities Related to, But Outside of, Land-Use Planning

The division into different responsibilities and compartments in the urban and regional administration results in fragmented perspectives on and views of the city. In addition to this sectoral split, there also seems to be a weakened regional approach in Latvia that acts as a barrier because it fails to provide a formal arena to meet; there is also no impetus for reaching agreement between local interests. In addition, impacts and further interdependencies and consequences cover wider areas than merely the urban level.

There is little incentive to prevent potential climate change impacts from occurring, to prepare for living with further change, and to integrate it into current land use. As these tasks are related to uncertainties and longer-term horizons, they have a weak position in the general policy domain. This is especially the case in comparison with other interests that initially promise short-term economic benefits, in Rostock, which has severe financial problems and is looking for tax income, and particularly in Riga, the capital of Latvia, which as we have said was the first country to be hit by the economic crisis in Europe in 2008. Second, other interest domains also provide "exact or concrete" and seemingly secure data, even if they are based on scenarios of future development, too. Thus the potential impacts of climate change do not lead to a search for alternative integrated solutions that are sustainable. Rather, other interests prevail and nonintegrated building measures are optimized, also in current flood-prone areas, as is the case in Rostock.

12.3 CONCLUSIONS AND OUTLOOK

There are many challenges to land-use planning associated with the potential impacts of climate change. At present, land-use planning is embedded in local and regional power structures, and is part of the established institutional system (which potentially has less interest in change). It is therefore difficult for land-use planners to start their own initiatives or act as stewards for any transformations that may be necessary. However, if land-use planning is attributed a central role in climate change adaptation, also in local and regional practice, it also needs the respective powers required to deal with land as a limited collective resource. It should also be fully established as a truly integrating and all-encompassing territorial planning. To achieve the latter objective (or both), it is necessary that planning relies not only on sectoral (and interest-driven) data. It must be able to (1) relate this data to other aspects more effectively and (2) extract it from this combination or integrate new results (probably generated by itself) focusing on the entire urban regional system. Land-use planners must therefore be able to request additional data to be identified and brought up by themselves or to collect their own integrated socioecological data. An ideal, yet unrealistic, change would be to implement a real cross-sectoral approach by overcoming the sectoral split at the institutional and government level. This would have resource implications that may not be brought up by the local communes or regions but would require support at the subnational and national scales.

Even now, land-use planning could use more integrated models or interlinking analyzing tools to identify socioecological and any possible technical interdependencies and to influence future situations. One explicit strategy in pursuit of this integrated path would be to apply the ecosystem services approach also in cross-cutting land-use planning for the entire planning region. Ecosystem services can be interlinked with climate change effects, and their vulnerabilities can be interlinked with social groups that may be affected to a different extent.

Tackling sectoral fragmentation also means assuming responsibility not only for part of the system but for the city or urban region as a whole. This cannot be managed by planning alone. However, planners can propose the integration of spheres, sectoral perceptions, perspectives, and data, and can bring together all of the relevant actors with their different perspectives as well as knowledge forms (see e.g., Galderisi, 2016). Bringing together all of the stakeholders would be a step toward tackling the discrepancy between practice and science and the knowledge requested by practitioners and that provided by scientists (Van de Ven, 2007), as was shown in the Rostock science-practice collaboration, which accelerated the urban process on adaptation.

Not all challenges can be resolved by land-use planning alone. However, the catalogue of opportunities ranges from choosing one's own integrated perspective and initiating governance-related collaborative processes (see Innes and Booher, 2010) to learning from other planning systems to change one's own planning instruments. As far as the cited case studies are concerned, planning practitioners, who are embedded in their established institutional contexts, are merely expected to serve as stewards of sustainably transforming existing urban and regional structures and functions in order to achieve socioecological resilience.

According to findings of others (Measham et al., 2011; Hurlimann and March, 2012), climate change impacts and adaptation are not predominant issues in political practice at the local level but are competing with other priorities, which also promise short-term benefits. This is a severe constraint that is difficult to overcome. To do so, political leadership or at least support for practical land-use planning is needed. This was an issue in Rostock after an agreement was found to start a collaborative science-practice process on adaptation with the backing of a political leader in the city. Also supportive would be a self-conception of planning that takes up a transformative or more proactive notion (Albrechts, 2010), which goes further than the perceived passive and controlling role.

Acknowledgments

This research was substantially supported by a grant from the German Federal Ministry of Education and Research from 2009 to 2014 (grant number: FKZ 01UU0909). An earlier version of parts of this book chapter was an ACSP Conference 2014 paper under the ID 5383_186.

References

Agard, J., Schipper, E.L.F., Birkmann, J., et al., 2014. Annex II: glossary. In: Barros, V.R., Field, C.B., Dokken, D.J., et al. (Eds.), Climate Change 2014. Impacts, Adaptation, and Vulnerability. Part B: Regional Aspects. Contribution of Working Group II to the Fifth Assessment Report of the Intergovernmental Panel on Climate Change. Cambridge University Press, Cambridge, UK and New York, NY, USA, pp. 1757–1776.

Albers, M., Hasibovic, S., Deppisch, S., 2013. Klimawandel und räumliche Planung: Rahmenbedingungen, Herausforderungen und Anpassungsstrategien. In: Stadtregionen im Ostseeraum. HafenCity University Hamburg, Hamburg. Neopolis Working Papers: Urban and Regional Studies, 13.

Albrechts, L., 2010. More of the same is not enough! How could strategic spatial planning be instrumental in dealing with the challenges ahead? Environment and Planning B: Planning and Design 37, 1115–1127.

Bulkeley, H., 2013. Cities and Climate Change. Routledge, New York.

Deppisch, S., Hasibovic, S., 2013. Social-ecological resilience thinking as a bridging concept in transdisciplinary research on climate-change adaptation. Natural Hazards 67, 117–127.

Galderisi, A., 2016. Nexus approach to disaster risk reduction, climate adaptation and ecosystems' management: new paths for a sustainable and resilient urban development. In: Colucci, A., Magoni, F., Menoni, S. (Eds.), Peri-urban Areas and Food-energy-water Nexus. Sustainability and Resilience Strategies in the Age of Climate Change. Springer, pp. 11–21.

Hagemeier-Klose, M., Albers, M., Richter, M., Deppisch, S., 2013. Szenario-Planung als Instrument einer klimawandelangepassten" Stadt- und Regionalplanung — Einflussfaktorenanalyse und Szenarienkonstruktion im Stadt-Umland-Raum Rostock. Raumordnung und Raumforschung 71 (5), 413–426.

Hurlimann, A.C., March, A.P., 2012. The role of spatial planning in adapting to climate change. WIREs Climate Change 3 (5), 477–488.

Innes, J.E., Booher, D.E., 2010. Planning with Complexity: An Introduction to Collaborative Rationality for Public Policy. Routledge, London and New York.

Kovats, R.S., Valentini, R., Bouwer, L.M., et al., 2014. Europe. In: Barros, V.R., Field, C.B., Dokken, D.J., et al. (Eds.), Climate Change 2014: Impacts, Adaptation, and Vulnerability. Part B: Regional Aspects. Contribution of Working Group II to the Fifth Assessment Report of the Intergovernmental Panel on Climate Change. Cambridge University Press, Cambridge, UK and New York, NY, USA, pp. 1267–1326.

Krämer, I., Borenäs, K., Daschkeit, A., et al., 2012. Climate Change Impacts on Infrastructure in the Baltic Sea Region. Baltadapt Report # 5. Danish Meteorological Institute, Copenhagen.

Latvian Ministry of the Environment, 2009. Vides aizsardzibas un regionalas attistribas ministrija: Informativais zinojums par neformalo vides ministru snaksmi, kas 2009. Gada 14–15. Aprili notiks Praga (Cehija). Riga.

Measham, T.G., Preston, B.L., Smith, T.F., et al., 2011. Adapting to climate change through local municipal planning: barriers and challenges. Mitigation and Adaptation Strategies for Global Change 16 (8), 889–909.

Othengrafen, M., 2014. Anpassung an den Klimawandel: Das formelle Instrumentarium der Stadt- und Regionalplanung. Verlag Dr. Kovač, Hamburg.

Revi, A., Satterthwaite, D.E., Aragón-Durand, F., et al., 2014. Urban areas. In: Field, C.B., Barros, V.R., Dokken, D.J., et al. (Eds.), Climate Change 2014. Impacts, Adaptation, and Vulnerability. Part A: Global and Sectoral Aspects. Contribution of Working Group II to the Fifth Assessment Report of the Intergovernmental Panel on Climate Change. Cambridge University Press, Cambridge, UK and New York, NY, USA, pp. 535–612.

Richter, M., Deppisch, S., Storch, H., 2013. Observed changes in long-term climatic conditions and inner-regional differences in urban regions of the Baltic Sea coast. Atmospheric and Climate Sciences 3 (2), 165–176.

RPR-Riga Planning Region Development Council, Riga Region Development Agency, 2005. Spatial (Territorial) Plan of Riga Planning Region. Part I: Current Situation. Riga.

SUR-Arbeitskreis, 2011. Stadt-Umland-Raum-Rostock. Entwicklungsrahmen Stadt-Umland-Raum Rostock. Amt für Raumordnung und Landesplanung, Rostock.

Underdal, A., 2010. Complexity and challenges of long-term environmental governance. Global Environmental Change 20, 386–393.

Van de Ven, A.H., 2007. Engaged Scholarship, a Guide for Organizational and Social Research. Oxford University Press, Oxford.

Wilkinson, C., 2012. Socioecological resilience: insights and issues for planning theory. Planning Theory 11 (2), 148–169.

CHAPTER 13

Importance of Multisector Collaboration in Dealing With Climate Change Adaptation: The Case of Belgrade

Ratka Čolić, Marija Maruna
University of Belgrade, Belgrade, Serbia

13.1 THE BELGRADE LOCAL CONTEXT

Belgrade is the capital of the Republic of Serbia, a city of 1,639,120 inhabitants (according to the 2011 census). Its territory covers an area of 3222 km². The city is a metropolitan area composed of 17 municipalities, differing in sizes and in spatial, natural, and economic characteristics, from the central core of the city to peripheral suburban areas containing land of either areas devoted to agriculture and/or undeveloped. Belgrade has a broader set of powers than other municipalities and towns of Serbia; the city is responsible for governing and ensuring protection of water resources, governing local and state roads, organizing its municipal police force, ensuring fire protection, etc. Belgrade is the hub of Serbia's education, health care, and culture, and accounts for 35% of the country's gross domestic product.

The territory of Belgrade has been subjected in the past to extreme weather events; the most severe were the 2014 floods, a disaster with consequences never recorded in the previous 200 years. At high water, the Sava and Danube rivers threaten the areas of the city lying along their banks. About 24 km of the Sava's course are located in the territory of Belgrade, as are about 50 km of the Danube. In this area, both rivers receive water from extensive networks of tributaries. About 29 km of embankments have been built in the city's central districts, and these cover about one-fifth of the entire riverbank length. Over 100 smaller torrential streams are active in Belgrade and pose risks to flooding of parts of the city, as their flash floods can be highly dangerous.

Belgrade is also faced with numerous environmental issues. These include sanitation and hygiene, from elementary issues with housing, water supply, and wastewater disposal and waste management, to modern-day environmental problems, such as natural and rural ecosystems' fragmentation, air pollution, CO_2 emissions, etc.

According to climate change forecasts, Belgrade is expected to be affected by more severe heat waves, greater precipitation, and more powerful storms, as well as less pronounced effects of extreme cold (CCAPVA, 2015).

Belgrade's urban development has been driven over the past 2 decades by both local and broader political and socioeconomic circumstances that the country has found itself in, including economic transition. Services dominate the city's economy (accounting for 69% of output), followed by manufacturing (21%), agriculture (3%), tourism (2%), etc. Greater investment, in particular into commercial developments and office space, has accelerated the growth of Belgrade since 2000. Belgrade is home to 21% of the total population of Serbia. The conflicts in the region and the NATO bombing campaign resulted in the influx of some 140,000 refugees (1996) and about 230,000 internally displaced persons (1999). In a more recent development, Belgrade has become a major focal point for the latest wave of migrations in Europe.

13.2 CURRENT INITIATIVES/PRACTICES

Belgrade has made efforts to define plans, strategies, and programs aiming to reduce vulnerability and improve resilience to climate risks. The formulation of policies is led by city institutions, mainly through multisectoral and multidisciplinary task forces, and in some cases, supported by international donor projects. These policies are described next in chronological order (Fig. 13.1).

13.2.1 Environmental Atlas of Belgrade (1998–2002)

At the initiative of the City Public Health Agency, a project was launched in 1998 to evaluate the City of Belgrade General Urban Plan from an environmental perspective. This environmental evaluation was designed to feed into the overall evaluation of the land-use plan as the starting point for striking a balance between the quality of the environment and the city's spatial and functional structure. The primary aim here was to develop and operationalize an environmental map model that could be used to enhance environmental management and be integrated into the planning, construction, and land-use process. Twelve sectoral data groups covered issues of land use, geology and hydrogeology, climate, renewable energy, polluters and geotechnical factors, hazards, air, water, soil, noise, and health. A total of 56 maps were produced that displayed both natural and man-made features of the Belgrade area. Data collected in the Environmental Atlas were also used in the development of the 2015 Climate Change Adaptation Plan and Vulnerability Assessment.

13.2.2 Green Regulation of Belgrade (2002–07)

The Green Regulation project was initiated in 2002 by the City Secretariat for Environmental Protection. The initial assessment here found that the key issues were lack of data; negligible percentage of green areas actually introduced in comparison to area planned; uncontrolled rezoning of green areas to allow development; lack of standards; inefficient systems management; and no monitoring of the state of play in this field (Cvejić, 2011).

This project was implemented in a number of stages: (1) situation analysis and proposal of a decision to protect and enhance green areas (2002–03); (2) preparation of a geographic information system for green areas (2003–04); (3) mapping and evaluation of Belgrade's biotopes (2005–07); and (4) development of the General Regulation Plan covering the system of green areas in Belgrade. The 2007 planning initiative constituted a key step in the management of green areas. A nearly 2.5-fold increase in green areas was envisaged, to 20,146 ha (25.96% of total area) from the then-current figure of 10,541 ha (or 14.96% of total area). The funds to develop the plan were allocated in the city budget, and the plan was prepared by the Belgrade Urban Planning Institute. The 2015 Climate Change Adaptation Plan defined green infrastructure as a priority initiative for the city.

13.2.3 Regional Spatial Plan of Belgrade (2003, 2011)

The Regional Spatial Plan (RSP) of Belgrade covers the administrative territory of the city, i.e., all of its 17 municipalities. It aims to incorporate climate change into sectoral strategies and ensure the development of a sustainable

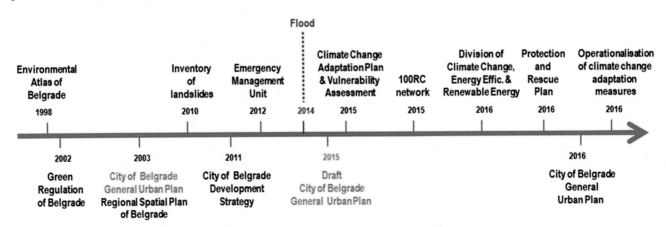

FIGURE 13.1 Belgrade resilience initiatives. *Authors' elaboration.*

system to manage the risk of climate change. The RSP sets the following strategic priorities: (1) development of a climate monitoring system and database of spatial information on local climate change; (2) implementation of a multidisciplinary research program to examine local climate change and its impact on agriculture, forestry, water management, energy, biodiversity and ecosystems, infrastructure, and human health; (3) introduction of environmentally friendly technologies, including use of renewable energy; and (4) creation of a National Climate Change Centre. The RSP was developed by the Belgrade Urban Planning Institute; it was enacted by the Belgrade City Assembly in 2003 and amended in 2011.

13.2.4 City of Belgrade Development Strategy (2007–11)

One of the principles underpinning the City of Belgrade Development Strategy is the introduction of environmental awareness and understanding of environmental sustainability as preconditions for growth. One project in particular can be highlighted from amongst the set of measures envisaged in this Strategy; this involves protecting and developing the environment and ensuring control of climate change impact, including monitoring of water quality and use, disposal and processing of both solid and liquid waste, natural disaster risk management, and elimination of risk flash points.

The first draft of the Strategy was developed in 2008 by the Belgrade-based nongovernmental organization (NGO) Palgo Centar together with a crosscutting team composed of both national and international experts. The preparation of the second, expanded and amended, version was led by a team of the City of Belgrade Urban Planning Institute; this iteration was enacted by the Belgrade City Assembly in 2011.

13.2.5 Emergency Management Unit (2012–16)

The City Assembly established the Emergency Management Unit in 2012. The Unit is headed by the mayor and made up of officers of city directorates, public authorities, and enterprises (for a total of 33 members). City bodies and public authorities represented include: information delivery; municipal police; inspectorates; business governance; urban planning and construction; utilities and housing; environmental protection; healthcare; social services; transportation; education and child welfare; emergency management at the national Ministry of Interior; water management; National Meteorological Office; fire and rescue brigades; energy; power plants; road maintenance; parking; public transportation; parks; public sanitation; water supply and sewage; public health; emergency medical response; Red Cross; national Institute for Biocides and Environmental Medicine; and veterinary medicine.

The Emergency Management Unit is responsible for coordinating protection and rescue efforts in the event of flooding, hail, and gale-force winds; earthquakes; building collapses; landslides, and erosion events; droughts and elevated temperatures; and fires and explosions. It is also tasked with civil defense duties, including evacuating members of the public at risk and providing them with shelter and care.

In 2014 a separate coordinating body was established to manage landslides in Belgrade; this authority has monitored the recovery phase following damage caused by landslides and reviewed the city's landslide inventory. Landslides had previously caused major damage and led to the resettlement of parts of a number of neighborhoods.

13.2.6 Inventory of Landslides (2010–16)

The city of Belgrade initiated the creation of an inventory of landslides in 2010, and this database was updated after the 2014 flooding. The critical parts of the urban area of the city lie on the right banks of the Sava and the Danube, with the neighborhoods of Karaburma, Mirijevo, and parts of Zvezdara facing the greatest risk. On the outskirts of the city, also affected are parts of the suburban municipalities of Obrenovac, Grocka, and Lazarevac, with the neighborhood of Umka threatened by one of the largest potential landslides in the Balkans. A total of 1155 critical locations have been inventoried throughout Belgrade; of these, 602 are active landslides.

Funds to rehabilitate areas prone to landslides are provided by the city itself, by the Ministry of Emergency Situations (which funded the development of the landslides inventory in Belgrade, as in other Serbian municipalities), by the European Union, etc.

13.2.7 Climate Change Adaptation Plan and Vulnerability Assessment (2015)

In 2015, the city of Belgrade adopted the Climate Change Adaptation Plan and Vulnerability Assessment (CCAPVA). This initiative was undertaken as part of a regional project, Climate Change Adaptation in the Western

TABLE 13.1 List of "High Priority" and "Very High Priority" Climate Change Adaptation Measures

Type of Measure	Measure	Priority
Urban green structures	Green infrastructure	++
	Green open spaces	+
	Green alleys	+
Water systems	Flood protection	++
	Retention basins	+
	Water saving and reuse	+
	Water drainage (rainwater and watercourse regulation)	+
Urban planning	Urban planning to avoid flood risk	++
Nonstructural measures	Awareness raising and behavior change	+
	Informing the public about adaptation to extreme events	+
	Institutional and organizational measures	+
	Warning system	+

+, high priority; ++, very high priority.
Skupština Grada Beograda (Belgrade City Assembly), 2015. Climate Change Adaptation Plan and Vulnerability Assessment.

Balkans, implemented by the German Corporation for International Cooperation, with the city's Secretariat for Environmental Protection serving as the primary counterpart in Belgrade. City secretariats, public enterprises, and institutions took part in drafting the plan, as did NGOs; these entities are also envisaged to take part in its implementation. The plan contains a vulnerability assessment, assessment of future risks and opportunities, and a list of priority measures and activities.

The vulnerability assessment was based on information on extreme weather events between 1995 and 2014 and spatial aspects of receptor vulnerability. Climate change trends for the area of Belgrade were taken from the Initial National Communication of the Republic of Serbia (2010) under the United Nations Framework Convention on Climate Change (UNFCCC).

Flood protection, green infrastructure, and urban planning are measures accorded the highest priority for the city. In addition, high priority is given to improving the warning system, informing the public, awareness-raising, and other institutional and organizational measures, as well as to constructing retention basins, water drainage, water saving and reuse, and creation of rehabilitation of green open spaces and streets (Table 13.1).

Adoption of the Climate Change Adaptation Plan and Vulnerability Assessment has made Belgrade the first city in Serbia to launch activities to implement prioritized climate change adaptation measures. All measures include an indication of the implementing agency, time frame, and means to monitor implementation of the plan.

13.2.8 Belgrade as Part of the 100 Resilient Cities (100RC) Network (2015)

Belgrade is part of the 100RC Network. Under the mandate of this association, which supports the adoption and incorporation of a view of resilience that includes not just the shocks but also the stresses, the following "resilience challenges" have been defined for Belgrade: aging infrastructure, chronic energy shortages, coastal flooding, lack of affordable housing, landslide, rainfall flooding, refugees, and social inequity. Among the Network's key pathways is the membership of Belgrade in a global network of member cities that can exchange their experiences and help one another, as well as guidelines to institutionalize the position of a chief resilience officer in city government.

13.2.9 Establishment of the Division of Climate Change, Energy Efficiency, and Renewable Energy (2016)

According to the city government's new organizational structure, enacted in 2016, a Division of Climate Change, Energy Efficiency, and Renewable Energy was established as part of the existing Department of Strategic Planning and Intersectoral Coordination at the Secretariat for Environmental Protection (Belgrade City Assembly, 2016).

13.2.10 Protection and Rescue Plan and Risk and Vulnerability Assessment

In 2016 the city of Belgrade adopted a Protection and Rescue Plan and a Risk and Vulnerability Assessment. The Plan deals not only with climate change risks, but also with landslides, and erosion events, fires and explosions, earthquakes, building collapses, evacuation of people, etc.

13.2.11 City of Belgrade General Urban Plan (2016)

The new General Urban Plan for Belgrade (GUP) was adopted in 2016. The Plan covers an area of some 56,540 ha zoned for development. In addition to defining strategic priorities through new "projects of interest for the City," the Plan provides new solutions and defines new objectives of development and primary program elements that pertain to agricultural areas; afforestation; the city's riverbanks; environmental protection; renewable energy; and energy efficiency of built structures.

In addition to objectives, the Plan also defines climate change risk protection and adaptation features such as environmental protection measures; measures to protect agricultural land; erosion and landslide protection; flood protection; local heat island impact mitigation; wind impact mitigation; reduction of pollution, emissions of volatile organic compounds, and noise; rainwater regulation; and measures to address and regulate torrential streams. Specific measures to mitigate climate change include greater efficiency in the use of Belgrade's own energy generation potentials; reduced emissions of greenhouse gases; and reduced imports of fossil fuels.

These measures are detailed in the narrative section of the GUP, primary land-use maps, and other thematic maps. Particularly significant in this regard is a Map of Limitations to Urban Growth, attached to the Plan as a separate appendix, which shows areas with limited use due to their geological features (such as landslides, ravines, and areas of tectonic weakness); divides the city into zones by feasibility for development according to engineering and geological criteria (reclamation and rehabilitation measures and seismicity); indicates water catchment areas with their immediate protected areas; shows hydrological features (watercourses; wells; thermal and mineral water; and stagnant water); and shows facilities with elevated risk of chemical accidents and special restricted areas (airports, broadcasting ranges, etc.).

Environmental and natural resource protection measures include a detailed definition of steps to mitigate the adverse impact of climate change (Table 13.2). These have been aligned with urban development considerations and incorporated into planning documents.

The GUP does not include any direct implementation mechanisms. Unlike the CCAPVA, which is an action plan with prioritized climate change adaptation activities, all measures envisaged under the GUP have identical priority rankings due to its strategic development nature. The manner and degree of its implementation can be monitored through programs and activities of the various responsible authorities.

13.2.12 Operationalization of Climate Change Adaptation Measures (2016)

Reviewing priority measures under the Climate Change Adaptation Plan and the City of Belgrade GUP, we analyzed local sectoral development programs that are primarily funded by the city. At the annual level, the

TABLE 13.2 List of Measures to Mitigate Adverse Impacts of Climate Change

Type of Measure	Measure	Identical Priority Ranking
Measures to mitigate adverse impacts of climate change	Improved maintenance of existing wellheads	-
	Expansion of existing systems to protect from flooding, erosion, and torrential streams; enhanced protection from high water	-
	Revitalization of forests in the Danube region	-
	Dynamic and purpose-oriented conservation of biodiversity of forests and forested and urban areas	-
	Adaptive management and enhanced sustainability and multifunctionality of forests under conditions of climate change	-
	Expansion of ecologically functional areas/biotopes (Biotope Area Factor or Green Factor)	-

Generalni urbanistički plana Beograda (City of Belgrade General Urban Plan), 2016. Službeni list grada Beograda, br. November 2016.

Belgrade City Assembly enacts the Construction Land Development and Allocation Program, which covers the development of land zoned for construction and investment into preparation and construction of major facilities of significance for the city. The 2016 Program has earmarked substantial funds for construction of water supply and sewage networks. In addition, a large number of detailed zoning plans have been under development since 2015 that deal with construction of water supply facilities, sewage systems, wastewater collectors, and wastewater treatment facilities; regulation of retention basins; construction and reconstruction of streets; and regulation of watercourses.

Belgrade's suburban municipalities at risk of flooding and other natural disasters and emergencies have prioritized the construction of roads; rehabilitation of existing Ranney Collectors; and construction of regional water supply mains, facilities, and networks. Both urban and suburban municipalities have set the construction of rainwater and wastewater collectors, sewerage, and wastewater collection facilities as their priorities. Funding has been provided for these purposes by the Belgrade Land Development Agency, from the city budget, and from special-purpose loans used to finance road building. The Secretariat for Environmental Protection has allocated funds for afforestation in its 2016 budget.

13.3 CRITICAL ANALYSIS OF CURRENT INITIATIVES/PRACTICES

The example of Belgrade has been used to analyze local climate change mitigation and adaptation policies in relation to the city's development strategies. Of particular significance is the integration of climate change adaptation policies into planning policies, contributed to by the multidisciplinary approach and intersectoral cooperation.

13.3.1 Transition and the Resilient City

The example of Belgrade has been used to demonstrate a chronological sequence of local climate change planning, mitigation, and adaptation policies. These, however, are not confined to a single unified plan or initiative, but rather constitute a complex set of individual efforts made by various local (and involved national) institutions and fields of activity, from public health, environmental protection, land-use planning, infrastructure, to water management, agriculture, forest protection, etc. These policies can be implemented only if there is intersectoral cooperation and integration of priority measures that provides a joint framework for action.

According to Lazarević-Bajec (2011), it is of importance for the city in transition and the postsocialist urban context to "strengthen links between climate change adaptation policies and sustainable development, which indicate the need for integration." Serbia is characterized by a specific postsocialist/transition context of urban restructuring. Political primacy has been accorded to economic development, attraction of investment, and reduction of unemployment. Transition changes have been accompanied by deregulation and reduced control, as well as a lack of funds. In addition, Serbia's institutions are underdeveloped and lack continuity; moreover, appropriate procedures and capacities are also lacking, although these are considered key factors in adaptation to climate change (Maruna, 2012).

13.3.2 Level of Integration Between Mitigation and Adaptation Initiatives and Plans

It is at the level of integration between mitigation and adaptation initiatives that all advantages of intersectoral cooperation are revealed. An example of this is the Emergency Management Unit, which has both spearheaded urgent flood relief actions and initiated political, organizational, and technical support for city institutions in developing strategies and plans that deal with cumulative effects and risks of climate change and its long-term mitigation. The development of the CCAPVA (2015) and GUP (2016), as well as Belgrade's inclusion into the 100 RC Network, are examples of efforts to consider these issues from the perspectives of both short- and long-term action.

13.3.3 Mainstreaming of Mitigation and Adaptation Practices into Urban Planning Processes

The Climate Change Adaptation Plan aims at "integrating climate change adaptation mechanisms into the process of managing urban development and urban planning." This, however, is nothing new, since planning has in Serbia traditionally included considerations designed to minimize the risk of flooding, landslides, fire, etc. (as is the case of the RSP for the Administrative Area of Belgrade, Belgrade General Urban Plan, numerous zoning plans, and the City Development Strategy). What does constitute an innovation is that climate change mitigation and adaptation measures have been incorporated into urban development in the Belgrade General Urban Plan. Yet, being a strategic plan, the GUP does not provide for operational implementation of these measures as it does not envisage mechanisms for putting them into effect. The GUP does allow the definition of context-adapted policies, rather than application of generic solutions. It also permits the definition of priority projects for implementation of measures by sector.

Integration of climate change adaptation into local development strategies and plans depends on sectoral coordination, knowledge, and ability to appropriately incorporate these considerations, but also relies on political support in an environment fraught with insufficient funding. Why, in addition to the development of strategies and plans, do we highlight the ability to implement them? Because this is the locus of key weaknesses in postsocialist/transition context of spatial planning and development, as well as in efforts to enhance urban resilience. The 2014 flooding that struck Serbia, and Belgrade in particular, additionally underscored all the weaknesses of the system but also provided better insight into critical areas and the importance of intersectoral work and joint action to address problems. The floods also underlined the significance of coordination and cooperation, not just in developing strategies and plans but also in implementing policies and pursuing concrete projects.

13.3.4 Existing/Potential Barriers or Criticalities that Might Be Obstacles to Achievement of the Expected Goals

Although climate change mitigation is often said to depend primarily on activities of the state (Maruna, 2012), bottom-up initiatives are particularly important for the local context. Efforts have been made since 2008 in Serbia to establish a legislative, institutional, and political framework to deal with climate change; these can be identified in parts of national strategies for various areas and sections of individual laws (Maruna, 2012). The principles that underpin Serbia's climate change legislation are based on international treaties that include reporting requirements with regard to implementation of multilateral treaties, including the UNFCCC. In 2010, Serbia adopted its Initial National Communication, the country's first report under the UNFCCC. More recent initiatives include the development of the national Climate Change Adaptation Strategy and Action Plan. The first draft of this Strategy was prepared in 2015 as part of a project to develop Serbia's Second National Communication under the UNFCCC in collaboration with the UN Development Program and the Global Environment Facility.

The level of coordination between climate change policies and other developmental policies has been assessed as low (Lazarević-Bajec, 2012). In these circumstances, decision-making at the local level is crucial for taking adaptation into account and linking it with sectoral and intersectoral policies. The peculiar nature of adaptation requires sound information, local knowledge, and dedicated actors that are prepared to develop and implement appropriate strategies and projects aimed at reducing vulnerability and enhancing resilience to climate risk (Lazarević-Bajec, 2012). The framework for action is here provided by the powers of local administrative authorities built up through several decades of decentralized land-use planning, as well as by the new requirements to monitor implementation of plans and respect municipal and urban planning order.

And yet, to be able to take a more realistic look at these critical aspects, understanding of weaknesses in implementing policies is also needed. These primarily entail poor implementation of and insufficient respect for current plans and decisions, as borne out by government officials: "[We] have now compiled an inventory of landslides for many towns, for instance the inventory for Belgrade is complete, but we have recently been seeing cases where these decisions have not been adhered to and where building permits have been issued for areas prone to landslides or at risk of flooding" (interview with Minister for Emergency Situations, April 5, 2016). Measures defined in urban plans do not entail automatic implementation. The degree to which plans, strategies, and programs are implemented is often highlighted as a weakness of the planning system. Coordination and cooperation between institutions was seen as crucial in more complex situations involving flood relief, rehabilitation of landslides and torrential streams, unpermitted construction, etc. Maintaining awareness of the importance of adaptation to climate change is not just an issue for the professional public and cannot be placed high on the priority ladder only at times of crisis. It requires continuity and perseverance in practical implementation and a high degree of awareness on the part of the general public. Findings of public opinion surveys show that members of the public are by and large dissatisfied with performance of decision makers and the media when it comes to climate change.

The example of local policies of Belgrade underscores the importance of local initiatives, underpinned by the specific logic of the local context, institutions, human resources, and willingness to make and integrate efforts in tackling problems and proposing solutions.

References

Cvejić, J., 2011. Zelena Regulativa Beograda- Koncept i Realizacija [Green Regulation of Belgrade – Concept and Implementation] Travanj- svjetski dan krajobrazne arhitekture, Park Maksimir 17.04.2011. Zagreb.

Generalni urbanistički plana Beograda [City of Belgrade General Urban Plan], 2016. Službeni list grada Beograda, br. 11/2016.

Lazarević-Bajec, N., March 2011. Integrating climate change adaptation policies in spatial development planning in Serbia. A challenging task ahead. SPATIUM International Review 24, 1–8.

Lazarević-Bajec, N., 2012. Integracija politika adaptacije na klimatske promene" ['Integration of climate change adaptation policies']. Uticaj klimatskih promena na planiranje i prjektovanje. Razvijanje optimalnih modela [Impact of Climate Change on Planning and Design. Development of Optimal Models] (ur. V.Đokić i Z.Lazović). Univerzitet u Beogradu, Arhitektonski fakultet, Beograd, pp. 58–82.

Maruna, M., 2012. Regionalne strategije prilagođavanja klimatskim promenama: smernice za urbanističko planiranje u Srbiji (Regional climate change adaptation strategies: Guidelines for urban planning in Serbia). Arhitektura i urbanizam 36, 50–55.

Skupština Grada Beograda [Belgrade City Assembly], 2015. Climate Change Adaptation Plan and Vulnerability Assessment (CCAPVA).

Skupština Grada Beograda [Belgrade City Assembly], 2016a. Rešenje o obrazovanju Gradskog štaba za vanredne situacije na teritoriji grada Beograda [Decision Establishing the City of Belgrade Emergency Management Unit] Službeni list grada Beograda, br. 64/2016.

Skupština Grada Beograda [Belgrade City Assembly], 2016b. Druge izmene i dopune Programa uređivanja i dodele građevinskog zemljišta za 2016.godinu [Second Set of Amendments to the Construction Land Development and Allocation Programme for 2016], Službeni list grada Beograda, br. 64–20.

Skupština Grada Beograda [Belgrade City Assembly], 2016c. Informator o Organizaciji i Radu Organa Grada Beograda [Information Bulletin on the Organisation and Operation of Bodies of the City of Belgrade], Beograd, May.

Further Reading

Čolić, R., Lalović, K., Maruna, M., Milovanović-Rodić, D., 2015. Disaster risk management in Serbia and flooding in obrenovac municipality. In: Fokdal, J., Zehner, C. (Eds.), Resilient Cities: Urban Disaster Risk Management in Serbia (Report on the Results of a Case Study Research Project (2015). Berlin University of Technology, Berlin.

Fokdal, J., Zehner, C. (Eds.), 2015. Resilient Cities: Urban Disaster Risk Management in Serbia (Report on the Results of a Case Study Research Project). Berlin University of Technology, Berlin.

Institute of Public Health of Belgrade, Agency for City Building Land and Development of Belgrade, 2002.). Environmental Evaluation of the Area of Belgrade Master Plan, Vol. B: Spatial Presentation (maps) of Sectoral analyses, 'Environmental Atlas of Belgrade', Belgrade. Available at: zdravlje.org.rs/ekoatlas/indexe.html.

Interview With the Minister for Emergency Situations, April 5, 2016. Available at: lajkovacnadlanu.rs/sedmica-u-znaku-posete-ministra-ilica/.

Jovčić, I., Šćekić, V., 2016. Research on Opinion of General Public Towards Climate Change in General and Climate Change Adaptation. Environment Improvement Centre, Belgrade. Available at: cuzs.org/files/Analysis_on_Public_opinion_towards_Climate_Change_in_general_and_Climate.pdf.

Maruna, M., Čolić, R. (Eds.), 2015. Inovativni metodološki pristup izradi master rada: doprinos edukaciji profila urbaniste [Innovative Methodological Approach to Master's Thesis Writing: A Contribution to Planner Education]. Arhitektonski fakultet, Beograd.

Maruna, M., Čolić, R., Fokdal, J., Zehner, C., Milovanović Rodić, D., Lalović, K., 2015. Collaborative and practice oriented learning of disaster risk management in post socialist transition countries. In: XVI N-AERUS Conference: Who Wins and Who Loses? Exploring and Learning From Transformations and Actors in the Cities of the South, Dortmund.

Ministarstvo poljoprivrede i zaštite životne sredine (MPZŽS) [Serbian Ministry of Agriculture and Environmental Protection, MAEP], 2015. Treći pregled stanja životne sredine u Srbiji, Prioriteti u oblasti životne sredine/Ključni izazovi koji nam predstoje [Third Environmental Performance Review of Serbia, Environmental Priorities/Key Challenges], Beograd, December.

Ministry of Environment and Spatial Planning, 2010. Initial National Communication of the Republic of Serbia Under the United Nations Framework Convention on Climate Change. Available at: unfccc.int/resource/docs/natc/srbnc1.pdf.

Regionalni prostorni plan administrativnog područja grada Beograda [Regional Spatial Plan for the Administrative Area of the City of Belgrade], 2011. Službeni list grada Beograda, br. 10/04, 38/11.

Strategija razvoja grada Beograda [City of Belgrade Development Strategy], 2011.

CHAPTER 14

Genoa and Climate Change: Mitigation and Adaptation Policies

Daniele F. Bignami, Emanuele Biagi

Fondazione Politecnico di Milano, Milan, Italy

14.1 THE GENOA CASE STUDY: CLIMATIC AND GEOGRAPHICAL FEATURES

Genoa, which is the capital of the province and the region of Liguria, is one of the 15 Italian metropolitan cities with an urban area of more than 800,000 inhabitants. It represents one of the several typical urban areas where climate change is projected to increase risks for people, assets, economies, and ecosystems, including risks from heat stress, storms and extreme precipitation, inland and coastal flooding, landslides, air pollution, drought, water scarcity, sea level rise and storm surges, all of which were recently indicated by the Intergovernmental Panel on Climate Change (IPCC) as a scenario with "very high confidence" (IPCC, 2014). These risks are amplified for those lacking essential infrastructure and services or living in exposed areas; sadly, we have to acknowledge that over the years Genoa has gone through a ruthless urbanization process, which has been careless of the territory and of the consequences that it could generate (Rosso, 2014).

As further stated by the IPCC (2014), in the case of Genoa too, effective adaptation and mitigation responses depend on policies and measures across multiple scales: international, regional, national, and local. Coherently, even if in Genoa adaptation options exist in all sectors, the context of implementation and the potential to reduce climate-related risks have to be carefully and locally evaluated. In the same way, mitigation options are available in every major sector, but they can prove themselves more cost-effective if combining the right mix of measures, for instance, to reduce energy use and decarbonize energy supply of the city.

As in many other Euro-Mediterranean sites, the urbanization in the Genoa coastal areas is mainly due to the lack of space needed for the development of the harbor and growing industrial activities. Since the beginning of the 20th century there has been a gradual expansion toward the sea: between the Voltri area of Genoa and the mouth of the Bisagno river, large areas with embankments and coastal defenses have been created, which radically transformed the coastline landscape. Furthermore, the Genoese streams have not been spared by such an invasive urbanization; they have suffered major changes, especially in their terminal sections, being forced most of the time to flow into small artificial channels, most of them covered and in most cases insufficient to discharge the design flow (Fig. 14.1).

It is no coincidence that hydraulic tests and hydrological studies (in addition to the recent floods of 2011, 2014, and 2015) show how the most critical areas of the basin, with respect to flood risk, are the ones situated along the principal axis of the Bisagno, in the final covered section (which goes from Brignole railway bridge right to its opening into the sea) and the uncovered one, located between the railway bridge and the confluence between the Bisagno and the Fereggiano stream.

14.1.1 Climate and Climate Change in Genoa

Generally, as a consequence of climate change, the metropolitan cities' territories, as it happens in most of the Mediterranean areas, are likely to be exposed to a higher intensity and/or frequency of phenomena such as summer heat waves, prolonged periods of drought, but also heavy rainfalls, winter frosts, and storms. These phenomena can have dramatic consequences on the territory, such as potential increase in the risk of fire, reduction in welfare

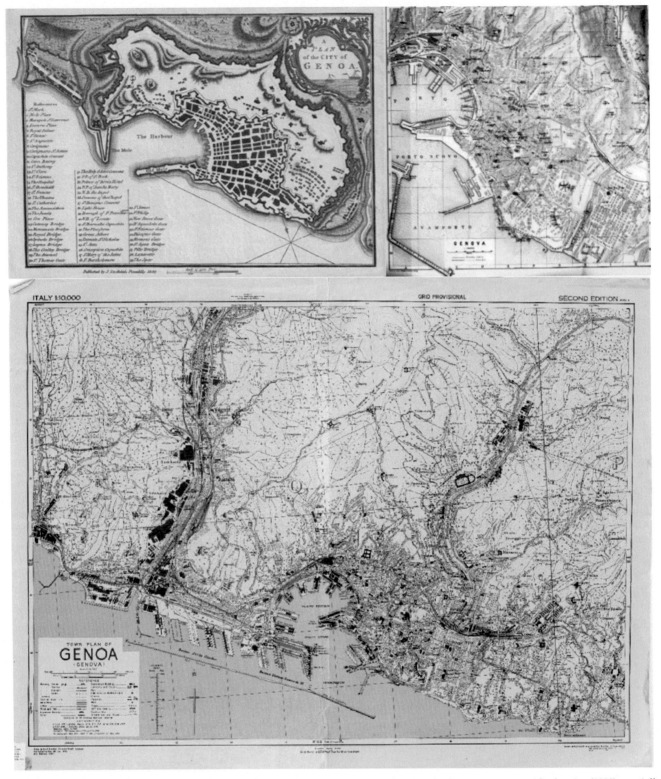

FIGURE 14.1 Genoa maps of 1800 (*top-left*), 1913 (*top-right*), and 1943 (*bottom*) showing the Bisagno river outside the city (1800), partially urbanized (1913), and covered (1943), as it is today.

(mainly due to heat waves), decrease of biodiversity, increase of hydrogeological risk, and reduction in water availability in summer.

For all these reasons it is necessary to accurately guide planning decisions in order to reduce the vulnerability of the territory.

Genoa's climate is generally very favorable to human activities; it is characterized by mild temperatures, which occur within a limited range, abundant rainfall, high solar radiation, and lively ventilation. These features, however, may be considered valid only in the coastal areas and in the central amphitheater of the city. The climate of the inner suburbs, where urbanization mostly expands toward inland Apennine valleys, may instead differ significantly, due to the rugged topography of the area and the increasing distance from the natural temperature mitigation provided by the sea.

In fact, the climate of the most urbanized parts of Genoa, even if these areas directly overlook the sea, have Mediterranean characteristics only along the coast, especially the western one (according to the best known, and still among the most frequently used systems, of Köppen (1936)). In fact, this small area is the only one that falls in the Csa area (hot dry-summer climate), which corresponds to a subtropical climate with dry and hot summer, widely known as Mediterranean climate. Moving eastward, toward the inland valleys, there is a shift from the Csa area to the transition area Cfsa (temperate climate transition to the Mediterranean), in which most of the municipal area can be found. Lastly, higher along the slopes, there is a shift to climate zone Cfb (temperate climate with warm summer).

This general framework seems to assimilate Genoa's climate with the one of many other Italian cities. However, there are some peculiar factors that, albeit sporadic and limited extension of time/space, strongly mark the meteorology of the city, sometimes affecting human activities. One factor that stands out among all the others is the distribution of rainfall, which can reach really high intensities, incomparable to the ones of other European cities (Genoa Municipality, 2014).

In such context, it is interesting to study the potential impact of climate change, as pointed out in a study carried out in 2012 regarding a Genoa case study (Brandolini et al., 2012), in which the time series of annual precipitation totals were crossed with the number of rainy days. This showed how, in recent decades, although the amount of total annual rainfall was basically the same, the rainy days have substantially decreased. This means that a net increase in the intensity of rainfall occurred, which is a possible indicator of a shift toward a more extreme climate regime. Experts nowadays are monitoring and evaluating climate classifications on the earth's surface; they can represent a convenient and still quite simple tool for the validation of climate models and for the analysis of simulated future climate changes (Belda et al., 2014). In addition, among observed and projected impacts from climate change for the Mediterranean region of Europe (EEA, 2012), other potentially relevant impacts involving the Genoa urban area are: rise in temperatures higher than European average; decrease in annual precipitation; decrease in annual river flow; increasing risk of biodiversity loss; increasing risk of forest fires; increase of mortality from heat waves; and decrease in hydropower potential.

14.2 DESCRIPTION OF CURRENT CLIMATE INITIATIVES IN GENOA

At the local level, to face the effects of climate change, first of all it is necessary to define effective adaptation policies in order to minimize the negative impacts on the territory. In fact, the adaptation process includes the innovation of urban areas' design criteria, with the purpose of harmonizing the land development needs with the principles of environmental sustainability.

In order to cope effectively with the problem of climate change, the province of Genoa has launched a number of initiatives, also in connection with major European projects.

Among these, a first line of action consists of the adaptation of the territory to the effects of climate change. For this purpose, a significant project is the one cofunded by Genoa province (today, Metropolitan City of Genoa), under the European Union (EU) program INTERREG IVC, the program of interregional cooperation across Europe, financed through the European Regional Development Fund (ERDF), the project "GRaBS: Green and Blue Space adaptation for urban areas and eco towns" (2008—11). The purpose of the project was to integrate adaptation strategies to climate change into regional planning and development, promoting the adoption of adaptation action plans (AAPs), through the development of a database of good practices and studies. In particular, the GRaBS project concerned adaptation strategies focused on planning, specifically related to urban areas and linked to green areas and drainage systems, playing a central role in local adaptation strategies. The project included the deployment of "green and blue" infrastructure (i.e., the use of water and vegetation components as primary elements of thermoregulation and ecological continuity). To meet the specific needs of the territory, the province of Genoa has carried out many activities, such as preparatory studies for local evaluation, pilot projects, divulgation activities in the area, and drafting of local Adaptation Action Planning Toolkit instruments for the municipalities that belong to the provincial area (see Cartogis website, GRaBS page - Città Metropolitana di Genova, 2016).

In parallel, another important local action introduced to face the effects of global warming was the signing in February 2009 of the European Covenant of Mayors (COM) by the municipality of Genoa. This initiative, mainly addressed to mitigation, was proposed by the European Commission as "Action Plan for Energy Efficiency: Realizing the Potential" (EU, 2006). The initiative aimed at actively involving as many as possible European cities in significantly improving energy efficiency and the use of renewable energy sources in urban environments, where policies and measures relating to certain key sectors, such as transport and construction, are more promising in reducing the production of CO_2. European cities joining the COM were requested to draft a Sustainable Energy Action Plan (SEAP), in order to fulfill the objectives set by the European Union. In particular, the main objective was to reach a reduction of 20% (at least) by 2020 of their own CO_2 emissions (but now by at least 40% by 2030), through local policies and measures aimed to increase the use of renewable energy and to improve energy efficiency, implementing specific programs on energy conservation, using appropriate promotion and communication actions.

By signing the COM, Genoa's mayor made an explicit, but nonbinding, commitment to reduce CO_2 emissions by 23.7% (compared to the baseline year 2005) by the year 2020. Genoa's SEAP, prepared in collaboration with the Liguria Regional Energy Agency and the Research Center in Town Planning and Ecological Engineering of the University of Genoa, is, meanwhile, still in the process of implementation; it is not intended to be part of ordinary planning or the initiatives already launched, but rather it is a set of guidelines and methodology for the local authorities' approach to government of the city in the short- and long-term.

Moreover, still in the field of mitigation initiatives, in 2008 the province of Genoa promoted, through the "Provincia Energia" group of integrated actions, the use of renewable energy, energy saving and energy efficiency in its own territory, as a necessary means to face climate change. This initiative called for an integrated approach, involving many different provincial departments and the main institutional and economic entities of the territory. In the context of "Provincia Energia," the province of Genoa has:

- Promoted and created a participated approach in drafting a Provincial Plan for Renewable Energy (with the financial resources of EU project "Res Publica" of 2007–09);
- Created a structure, the Science Center "Muvita," aimed to communicate, inform, and disseminate knowledge regarding climate change, renewable resources, and energy saving and efficiency;
- Accomplished, in 2009–10, the inventory of CO_2 emissions of the entire province and of all 67 municipalities of the territory. This inventory is available to municipalities that have signed the COM, so that it can be used as a "baseline" for the SEAP of each municipality.

Not long after, the province of Genoa and the Chamber of Commerce of Genoa, through their development agency *Gal Genovese*, participated in the EnSURE (Energy Savings in Urban Quarters Through Rehabilitation and New Ways of Energy Supply) project (2010–13) under INTERREG Central Europe. The project implemented strategies for the energetic rehabilitation of the building stock and energy efficiency in urban development (e.g., refurbishment measures and new concepts for the energy supply such as local and district heating, combined heat and power, and the use of renewable energy resources).

Within the "Provincia Energia" actions, the COM has been identified as the most suitable instrument to convey the provincial policies to face and adapt to climate change and to address the territory development toward "green economy" issues.

Moreover, the province of Genoa and Muvita (meanwhile transformed in a foundation) provide methodological support to the SEAPs' implementation. In detail, they support the municipalities both in the actions aimed at involving citizens and in the different phases for carrying out the SEAPs.

With respect to the first point, the involvement of all the local stakeholders is the starting point for encouraging the behavioral changes needed to support the actions promoted by the SEAP. The SEAP is developed through a consultation process that provides the basic knowledge of the urban system at stake, supported by transverse communication actions to be planned in each of the three phases of the project:

- Planning: detailed design of interventions, coordination with internal structure, creation of support materials, implementation of the mapping of local stakeholders of each involved municipality;
- Participation: starting of activities within the local community and with groups of interest in order to define the future scenario (to be pursued through the actions of the plan);
- Results: presentation and setting of the future scenario, definition of priorities, and public communication of the results obtained.

For what concerns the drafting of the SEAP, the province of Genoa entrusted Muvita with drafting all of the plans for the different municipalities.

Muvita reported the stages and the results of the participation processes, and drafted the profile of each of the actions planned for the implementation of the SEAPs, including part of the proposed strategies into the lines of activity of the "Genova Smart City" Association, which were then promoted by the municipality of Genoa to create synergies among local public and private actors. Thus, the work done within the numerous SEAPs became part of a larger strategy aimed at developing a green economy.

Meanwhile, the Liguria region, through the IRE SpA (the regional agency of Liguria in the field of infrastructures, building refurbishment, and energy, created in 2014 from the merger of three regional bodies including the energy agency (former ARE Liguria), which is the energy operator for the Liguria region) participated into COOPENERGY, a 3-year European-funded project (cofunded through the Intelligent Energy Europe Program) with the aim to help regional and local public authorities develop joint action plans by using multilevel governance agreements, deeming those plans fundamental in supporting the successful implementation of SEAPs.

14.3 CONCLUSIONS AND CRITICAL ANALYSIS

The goal of the aforementioned projects and initiatives is to encourage local authorities to adopt a new approach in the way they deal with climate change. Adaptation policies regarding urban areas' design criteria have now started, even if they still have a long way to go, just as energy policies (mitigation) have likewise started. These initiatives are no longer pursued by sporadic interventions but are designed to set in motion a rigorous process (particularly SEAPs), which, starting from the definition of quantifiable goals, are carefully planned until the final monitoring of the obtained results (which are all overseen by the EU). The Stern Review on the Economics of Climate Changes (2006) showed the economic consequences of passiveness in facing the challenge of climate change and also highlighted that the concern regarding the natural environment will not compromise economic growth (green economy). A proactive urban approach, like the one of the green economy, is actually even more justified in the cities that host the majority of the population of the Mediterranean area; it is, in fact, a trend that keeps on growing, as in the case of Genoa. The adoption of these strategies implies giving equal importance to natural environment, economy, social issues, and territory.

Hence, in addition, we emphasize the need for giving priority to urban redevelopment of brownfield sites rather than calling for new expansions, a need that we consider necessary also as far as Genoa is concerned. This choice may be coupled with the processes of modernization of municipal public lighting and substitution of nonhistoric housing stock, in order to relaunch their functional and energetic efficiency and their microeffects on climate. In a perspective of more integrated and cooperating outcomes (nowadays this goal is rather unexploited), actions toward environmental sustainability could be implemented in Genoa also for the mobility sector, as it has been done in the case of the Port Environment Energy Plan (Genoa Port Authority, 2011). In this regard, the primary objective of the Urban Mobility Plan issued in 2010 was to give priority to public transportation (Genoa Municipality, 2010), discouraging private transportation. Other actions should involve the preservation and enhancement of the ecological network, the green areas of the city and the hilly areas, the spread of flood control, and emergency management activities, which have a fundamental importance in facing the increasingly alarming issue of climate change. All such actions, together with mitigation and adaptation initiatives adopted by Genoa's institutions, should finally bring some benefits, in a midterm range. These will actually be strictly connected and achievable by an efficient coordination of the aforementioned plans, and should be integrated with ordinary planning devices and bottom-up practices. The latter are still not frequent in Genoa, due to the building characteristics of Genoa itself, even if, as shown by the illustrated projects, stakeholders, if involved, are very interested in cooperating with the local authorities on issues of common interest, as energy and planning.

Thanks to the numerous initiatives already undertaken, Genoa, or at least part of it, demonstrates to be on the right path toward an effective contribution both to the reduction of greenhouse gas emissions and to the increase of urban resilience in the face of climate impacts. Nevertheless, a permanent and stronger involvement of political local leaders in the path to achieve these goals is indeed highly desirable.

References

Brandolini, P., Cevasco, A., Firpo, M., 2012. Geo-hydrological risk management for civil protection purposes in the urban area of Genoa (Liguria, NW Italy). Natural Hazards and Earth System Sciences 12, 943–959. https://doi.org/10.5194/nhess-12-943-2012.

Belda, M., Holtanová, E., Halenka, T., Kalvová, J., 2014. Climate classification revisited: from Köppen to Trewartha. Clim Res 59, 1–13. https://doi.org/10.3354/cr01204.

Città Metropolitana di Genova, Cartografia Tematica On-line. Archivio Cartografico. Available at: http://cartogis.provincia.genova.it/cartogis/grabs/progetto.htm.

EEA (European Environmental Agency), 2012. Climate Change, Impacts and Vulnerability in Europe 2012. An Indicator-based Report. EEA Report, 12/2012, Copenhagen, Denmark.
EU (European Union), Action Plan for Energy Efficiency: Realising the Potential, Brussels, 19.10.2006, Communication from the Commission - COM(2006)545 final.
Genoa Municipality, 2010. Piano Urbano della Mobilità Genovese. Documento finale. Available at: http://www.comune.genova.it/sites/default/files/pum_-documento_finale_allegato_alla_delibera_n._1_2010_0.pdf.
Genoa Municipality, 2014. Norme Di Attuazione. Piano Urbanistico Comunale. Available at: http://puc.comune.genova.it/Aggiornamento2014/Norme_Attuazione_agg_02_2014.pdf.
Genoa Port Authority, 2011. Port Environment Energy Plan. Available at: http://servizi.porto.genova.it/en/cosa_mi_serve_per/impianti_produzione_energia_fonti_rinnovabili.aspx.
IPCC, 2014: Climate Change 2014: Synthesis Report. Contribution of Working Groups I, II and III to the Fifth Assessment Report of the Intergovernmental Panel on Climate Change [Core Writing Team, R.K. Pachauri and L.A. Meyer (eds.)]. IPCC, Geneva, Switzerland, 151 pp.
Köppen, W., 1936. Das geographische System der Klimate. Handbuch der Klimatologie. Geiger (publisher), Rudolf. Berlin, Borntraeger.
Rosso, R., 2014. Bisagno. Il fiume nascosto, Marsilio.
Stern, N., 2006. Stern Review on the Economics of Climate Change. Cambridge University Press.

Websites

Covenant of Mayors. www.eumayors.eu.
EU "Coopenergy" Project Website, Funded by the Intelligent Energy Europe Programme. www.coopenergy.eu/.
EU "Ensure - Energy Savings in Urban Quarters Through Rehabilitation and New Ways of Energy Supply" project Website, INTERREG Central Europe. www.ensure-project.eu/about-ensure/.
EU "GrABS - Green and Blue Space Adaptation for Urban Areas and Eco Towns" Project Website, INTERREG IVC. www.grabs-eu.org.
EU "Res Publica - Renewable Energy Sustainable Planning and Use Within Public Bodies in Liaison With Involved Community Actors", project website, funded by the Intelligent Energy Europe Programme. https://ec.europa.eu/energy/intelligent/projects/en/projects/res-publica.
Fondazione Muvita. www.muvita.it.

Further Reading

Genoa Municipality ARE, CRUIE, 2010. Sustainable Energy Action Plan. Available at: http://www.urbancenter.comune.genova.it/sites/default/files/archivio/allegati/SEAP%20summary_0.pdf.
ISAC-CNR - Istituto di Scienze dell'Atmosfera e del Clima del Consiglio Nazionale delle Ricerche, 2016. Climate Monitoring for Italy. Available at: http://www.isac.cnr.it/climstor/climate_news.html.

CHAPTER 15

The Evolution of Flooding Resilience: The Case of Barcelona

Andrea Favaro[1], Lorenzo Chelleri[2,3]

[1]Universidad Politécnica de Madrid, Madrid, Spain; [2]Gran Sasso Science Institute (GSSI), L'Aquila, Italy; [3]Universitat Internacional de Catalunya (UCI), Barcelona, Spain

15.1 INTRODUCTION: FLOODING RESILIENCE AND BARCELONA

Flooding is one of the most threatening hazards affecting coastal cities nowadays (Fuchs et al., 2011), related to climate change but also to past and current patterns of urban development, which have been increasing soils sealing, reducing the drainage potential (Muller, 2007). Because of the relevance of the issue, and the limitation of mitigation measures, adaptation (and therefore the concept of resilience) to flooding is playing a key role among climate plans and resilience theory (Liao, 2012). Challenges related to flooding can be addressed both with top-down hard infrastructure solutions or more soft, bottom-up approaches, in which people behave and practices can leverage a reduction in the vulnerability to flood impacts (Chatterjee, 2010; Odemerho, 2015). In both cases, opportunities for the implementation of a variety of solutions are not threatened anymore by physical and technical limits, but relaying of the political commitment in shaping sustainable approaches, setting integrated design strategies and policies for the long-term transformation (Schuetze and Chelleri, 2011, 2013; Chelleri et al., 2015), rather than aiming at specific and short term projects. Indeed, flooding issues are part of long-term dynamics, in which urbanization patterns and natural processes evolve together, as this chapter addresses.

Barcelona has always struggled with floods, since the medieval ages in which the city was protected by walls and surrounded by rivers. Indeed, in the second half of the 19th century, Barcelona suffered from 21 floods of which 15 were of extreme intensity and 6 catastrophic (Barrera et al., 2006).

At that time, the major issue with flooding was related to health. In the historical city there were diseases such as cholera (Buj, 2001), typhus (Buj, 2003), and gastroenteritis (Recaño and Esteve, 2006). The *Ensanche* (first city expansion of the 19th century), devoid of a sewage and drainage system (Capel and Tatjer, 1991), turned Barcelona into a swampy and unhealthy area due to the appearance of malaria (Buj, 2000), with a mortality rate about of 27.69% in early 1900, one of the highest in Europe (Ajuntament de Barcelona, 1991).

While health and hygiene related issues were resolved over time (Urteaga, 1980), the problem of flooding persisted. Flooding risk areas were identified in the *Plan Especial de Alcantarillado de Barcelona* (PECB) in 1988, covering an area of 1052.9 ha (Fig. 15.1).

According to Llebot (2010), flooding is the most common natural hazard affecting the region of Catalonia. The evolution of rainfall events affecting Barcelona shows an uncertain trend between 1854 and 2005 (Barrera et al., 2006) and an absent trend for the hundred years prior to the PECB (Clavegueram de Barcelona, 1997). Conversely, in a shorter period of time, between 1982 and 2006, there was a decrease in annual rainfall (Llasat et al., 2009). The lack of a trend in the evolution of rainfall therefore doesn't explain the increase of flood events. The likely explanation is then the augmented vulnerability due to the increase of city exposure caused by changes in land use, rather than weather hazard (Llasat et al., 2012). Impermeable soil in the 1980s reaches indeed the order of 90% in the municipal area (excluding the natural park of Collserola standing on the edge of the municipality). Flooding became more frequent because of the unplanned growth of the drainage network, of which the *Bogatell* catchment, the main part of the city, is an example. In fact, the outlet sewer, initially designed by Garcia Fària for a 1582 ha catchment surface, in

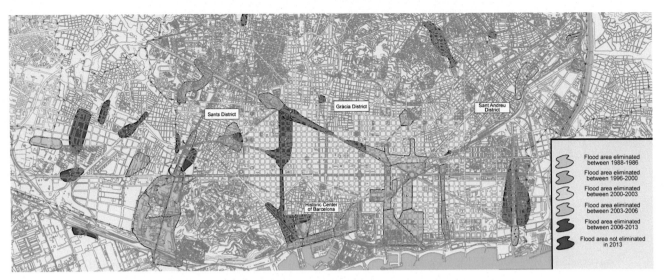

FIGURE 15.1 Evolution of flooding areas. *Modified from Clavegueram de Barcelona S., 2006. In: Barcelona, A.D., Clavegueram de Barcelona, S. (Eds.), Plan Integral de Clavegueram de Barcelona 2006 (PICBA06). Master Drainage Plan of Barcelona, Planning and Project Department.*

1962 collected water from 3600 ha (Ajuntament de Barcelona, 1991). Only with the PECB of 1988 was the catchment brought back to its starting size and a century-long issue was finally solved.

The results of these underdesigned water management areas caused enchained (and growing) economic consequences. Indeed insurance companies paid 1,574,530,945 euros for flooding damages between 1971 and 2002 in the region, a sum equal to 78.86% of the total refunds paid in that period (Barredo et al., 2012).

The increasing risk of flooding for Barcelona also comes from some major infrastructure planning choices, postponed for decades, like the deviation of the Llobregat riverbed. In 1962, a heavy rainfall—250 mm of water fell in just 6h (Servei Meteorològic de Catalunya)—provoked the flooding of the Llobregat River, causing 441 deaths, 374 missing persons, 213 injured, and a damage amount of 2.65 billion pesetas (Parlament de Catalunya, 2000). A parallel issue to the flooding one is the consequent processing of the water body contamination management. Of the average 84 rainy events each year, 75% have less than 15 mm volume. Only 40% of these are brought to the wastewater treatment plants, while the remaining 60% goes directly into the receiving water body, with a contamination rate that can rise up to 30% of entire contamination present in the receiving water body (Malgrat Bregolat, 1995).

Despite the reduction of the flooding exposed areas to 345.4 ha in 2006 (Clavegueram de Barcelona, 2006), the problems of flooding and water body contamination still represent a current challenge.

15.2 FIRST PATTERNS OF FLOODING RESILIENCE: LINKS BETWEEN DRAINAGE SYSTEMS DESIGN AND URBAN PLANS

Idelfonso Cerdà was the chief civil engineer from 1841 to 1848 and conceptualized the first real expansion of Barcelona in 1859. In his plan (Cerdá, 1859), Cerdà made storm water management one of the key strategic elements of urban development (Magrinyà, 2009). The *Ramblar Colector* is better known as the pipe drain that intercepts the torrential waters coming from the mountains to protect the new urban expansion; it became an essential pillar underpinning the urban expansion plan (Fig. 15.2). From another side, the organization of the sewer system Cerdà's plan uses an innovation consisting of separating the networks for storm water and wastewater. The former network was built according to the maximum slope, that is the vertical axis of the *Ensanche*. The latter waters instead are accumulated in septic tanks.

Unfortunately, not all these innovations were operationalized. Indeed, despite its strategic importance, the *Ramblar Colector* were not realized, and it has been replaced with the "deviation of *Riera Malla*" (Fig. 15.2), an intermediate cost-effective solution, that solved the most difficult problems of this torrent, but wouldn't address future expansion pressures (Magrinyà, 1995). At the same time, the demolition of the city walls led to the necessity of

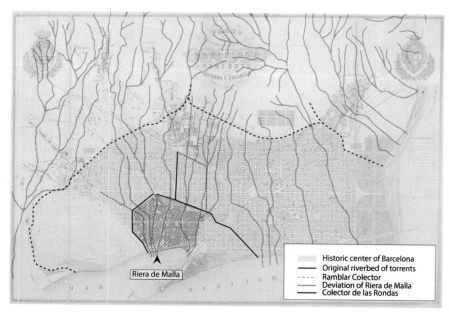

FIGURE 15.2 Localization of the three projects: Ramblar Colector, Colector de las Rondas, and Deviation of Riera Malla and as background the Cerdà Plan. *Authors' elaboration from Cerdá I., 1859. Teoría de la construcción de las ciudades aplicada al proyecto de Reforma y Ensanche de Barcelona. Cerda y Barcelona, 107—450.*

framing some kind of water protection strategy for the historical city center. Hence, after the flooding of 1862, the *Colector de las Rondas* was built in 1863 in the old location of walls; it was a drain pipe that acted like an "underground wall" (Da Costa, 1999) to protect the historical city center (Fig. 15.2).

We could therefore conclude that if the Cerdà plan anticipated by many years the idea of the "drained city" (Brown et al., 2008), in contrast it failed to grasp the hygiene-related problems. The use of septic tanks did aggravate the underground water presence of infections with a related increase in typhoid and cholera diseases (Garcia Fária, 1894). Also the *Ensanche* urban consolidation actualized Cerdà's urban plan but without upgrading its related sewer system, which was only 31 km long (Ajuntament de Barcelona, 1991; Suriol, 2002).

Because of this, the next urban development of the city, designed in 1891 by Pere Garcia Fária, was a turning point for urban drainage of the city, since it was centered around health principles (Vilalta, 1997). The plan was conceived around a combined system for both storm water and wastewater, focusing on the concept of water utilization and treatment, in order to close the water cycle and defuse water infection (Gómez Ordóñez, 1987). Wastewater was subsequently sent to the Llobregat river mouth for crop watering and fertilizing purposes through a single sewer crossing the whole city under the main avenue. Unfortunately, also in this case, the forward-looking plan didn't find the necessary support from the political patronage (Ajuntament de Barcelona, 1991), which built only the sewer part of the plan, changing all the other features, and leading to the design of different socially homogeneous areas, thus restructuring the social order (Da Costa, 1999). Fortunately, the next plan known as *Plan de Saneamiento y alcantarillado* (Vilalta, 1969), was focused on enhancing seawater quality, in a time of increasing importance of the seafront. In order to reach this objective, Vilalta obtained a detailed assessment of the underground networks and though this information emerged an infrastructure deficit of 49 km of collector drains, in addition to other critical issues (like inverted siphons and low-slope sewer stretches). This was the first time an accurate methodology was introduced for sizing underground networks (besides, at that time the planned return period for flooding was only 10 years of rains).

The upgrading of the drainage network during and since the Vilalta plan has been impressive, with the underground networks increasing to 860 km in 1975, thanks to an annual investment of 400 million pesetas.

Although its actualization is incomplete, this plan's guidelines, methodology, and solutions will be used for future plans, which will characterize Barcelona's modern flooding resilience.

One of the biggest failures related to flooding resilience from those three introduced plans, however, has been the decision to abandon the decision of Cerdà of introducing from the very first stage of urban expansion separate systems for rain and wastewaters, instead merging them into a combined system (Jara Urbano, 1954; Ajuntament de Barcelona, 1991).

15.3 BARCELONA OLYMPIC MODEL FRAMING THE CONTEMPORARY CHALLENGES IN FLOODING RESILIENCE

The 1992 Olympic Games were the perfect opportunity for the next step in Barcelona city's evolution toward a global city (Maragall, 1999). The definition of a new city drainage plan (Calavita and Ferrer, 2000) finally fully expressed the strict relationship between water management and urban development. In 1988 the PECB was approved, with the goal of finally resolving the city's chronic flooding problems (Arandes i Renù et al., 1988).

Despite an usual actualization rate of only 33% of the proposed drainage works, this plan gave rise to a 580.4 ha reduction of flood areas. PECB ushered in historical changes in drainage management, such as network decentralization and major flexibility of the network functioning. In line with the emerging of the well-known Barcelona Model, consisting of a governance model based on the framing of public-private joint ventures to achieve city improvement goals (Marshall, 2000), in 1992 the *Clavegueram de Barcelona, S.A.* (CLABSA) was established as an authority with the goal to "carry out the technical management of urban drainage: planning, control and exploitation" (Martí, 1995; Cabot and Bregolat, 2013).

CLABSA conceived and implemented in a single philosophy PECB innovations, proposing the Advanced Urban Drainage Management (GADU as the original acronym), which is based on four pillars: (1) perfect system knowledge, (2) effective planning, (3) dynamic usage, and (4) integrated management (Martí, 1995). Therefore, thanks to IT systems and mathematical modeling, it has been possible to manage and develop much more effective drainage strategies, reducing mistakes to 10% from 50% (Ajuntament de Barcelona, 1991). Furthermore, IT systems and remote control procedures gave birth to an "active" drainage network, no longer gravity dependent, but able to be managed in order to modify intensity and direction of water flows according to real-time requirements. This feature improved network efficiency, reduced flood chance risks up to 75%, while saving 30% of operational costs.

PECB network efficiency hallmark was linked to a second key innovation: the introduction of underground storm water tanks. These huge underground spaces, retaining the rainwater that exceeds the current capacity of the drainage system, definitely increased the city flooding resilience (Liao, 2012). Fig. 15.3 and Table 15.1 show where they have been planned, and how many have been already built by 2017. Finally, these tanks could also avoid wastewater treatment plant overloads during peak rains thanks to the rainwater underground retention, concurrently

FIGURE 15.3 Localization of all storm water tanks and antidischarge unitary system tanks planned and realized. *Authors' elaboration from Clavegueram de Barcelona S., 2006. In: Barcelona, A.D., Clavegueram de Barcelona, S. (Eds.), Plan Integral de Clavegueram de Barcelona 2006 (PICBA06). Master Drainage Plan of Barcelona, Planning and Project Department; background from Apple Maps.*

TABLE 15.1 Storm Water Tanks and Antidischarge Unitary System Tanks Planned and Realized

Storm Water Tanks				Antidischarge Unitary System Tanks			
N°	Name	Volume (m3)	Phase	N°	Name	Volume (m³)	Phase
1	Zona Universitària	105,500	Realized	35	Bogatell	80,000	Not realized
2	Bori i Fontestà	71,000	Realized	24	Ciutadella-Barceloneta	50,000	Not realized
3	Parc Joan Mirò	55,000	Realized	25	Port Vell – Colon	15,000	Not realized
4	Doctors Dolsa	50,500	Realized	26	Port Vell- Passeig Montjuic	7,500	Not realized
5	Taulat	57,000	Realized	27	Cementiri Montjuic	5,000	Not realized
6	Escola Industrial	27,000	Realized	28	Motors	72,000	Planning
7	Parc Central Nou Barris	14,000	Realized	29	Amadeu Torner	22,000	Not realized
8	Diagonal Mar	17,500	Realized	30	Seat	16,000	Not realized
9	Fira M2	1,600	Realized	31	ZAL	32,000	Not realized
10	Parc del Poblenou	1,400	Realized	32	Bac de Roda	80,000	Planning
11	Placa Fòrum	800	Realized	34	Bassa IZF	188,000	Not realized
12	Mallorca – Urgell	16,000	Realized	33	Bassa IZF anti-DSU	28,000	Not realized
13	Carmel –Clota	75,000	Realized	35	Vallbona	2,000	Not realized
14	Av. Hospital Militar	27,000	Not realized	36	Torrent Tapioles-Torre Baró	30,000	Not realized
15	Navas de Tolosa	17,000	Not realized	37	Interceptor Estadella	23,000	Not realized
16	Artesania	12,100	Not realized	38	Torrent Estadella-Bon Pastor	41,000	Not realized
17	La Sagrera	90,000	Planning	39	Guipúscoa-Alarcón	10,000	Not realized
18	Les Planes	35,000	Not realized				
19	Can Boixeres	65,000	Not realized		TOTAL Planning	152,000	
20	Ciutat Judicial	60,000	Not realized		TOTAL Not Realized	549,500	
21	Can Batlló	35,000	Planning				
22	Ikea L'Hospitalet	4,370	Not realized		TOTAL	701,500	
23	Camp de l'Empedrat	17,000	Not realized				
	TOTAL Realized	492,300					
	TOTAL Planned	125,000					
	TOTAL Not Realized	237,470					
	TOTAL	854,770					

Authors' elaboration from Clavegueram de Barcelona S., 2006. In: Barcelona, A.D., Clavegueram de Barcelona, S. (Eds.), Plan Integral de Clavegueram de Barcelona 2006 (PICBA06). Master Drainage Plan of Barcelona, Planning and Project Department.

reducing the amount of contamination that the sewages would receive up to 30% with only one-tenth of the installation expense with respect to a treatment installation. This is possible since most of the pollution usually comes from the first flow of rainwater (first flush), which drags oils and pollution from the ground, which fall into the retention tanks; here there is time for the heavy metal pollution particles to precipitate from the stored water, remaining in the bottom of the tank when the water is pumped again to the sewage system, after the flash flood is finished.

Between 1990 and 2000, during the so-called Barcelona Olympic (or Barcelona Model) period, thanks to the recent solutions implemented, Barcelona's annual discharge unitary system in the receiving water body was decreased from 15% to 4% (Clavegueram de Barcelona, 2006).

Lastly, the Llobregat River Delta plan actualization entailed the deviation of the final tract of the riverbed and the building of El Prat wastewater treatment plant, at a cost of 13.3 million euros (Empresa Metropolitana de Sanejament SA).

15.4 SUSTAINABLE DRAINAGE: ENHANCING DECENTRALIZATION FOR THE FUTURE OF URBAN FLOODING RESILIENCE

Despite acting in a complex regulatory system (Valls and Perales, 2008) and by taking advantage of the political willingness to facilitate public-private relationships (Maragall, 1999), CLABSA was able to exploit the urban development in order to improve the drainage infrastructure (Clavegueram de Barcelona, 2006). The last step in the evolution of flooding resilience is represented by the *Plan Integrado of Alcantarillado de Barcelona* (PICBA).

European funds were used to further support the development of new storm water tanks (as in Fig. 15.3): *Urgell-Mallorca* and *Carmel-Clot* (total costs were, respectively, 6.383.253€ and 23.358.401€). One of the innovations and guiding principles of this plan was the Sustainable Urban Drainage System (henceforth TEDUS, as introduced within the plan), which "can bring back the catchment to a more natural state" (Jha et al., 2012) and that "works according to the ideas of sustainable development" (Woods-Ballard et al., 2007). There are different types of TEDUS: green roofs, porous/permeable paving, filter strips, soakaways, and infiltration trenches, filter drains, swales, infiltration basins, detention basins, retention ponds, and wetlands. All of these types, depending on the climatic, geological and urban conditions, have the goals of reducing peak flow rates, reducing the volumes of surface water flows and retained contamination. The volume of collected water is retained as much as possible through the use of microretention tanks, and then the waters by means of porous pavings with filter systems that allow the infiltration of water retaining the contaminants. The TEDUS provide a savings cost of 20% over the conventional techniques of drainage and purification, i.e., with the same efficacy, TEDUS are the cheapest solution.

Two pilot projects of TEDUS were carried out in Barcelona: *Torre Barò* (completed in 2005) and *La Marina de la Zona Franca* (still in the design stage).

The first project is localized in Torre Barò, a northern district of Barcelona, where the streets turn into torrents during the rains due to their considerable slope. The water, thanks to the favorable section of the road, is conveyed in the porous paving that collects, filters, and leads the water in a series of microretention deposits located intermittently along the entire route. At the end of the path the water is stored in a retention deposit for its reuse for local irrigation and street cleaning. The study of Llopart-Mascaró et al. (2010) has shown effectiveness in the use of TEDUS for the reduction of the surface flows (Fig. 15.4) and for the reduction of pollutants within the water.

The second project is located in La Marina de la Zona Franca and is the urban regeneration project of the area and the respective ground. The objective of TEDUS is to provide a separative management of rainwater: on the one hand to reduce the sliding surface, and on the other hand for water retention and groundwater recharge. The TEDUS consists in dividing the project area into four subbasins with independent catchment, transport, and infiltration structures, in order to recharge the aquifer, which in turn will be used for the irrigation of the local area (Fig. 15.5).

Linking and supporting the rise of such sustainability drainage projects demonstrates the recent trend in the city to promote a more integrated urban development, in which the coordination of different sectors and municipality departments work in synergy. This has also been fostered through the recent framing of an interdepartmental board called the city resilience office (Chelleri et al., 2013). Such institutional reframing is also allowing an integrated vision

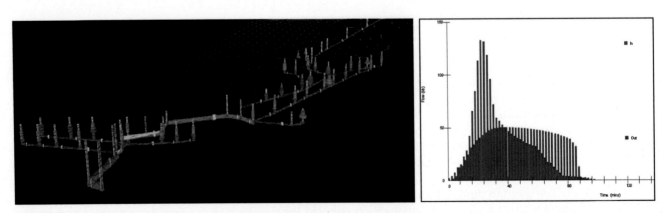

FIGURE 15.4 Simulation of the set of minideposits and hydrographs of entrance and exit and effect of lamination in a reservoir of urbanization. *Modified from Febles Domènech, M.D., Perales Momparler, S., Soto Fernàndez, R., Octubre 27–28, 2009. Innovación y Sostenibilidad en la Gestión del Drenaje Urbano: Primeras Experiencias de SuDS en la Ciudad de Barcelona. Jornadas de Ingeniería del Agua. CEDEX. Madrid.*

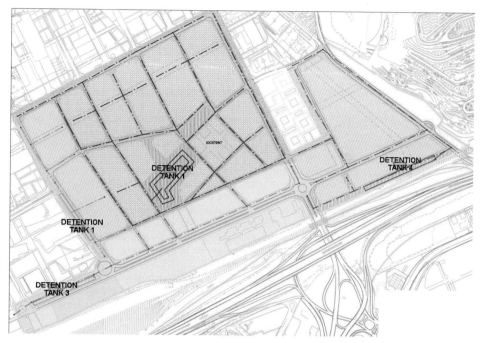

FIGURE 15.5 Proposal of sectorization of rainwater management by Suds in the La Marina de la Zona Franca urbanization. *Modified from Febles Domènech, M.D., Perales Momparler, S., Soto Fernàndez, R., Octubre 27–28, 2009. Innovación y Sostenibilidad en la Gestión del Drenaje Urbano: Primeras Experiencias de SuDS en la Ciudad de Barcelona. Jornadas de Ingeniería del Agua. CEDEX. Madrid.*

for water management, since CLABSA has become a public enterprise called *Barcelona Ciclo de l'Aigua SA*, integrating and managing the entire water cycle, from drinking water supply to sewer drainage networks and water treatment management, and thus re-proposing the drainage management vision proposed by Garcia Fária.

15.5 CONCLUSIONS

After presenting the three main periods of plans and related approaches to enhance flooding resilience through upgrading urban drainage, it is worth critically addressing which are the key insights from this evolution. A cautionary note, before discussing the details, is about the terminology used through this evolution. Indeed, the concept of resilience began emerging in Barcelona only in 2013, when through the "city resilience officer" the city implemented the Rockefeller 100 Resilient cities program. When presenting its resilience skills and experience, the "flooding resilience strategy" was represented as one of the key features of city resilience, thanks to the urban drainage plan's effectiveness. Indeed, as explored through the chapter, flooding resilience has always been one of the pillars of urban development in Barcelona, but the concept of resilience with respect to flooding has been the most recent "labeling" of those drainage plans.

Looking to the evolution of drainage plans, technologies proffered some key improvements toward the mitigation of flooding, while other past solutions only aimed at adapting to it. Looking at the different plans and strategies it is clear that there is a tension between short- and longer-term adaptations. Cerdà, when designing the first-ever urban expansion of Barcelona, was looking to a long-term sustainability, for urban drainage, and therefore proposed the never-realized "separate systems" for waste and rainwaters. This would have been the long-term best solution coping with future city expansion, which unfortunately, was not chosen at that time. The options chosen were from following plans, to mix waste- and rainwaters, which led to increasing flooding problems, and the need to create solutions, like the underground water-retention tanks, in order to cope with unmanageable water flows during peak rains. It is clear that a short-term adaptive response to flooding dominated the policy agendas of the past, and that only recently a more sustainable and long-term perspective has been introduced through the TEDUS approach. This should teach us that sometimes very inspiring and innovative solutions, promoting sustainability or resilience, are not put into practice because they cannot deal with the support of the politic parties, which are governing in that moment. Perceptions and cultural behaviors sometimes are simply not ready for accepting innovations (Davoudi et al., 2013), leading to not-so-convenient choices for the longer term.

References

Ajuntament de Barcelona, 1991. Sota la ciutat. Exposició al Collector del Passeig de Sant Joan (Barcelona).
Arandes i renù, R., Malgrat i Bregolat, P., Vàzquez, G., 1988. In: Barcelona, A.D. (Ed.), Pla Especial Clavegueram Barcelona (Barcelona).
Barredo, J., Sauri, D., Llasat, M., 2012. Assessing trends in insured losses from floods in Spain 1971–2008. Natural Hazards & Earth System Sciences 12 (5).
Barrera, A., Llasat, M., Barriendos, M., 2006. Estimation of extreme flash flood evolution in Barcelona County from 1351 to 2005. Natural Hazards & Earth System Sciences 6 (4).
Brown, R., Keath, N., Wong, T., 2008. Transitioning to water sensitive cities: historical, current and future transition states. 11th International Conference on Urban Drainage.
Buj, A.B., 2000. De los miasmas a Malaria. www. Permanencias e innovación en la lucha contra el paludismo. Scripta Nova. Revista Electrónica de Geografía y Ciencias Sociales 4.
Buj, A.B., 2001. Los riesgos epidémicos actuales desde una perspectiva geográfica. Universidad Autónoma del Estado de Mexico.
Buj, A.B., 2003. La vivienda salubre. El saneamiento de poblaciones (1908) en la obra del ingeniero militar Eduardo Gallego Ramos. Scripta Nova. Revista electrónica de geografía y ciencias sociales 7.
Cabot, J., Bregolat, P.M., 2013. La gestión de inundaciones urbanas: de la planificación tradicional a la gestión integral inteligente. Revista de Obras Públicas: Organo profesional de los ingenieros de caminos, canales y puertos 3542, 67–72.
Calavita, N., Ferrer, A., 2000. Behind Barcelona's success story: citizen movements and planners' power. Journal of Urban History 26 (6), 793–807.
Capel, H., Tatjer, M., 1991. Reforma social, servicios asistenciales e higienismo en la Barcelona de fines del siglo XIX (1876–1900). Ciudad Y Territorio 3 (89), 81–94.
Cerdá, I., 1859. Teoría de la construcción de las ciudades aplicada al proyecto de Reforma y Ensanche de Barcelona. Cerda y Barcelona 107–450.
Chatterjee, M., 2010. Slum dwellers response to flooding events in the megacities of India. Mitigation and Adaptation Strategies for Global Change 15 (4), 337–353.
Chelleri, L., Favaro, A., Lucchitta, B., Raventos, J., Fernandez, M., 2013. Dall'adattamento urbano al cambiamento climatico alla resilienza urbana: il caso di Barcellona, Spagna. In: National Conference "Climate Changing Cities" (Venice).
Chelleri, L., Schuetze, T., Salvati, L., 2015. Integrating resilience with urban sustainability in neglected neighborhoods: challenges and opportunities of transitioning to decentralized water management in Mexico City. Habitat International 48 (0), 122–130.
Clavegueram de Barcelona, S., 1997. In: Barcelona, A.D., Clavegueram de Barcelona, S. (Eds.), Plan Especial de Clavegueram de Barcelona 1997 (PECLAB97).
Clavegueram de Barcelona, S., 2006. In: Barcelona, A.D., Clavegueram de Barcelona, S. (Eds.), Plan Integral de Clavegueram de Barcelona 2006 (PICBA06). Master Drainage Plan of Barcelona, Planning and project department.
Da Costa, F.D.A., 1999. La Compulsión por lo limpio en la idealización y construcción de la ciudad contemporánea: salud y gestión residual en Barcelona, 1849–1936.
Davoudi, S., Brooks, E., Mehmood, A., 2013. Evolutionary resilience and strategies for climate adaptation. Planning Practice & Research 28 (3), 307–322.
Empresa Metropolitana de Sanejament SA. Available at: http://www.emssa.com/.
Febles Domènech, M.D., Perales Momparler, S., Soto Fernàndez, R., Octubre 27–28, 2009. Innovación y Sostenibilidad en la Gestión del Drenaje Urbano: Primeras Experiencias de SuDS en la Ciudad de Barcelona. Jornadas de Ingeniería del Agua. CEDEX. Madrid.
Fuchs, R., Conran, M., Louis, E., 2011. Climate change and Asia's coastal urban cities: can they meet the challenge? Environment and Urbanization Asia 2 (1), 13–28.
Garcia Fária, P., 1894. Medios de aminorar las enfermedades y mortalidad en Barcelona. Adm. Industria e Invenciones (Barcelona).
Gómez Ordóñez, J.L., 1987. García Fària i el seu projecte de sanejament. El naixement de la infraestructura sanitària a la ciutat de Barcelona 21–28.
Jara Urbano, L., 1954. Plan general de saneamiento y alcantarillado de la ciudad de Barcelona (Barcelona).
Jha, A.K., Bloch, R., Lamond, J., 2012. Cities and Flooding: A Guide to Integrated Urban Flood Risk Management for the 21st Century. World Bank Publications.
Liao, K.H., 2012. A theory on urban resilience to floods a basis for alternative planning practices. Ecology & Society 17 (4).
Llasat, M., Llasat-Botija, M., Barnolas, M., López, L., Altava-Ortiz, V., 2009. An analysis of the evolution of hydrometeorological extremes in newspapers: the case of Catalonia, 1982–2006. Natural Hazards & Earth System Sciences 9 (4).
Llasat, M.C., Llasat-Botija, M., Gilabert, J., Marcos, R., 2012. Treinta años de inundaciones en Cataluña: la importancia de lo cotidiano. Publicaciones de la Asociación Española de Climatología (AEC), Salamanca, Spain 799–807.
Llebot, J.E., 2010. Segon informe sobre el canvi climàtic a Catalunya.
Llopart-Mascaró, A., Gil, A., Martínez, M., Puertas, J., Suárez, J., del Río, H., Paraira, M., 2010. Caracterización analítica de las aguas pluviales y gestión de las aguas de tormenta en los sistemas de saneamiento.
Magrinyà, F., 1995. La propuesta de saneamiento de Cerdà para Barcelona. OP, Obra Pública.
Magrinyà, F., 2009. El Ensanche de Barcelona y la modernidad de las teorías urbanísticas de Cerdà. Ingeniería y Territorio 88, 68–75.
Malgrat Bregolat, P., 1995. Control de la Contaminación Producida en Tiempos de Lluvia por las Descargas de Sistemas Unitarios de Alcantarillado. OP-Revista del Colegio de Ingenieros de Caminos, Canales y Puertos. N° 3 Año.
Maragall, P., 1999. Aula Barcelona. Presentacion de la Colección Modelo Barcelona. Cuadernos de Gestion. http://aulabcn.com/catala/introduccio.htm.
Marshall, T., 2000. Urban planning and governance: is there a Barcelona model? International Planning Studies 5 (3), 299–319.
Martí, J., 1995. Explotación centralizada de redes de saneamiento. Revista Obra Pública. Saneamiento II, n° 33.
Muller, M., 2007. Adapting to climate change: water management for urban resilience. Environment and Urbanization 19 (1), 99–113.
Odemerho, F.O., 2015. Building climate change resilience through bottom-up adaptation to flood risk in Warri, Nigeria. Environment and Urbanization 27 (1), 139–160.
Parlament de Catalunya, 2000. Informe extraordinari del Síndic de Greuges al Parlament de Catalunya sobre l'actuació de l'Administració Pública en matèria de prevenció i intervenció davant el risc d'inundacions i avingudes. Butlletí Oficial: Tramitacions Generals.

REFERENCES

Recaño, J., Esteve, A., 2006. De los factores espaciales y medioambientales de la mortalidad en la barcelona de finales del siglo XIX. Centre d'Estudis Demogràfics.

Schuetze, T., Chelleri, L., 2011. Climate adaptive urban planning and design with water in Dutch polders. Water Science and Technology 64 (3), 722–730.

Schuetze, T., Chelleri, L., 2013. Integrating decentralized rainwater management in urban planning and design: flood resilient and sustainable water management using the example of coastal cities in The Netherlands and Taiwan. Water 5 (2), 593–616.

Servei Meteorològic de Catalunya, 2014. Available at: http://www20.gencat.cat/portal/site/meteocat/menuitem.0733ee5bfae8638c5c121577b0c0e1a0/?vgnextoid=d5eb5cef6aee2210VgnVCM1000000b0c1e0aRCRD&vgnextchannel=d5eb5cef6aee2210VgnVCM1000000b0c1e0aRCRD&vgnextfmt=default.

Suriol, J., 2002. Los ingenieros de caminos en la transformación urbana de las ciudades españolas a finales del siglo XIX. El caso de Barcelona. Scripta Nova. Revista Electrónica de Geografía y Ciencias Sociales 6, 120.

Urteaga, L., 1980. Miseria, miasmas y microbios. Las topografías médicas y estudio del medio ambiente en el siglo XIX. Geo Crítica: cuadernos críticos de geografía humana.

Valls, G., Perales, S., 2008. Integración de las Aguas Pluviales en el Paisaje Urbano: un Valor Social a Fomentar. I Congreso Nacional de Urbanismo y Ordenación del Territorio (Bilbao).

Vilalta, A., 1969. In: Proyecto de saneamiento y alcantarillado. S. D. S. Y. A. e. B. Ayuntamiento de Barcelona. Unidad de Vialidad.

Vilalta, A., 1997. Pere Garcia i Fària: sanejament i construcció de la ciutat. Aportacions catalanes en el camp de la urbanística i de l'ordenació del territori, des de Cerdà als nostres dies 53.

Woods-Ballard, B., Kellagher, R., Martin, P., Jefferies, C., Bray, R., Shaffer, P., 2007. The SUDS Manual. CIRIA.

CHAPTER

16

Athens Facing Climate Change: How Low Perceptions and the Economic Crisis Cancel Institutional Efforts

Kalliopi Sapountzaki

Harokopio University of Athens, Athens, Greece

16.1 THE PROFILE OF METROPOLITAN ATHENS: CURRENT VULNERABILITIES, EXPOSURE TO CLIMATE CHANGE AND RESILIENCE ASSETS

16.1.1 Demographic, Social, Economic, and Governance Features of the Metropolitan Area

Athens Metropolitan Area, geographically almost identical to the administrative region of Attica, covers 2,928,717 km² and consists of 58 municipalities, which are organized into seven regional units (Northern, Western, Central and Southern Athens, Eastern Attica, and Western Attica). According to the last population census (2011), the total population of metropolitan Athens amounts to 3,753,783 inhabitants. The city center rests within the Municipality of Athens with a population density of approximately 17,000 inh/km² (Fig. 16.1).

It was in the first three post-World War II decades that the rapid growth of Athens took place; in this period its population more than doubled (from 1,500,000 in 1951 to 3,500,000 in 1981). As Maloutas and Spyrellis (2015) point out:

> The city's increasing population was housed then in two ways: (a) individual privately-owned housing in the city's outskirts,… and (b) housing in modern apartments built through the flats-for-land system that mainly covered the needs of the middle and working-class social strata.

Currently, Athens is experiencing population decline, which is common across Greece due to an aging population and the economic crisis generating outward migration, especially among the young people. The population of Athens Municipality in particular dropped from 745,500 in 2001 to 664,000 in 2011 since it has suffered as well from abandonment, as medium- and high-income residents moved outwards to peripheral or suburban locations in an effort to escape from environmental degradation and the perceived or real conditions of "insecurity" in the city center. The predominant trend for years now is for economic migrants to occupy the degraded and abandoned neighborhoods of Athens and Piraeus while former residents relocate to the northern and eastern parts of the agglomeration.

Emmanuel (2015) refers to the housing model of southern Europe by Allen et al. (2004) to present in brief the home ownership pattern in Athens. This model is featured by (1) high percentage of home ownership, (2) high percentage and significance of second houses, and (3) the important role of extended family in supporting access to home ownership. In 2011, the rate of private-dwelling ownership in Attica was 68.4%, one of the highest in western Europe.

Social segregation in Athens is both horizontal and vertical. In horizontal terms there is a historical residential divide between the eastern and western parts of Greater Athens. On top of this came the division center-periphery, with a major part of the center occupied today by low-income and migrant groups. More specifically, higher (declared) incomes are concentrated in the northeastern, the southern, and southeastern parts of the

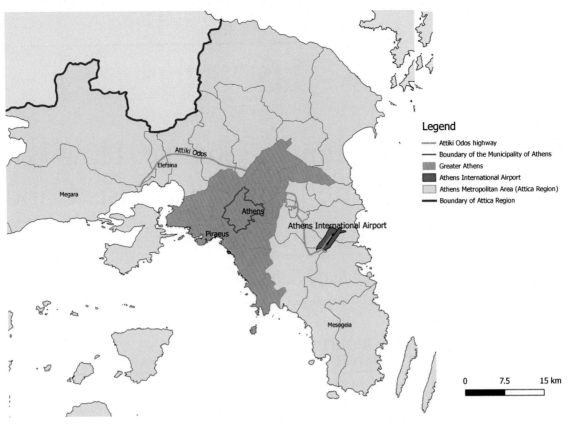

FIGURE 16.1 Boundaries and location of Attica Region, Greater Athens, the Municipality of Athens, and major transportation axes and nodes. *Author's elaboration.*

metropolitan area, and an enclave in the city center, while lower incomes are located in the western and northwestern parts of the agglomeration (Pantazis and Psycharis, 2015).

The Athens city center started to experience downward transformation in the 1990s. The flats-for-land system[1] that was extensively applied between 1950 and 1980 resulted in rapid proliferation of apartment buildings and rapid increase of the building density and traffic circulation. These changes, combined with a lack of infrastructure improvements, degraded living conditions, which in turn caused an exit of upper-middle strata from the declining central areas and an inflow of migrant groups attracted by the affordable cost of living in these areas (Maloutas and Spyrellis, 2015).

In vertical terms, social segregation in Athens resulted basically from the flat-for-land apartment block system per se. According to Maloutas and Spyrellis's (2015) arguments:

> The system allows for living conditions in densely built central areas to differ by floor. Building density did not affect the upper and lower apartment floors equally, contributing to create vertical social segregation.

Upper floor apartments are usually larger and enjoy better conditions (better view, less noise, more light, better ventilation, usable balconies). However, vertical social segregation has been facilitated only in buildings dated before 1980, which nevertheless still house the majority of people living in apartment buildings in the Municipality of Athens (75%). Higher social categories are overrepresented in upper floors, while lower categories, migrants, and vulnerable groups prevail in ground floors and basements (Maloutas and Spyrellis, 2015).

Regarding the economic profile of the capital region, Attica is the largest region in Greece, representing more than a third of the country's population and accounting for over 40% of the country's gross domestic product (GDP).

---

[1] The flats-for-land system is a barter system based on an agreement between a land owner and a builder-contractor to construct a building and split the ownership of the apartments and/or offices and shops built, as per an initial contract describing each side's level of participation in the relevant investment.

While the region presented satisfactory growth per capita up to 2008, since then the negative development of the country due to the economic crisis has been reflected in the indicators of the regional economy. In the period from 2008 to 2012, the regional GDP per capita (PPS) shows a downward trajectory at an average annual rate of −3.9%, which, however, is less than the country's respective rate (−4.7%) (Pavleas, 2015). The most severe socioeconomic impact of the crisis has been unemployment, which in Attica performed a trend worse than the national average as it started from 12.6% in 2010 (less than national average) to reach 27.3% in 2014 (more than national average). Certainly, Attica is the country's dominant economic and political center and a node for research infrastructure. Significant sectors are commerce, financial services, transport, information technologies, health and social services, as well as leisure and recreation. The sharp decline in private investment after 2008 has reduced the already low levels of private research and innovation expenditure in the region (Pavleas, 2015). Major development challenges for the capital region are (1) productivity growth through the introduction of new technologies in the public, social, and private sectors; (2) attracting private investments and supporting innovative enterprises; (3) increasing the capabilities and structures of cooperation between local authorities, educational and research institutions, and private SMEs, and (4) being prepared for threats and uncertainties (originating from environmental hazards and the economic crisis) and improving the quality of life, especially for the vulnerable and nonprivileged groups.

16.1.2 The Building Stock, Urban Form, and Green Infrastructure in the Municipality of Athens: Exposure Issues

Currently, the building stock in the city center (Athens Municipality) indicates signs of obsolescence. The value of this stock has been significantly reduced due to a combination of factors, among them old age, lack of maintenance, a degraded urban setting, and the economic crisis. In a study carried out in a wide central area by the University of Thessaly (2014), it was found that 55% of the building stock has had a lifetime of more than 50 years, while 20% was even older. Of the total stock only 7% was built later than 1990 (Triantafyllopoulos, 2015). Almost all buildings in the city center necessitate costly interventions in order to be restored and modernized and become energy efficient. Triantafyllopoulos (2015) argues that "the cost of restoration and modernization for over 60% of the buildings exceed their current market value…Besides, high plot coverage ratios and high building densities make it rather difficult to apply bioclimatic design." In general terms, the buildings in the city center are multiownership, energy intensive, and in a poor state of maintenance; they suffer from high rates of empty spaces due to the crisis and are unprofitable to modernize. These conditions are impossible to reverse without spatial planning interventions based on consensus/contracts among owners, investors/developers, and planning agencies.

Besides, urban form and pattern in the Municipality of Athens are accountable for high levels of exposure of central districts to the urban heat island (UHI) phenomenon. Of course, high density in the city center becomes an advantage when it comes to energy saving in winter, but at the same time it is a multiplying factor for UHI and exposure to heat waves in the summer. The prevailing construction materials within a densely built urban space radically change the energy balance as these absorb rather than reflect incident solar radiation. Additionally, atmospheric pollution, inadequate ventilation, man-made thermal energy (including the heat given out by numerous air-conditioning devices facilitating individualized adaptation to heat waves but undermining climate change [CC] mitigation), and the geometric features of the urban form intensify UHIs. On the other hand, sizeable greenery areas exist only in the historic precinct of Athens. Indeed, lack of or insufficient urban greenery is a crucial factor for configuration of UHIs in Athens. The phenomenon is eliminated only at the few sites with dense and high greenery, as is the case of the National Garden (Fig. 16.2). However, the relieving effect of the National Garden dissipates within a short distance due to the dense urban form and high urban traffic (Zoulia et al., 2008). Generally speaking, the center of the city during the day is featured by much higher temperatures than suburban areas (7°–8° warmer) (Bank of Greece, 2011).

Geomorphological factors and factors related to man-made interventions have been crucial for Athens' proneness to flooding. The rapid urbanization of the last decades, which left minimal space for greenery, deteriorated flood parameters (Koutsoyiannis, 2002). The natural stream network has been eliminated due to fill-in processes and on top street construction or building development. Sapountzaki and Chalkias (2014) argue that "the policy of streambed coverage that has been followed for decades or its unauthorized realization by private individuals with the connivance of the authorities dismantled Athens from its natural flood protection infrastructure." Generally, the old city center with combined sewers is less vulnerable to floods than the newer suburbs. Flood-prone areas are particularly those at the periphery of the urban complex, i.e., the municipalities along the Saronikos coast and

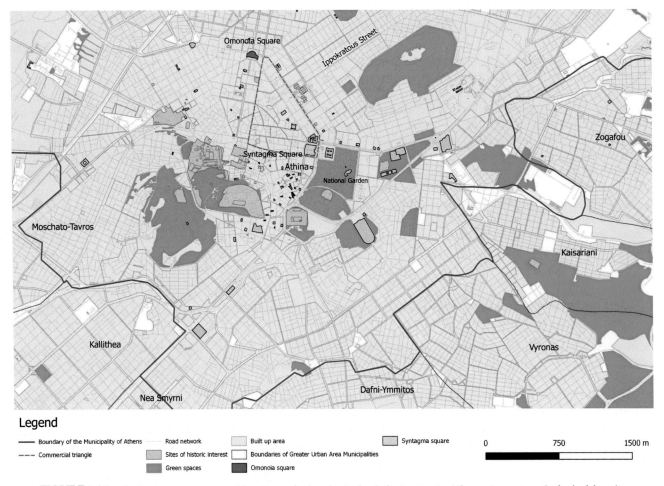

FIGURE 16.2 Basic greenery spaces and locations of urban heat island elimination in Athens city center. *Author's elaboration.*

those at foothill territories (Nikolaidou and Hatzichristou, 1995). Repeatedly, forest fires in the mountain ranges surrounding Athens basin have added to the exposure to flooding of the municipalities of the urban periphery.

16.1.3 Climate Change Trends in Athens Metropolitan Area—Catalysts and Impacts

In recent decades microclimate changes in Attica Region are the result of the combined influence of several man-made factors: expansive urbanization, loss of peri-urban greenery due to forest fires/deforestations, and global climate change.

The National Observatory of Athens (NOA) supplies research on CC in Athens with meteorological data time series covering a period longer than a century. According to NOA measurements, since the mid-1970s the mean annual temperature in Athens shows a continuous rising trend (1.3°C in 1976—2008). Similar are the trends for the average maximum and minimum annual temperatures. The increasing trend of the temperature is stronger during the summer, and in the last decades the mean temperature of the summer months (June, July, and August) has risen rapidly at a rate of 1°C per decade (Founda et al., 2004; Founda, 2011). The decade 2001—10 was the warmest up to then with respect to annual temperatures (mean, maximum, and minimum).

Except long-term trends of mean values of temperature, noteworthy are changes with regard to extreme weather events, particularly during the summer months:

- Increase of frequency of events of extreme temperature (both in terms of isolated extremely hot days and hot events lasting at least 3 days (Founda and Giannakopoulos, 2009);
- Increase of the intensity of the phenomena (absolute maximum temperatures);
- Prolongation and persistence of the phenomena;
- Shifting of manifestation to earlier seasons/periods of the year.

After the mid-1990s, the number of days with higher than 37 or 40°C has increased considerably. Similar has been the increase of the frequency of heat wave events (i.e., at least three successive days with temperatures higher than 37°C). The highest temperature ever recorded by NOA (according to recordings of 150 years) is 44.8°C (June 24, 2007), while the heat wave of June 2007 was the most severe that has ever been recorded. Indeed, summer 2007 seemed like a continuous heat wave. The impact of extremely high temperatures on the urban agglomeration of Athens is amplified by the combined effect of UHIs.

Predictions of the future climate of Athens in the decades to follow are rather worrying. Athens is located in the region of Eastern Mediterranean, which is considered one of the most vulnerable to the man-made component of CC (Giorgi and Lionello, 2008). In particular, predictions refer to a significantly hotter climate for the Mediterranean basin with long-lasting heat waves and lower precipitation but at the same time extreme precipitation events (Tolika et al., 2008).

According to three regional climate models (RCMs), Athens will experience an increase of mean maximum summer temperature by 2°C in the period 2021—50 and 4°C in the period 2071—2100 (Founda and Giannakopoulos, 2009). Predictions for the immediate future (2021—50) make reference to an additional number of 15 days (annually) with maximum temperature over 35°C, compared to the period 1961—90 (Hellas-W.W.F., 2008). While the summer of 2007 in Athens was considered then to be exceptional, it is predicted to become normal for Athens in the second half of the current century.

With regard to precipitation, the predictions for the next decades show reduction of the total annual rainfall in Athens and a simultaneous increase in the frequency of extreme rainfall events. In the case of Athens, a day's cumulative precipitation that is over 10 mm is an extreme precipitation event that may cause flooding (Vlahou, 2010). Attica is one of the regions of the country most frequently hit by extreme precipitation in the last 30 years (Vlahou, 2010). On the other hand, overall reduction of rainfall will deteriorate the quality of underground water reserves and increase the demand for water transfer from distant water basins.

All of the previous scenarios are a predominant topic of concern among experts but not an issue that typically concerns the general public. Heat waves in summer have always been a familiar phenomenon in Athens, as well as occasional floods, and which are difficult to attribute to CC. On the other hand, drought remains out of sight for as long as water authorities manage the water supply of Athens by attracting water resources from distant regions.

16.2 CURRENT RESPONSES TO CLIMATE CHANGE: POLICIES, INITIATIVES, PRACTICES

Recording of the greenhouse gas emissions in Greece is implemented only at the national level. This lack is due to the difficulties involved in measuring gas emissions within the boundaries of specific administrative units. As a result, elaboration of comprehensive plans for quantitatively defined gas emission reduction targets at local, even regional, levels is a difficult task. Consequently, plans follow only the national policy framework for gas emission reduction, including:

- The Second National Program for CC that was drafted and adopted in 2002; the basic objective of this program has been the satisfaction of national obligations, those arising from the implementation of the Kyoto Protocol for the period 2008—12.
- The Fourth National Report on CC (reviewing implementation of the above Program).
- The National Plan for the Allocation of gas emission allowances (NAP), 2008—12, which includes the "business as usual scenario."

According to the Second National Program for CC, the energy sector is the basic source of emissions in the country (76%—79% of the total emissions), rendering it a basic and critical target of the mitigation measures. Basic strategy of the plan for climate action is characterization of Athens as a climate neutral city, meaning that its emissions should not affect the limit of 2°C as the upper threshold of mean temperature increase (Special Unit for Climate Protection and Energy, 2014). According to Krommyda and Economou (2016), the changes and interventions for CC mitigation should target the following sectors: the energy supply system and respective infrastructure, energy efficiency of the building stock, social awareness and consumption patterns, urban planning and infrastructure, industry, transportation planning and mobility, economic prosperity and innovation, and waste management. Some of the respective interventions overlap or it is desirable to present synergies with CC adaptation measures.

The basic official document for CC adaptation at the national level is the "National Strategy for Adaptation to CC-NSACC" (Ministry of Environment and Energy, 2016). This includes sectoral policies for adaptation, among them the policy for the built environment and the cities. The actions provided refer to adaptation of urban planning to CCs and improvement of the microclimate of the built environment in urban centers; assessment of vulnerability of the building stock; architectural and urban redesign with appropriate arrangements of closed, semiclosed, and open spaces; updating the Building Construction Code and the Code for Energy Efficiency of Buildings; increasing the share of urban greenery in urban areas; utilizing appropriate energy-saving materials; combining the use of energy-saving technologies with the use of renewable energy sources (RESs); and training building users to achieve energy efficiency improvements by means of behavioral changes.

The most influential planning documents/policies for the socioeconomic and spatial future of the metropolitan area and Athens city center are: (1) The "Regional Operational Program of Attica (ROPA) 2014–2020" (Ministry of Economy and Development, 2016) (funded by the European Fund for Regional Development and the European Social Fund); (2) The "Smart Specialization Strategy" for the Region of Attica, part of which is funded by ROPA; (3) the "Master Plan of Athens/Attica 2014–2021" and the "General Urban Spatial Plans (GUSP)" of the municipalities of the metropolitan region, among them the GUSP of Athens Municipality.

As presented in the ROPA 2014–20, the Strategy for Regional Development of Attica puts high priority on improvement of the research infrastructure and the human capital of the region, development of entrepreneurship, completion of basic infrastructure (especially the projects for environmental protection), environmental risk mitigation, reduction of disparities, and combating poverty and social exclusion. Two of the 11 thematic objectives of the ROPA refer directly to CC mitigation and adaptation:

- Fourth Objective: "Promotion of energy efficiency, of RES usage and of co-generation and promotion of low-carbon emissions in urban areas" (European Regional Development Fund, 6.9% of the European Union [EU] allocation). The basic source of atmospheric pollution in Attica is vehicular traffic that, in contrast to the industrial sector, shows an increasing trend in pollutant emissions. In recent years, however, traffic loads have decreased due to the economic crisis, a trend that keeps pace with an increase in the use of means of massive transportation, principally tram and metro. On the other hand, the crisis added a new source of atmospheric pollution, i.e., woodstoves and fireplaces, which are used for heating instead of central heating equipment (necessitating costly heating oil). The thematic objective, aiming at CC mitigation, targets urban mass transportation and industry as well as the building sector (representing 39% of the total consumed energy). The actions/measures provided include energy saving and efficiency in the building sector (in public buildings and a limited number of private dwellings in the context of urban regeneration plans). Cogeneration of heat and power in public buildings, as well as management of urban liquid and solid waste for the production of energy, also enjoy high-funding priorities.
- Fifth Objective: "Promotion of CC Adaptation and of risk prevention and management" (European Regional Development Fund, 8.4% of the EU allocation). Attica Region is exposed to CC hazards, namely forest fires, heat waves, droughts, and flash floods. In addition, the region experiences rapid increase of energy demand for cooling. These risks have an impact on nature and biodiversity of the region, including the NATURA 2000 network. Therefore, the objective provides for enhancement and support of safety equipment and the civil protection mechanism, for antiflood programs and limited interventions in NATURA 2000 areas (adaptation and mitigation). The creation of green corridors to enhance and expand the network NATURA 2000 is anticipated to both improve air quality (mitigation) and limit runoff from extreme rainfall to protect downstream urban areas (adaptation). Antiflood programs are provided for specific high-flood-risk zones in conformity with the respective River Basin Districts (RBD) management plans (adaptation).
- On top of the aforementioned, objectives six and seven contribute indirectly to CC mitigation and adaptation.
- The sixth objective targets urban sprawl through gentrification interventions in degraded central neighborhoods (mitigation) and protection/preservation of open spaces in the city center (mitigation and adaptation), whereas the seventh provides for the extension of tramlines to achieve multimodality in urban mass transportation (mitigation).

The expected impacts after implementation of the ROPA include 270 public buildings and 1700 households to receive financial support for improving energy efficiency; reduction of CO_2 emissions by 20,600 equivalent tons; 385,000 more people protected by antiflood works; creation or renovation of 2,000,000 m^2 of open spaces in central urban areas; 5.5 km of new tram lines; 115,000 people covered by a Health Safety Net; and 150,000 people covered by improved health services.

In the former period of 2007–13 an important output of the ROPA for the purpose of reducing CO_2 emissions (i.e., CC mitigation) has resulted from the "Saving at Home" project (total budget 100 million euro) with 2691 completed projects of housing improvements for energy efficiency. An equally important output has been the "Upgrading of the terminal station of Liquefied Natural Gas (LNG) and the Station of Co-production of heat and electricity (13 MW)" covering among others Attica Region. Other projects and works relevant to CC adaptation and mitigation are improvement and extension of the water pipe network, localized urban regeneration projects in central urban areas through interventions in public open spaces (e.g., the project "green life in the city"), and extensions of fixed-track urban transport lines (Athens tram and metro, suburban railway, and electric rail).

The Attica Region's "Strategy for Smart Specialization" (RIS3) in the context of the EU Cohesion Policy is based on three fundamental objectives: (1) "Restructuring of the production base of Attica through technological enhancement of enterprises and innovation development"; (2) "addressing social needs and resolving the acute problems generated by the socio-economic crisis"; and (3) "promoting spatial investments in urban space and upgrading the social and environmental infrastructure for sustainability (urban transport, energy saving, combating urban pollution, integrating Information and Communication Technologies (ICT) in urban functions, upgrading social infrastructure, and urban waste management)". In particular, RIS3 supports and fosters developments and pilot implementation actions focusing on optimal environmental and energy management of urban functions, e.g., smart systems of transport, of living and governance in the city, smart buildings and neighborhoods, smart grids and smart-eco mobility, etc. In addition, soft entrepreneurship options in the form of smart city logistics that take advantage of green technologies in freight transport are also supported. All of the aforementioned options of research and development formulate jointly what RIS3 designates as "Sustainable Economy of Needs." Though RIS3 does not refer directly to CC mitigation and adaptation (with few exceptions, such as the e-environment for early warning in case of CCs and catastrophes), several of the aforementioned smart city applications are highly relevant.

The "Master Plan of Athens/Attica 2021 (MPA)" (L.4277/1.8.2014) for the period 2014–21, which is a crucial policy document for the future spatial profile of the metropolitan region, refers to CC adaptation as a strategic objective for sustainable spatial development, resource saving, and a more effective protection of the environment and cultural heritage. However, despite declarations, the MPA does not place specific attention on issues of CC in Athens and spatial interventions contributing to global CC mitigation and the reduction of vulnerability and exposure to CC hazards (i.e., CC adaptation). The MPA focuses basically on conventional and one-dimensional environmental sustainability issues, like natural resource conservation and biodiversity, preservation of high-productivity agricultural land, and formulation of a network of green corridors. The plan elevates vacant land as a crucial contributor to sustainability and provides for reuse of abandoned and empty properties in the city center in parallel with restriction of the out-of-plan building development[2] (in peri-urban areas) that has created a disorderly landscape damaging the countryside of Attica. Sustainable mobility (through extending fixed-track Mass Meansof Transportation (MMT)) is also one of the concerns of MPA but not clearly as a policy leading to CC mitigation. Lack of explicit reference to CC hazards in MPA is evidence of depreciation of the CC factor by the most important strategic spatial planning document for Athens.

The "General Urban Spatial Plans" (GUSPs) of the municipalities of Attica, among them the GUSP of Athens Municipality, are outdated plans as the majority of them have been elaborated and institutionalized before 2000 (their more recent modifications refer only to minor localized adjustments). Therefore, the prescriptions and objectives that these plans are based on do not take into account the CC adaptation and mitigation issues at stake. Finally, there have been several cases of bioclimatic renewal plans of open space precincts in Athens that originated from corporate initiatives of the Ministry of Environment or Athens Municipality and the private sector and being supported by European Commission (EC) funding, which however, did not find the way to implementation. The most important among them has been the plan "Rethink Athens," which started in 2010 and was cancelled in 2014 by the Council of State.

Currently, a new initiative by the Municipality of Athens for a resiliency strategy including adaptation to local CC impact is under way. This initiative originates actually from the program "100 Resilient Cities" (supported and funded by the Rockefeller Foundation) seeking out creative solutions to "increase employment, manage climate-change risks, and execute ambitious urban regeneration plans" (http://www.100resilientcities.org/cities/entry/athens#/-_/). The new resiliency plan is titled "Redefining the City: Launching a Resilience Strategy for Athens 2030" and has been developed in collaboration with the Athens "100 Resilient Cities" Office. The plan is founded on four pillars: the open, the green, the proactive, and the vibrant city. The goals connected to the green and proactive city are particularly relevant to CC mitigation and adaptation. The green city aims among others at (1) integrating

---

[2] A legal provision offering to land owners in out-of-plan areas the possibility of obtaining building development permits under certain conditions of plot size, geometrical features, and geographical location.

natural systems into the urban fabric, (2) promoting sustainable mobility, and cocreating public spaces, and (3) establishing a more versatile, sustainable, and equitable energy system, whereas the proactive city aims among others at streamlining and upscaling best "survival" skills within the municipality, and through planning and communication, to create trust and a safe environment for people (Municipality of Athens, 2017). With regard to adaptation, in particular, priority actions are:

- Pushing forward with green and blue infrastructure;
- Improving the built environment by using sustainable materials, such as cool pavements, materials with low-embodied energy, and bioclimatic design to reduce ambient air temperatures and improve microclimatic conditions;
- Intensification of public information and awareness campaign;
- Informing and raising awareness of citizens about the heat and flood risks in the city.

It is the municipality's hope that by restructuring the economy using environmental resilience strategies, they will reduce the impacts of environmental disasters and cut homelessness, poverty, and economic stagnation. It remains to be seen whether the plan will find the way to implementation after adoption and consensus on the part of the wider planning system of the country, the involved authorizing and funding institutions (including the Council of State), interested social movements, and civil society associations.

16.3 CRITICAL ANALYSIS OF CURRENT POLICIES, INITIATIVES, AND PRACTICES

As mentioned, CC in Athens as a result of both global and locally shaped man-made adverse conditions is a more or less invisible process to the general public. The urban society is featured by low levels of CC perception and awareness of the CC hazards to which Athens is exposed and the necessary mitigation and adaptation responses at both levels of the individual household and the institutions (national, regional, local). According to a public opinion survey carried out in the period 2007–08 (Public Issue, 2008), the responses/measures to face CC should basically originate from international-level organizations (64%—75% of the sample of 1052 persons) and much less from national (6%—9% of the respondents) or local level institutions (9%—14%). According to the same survey, people in Greece believe that CC impacts concern basically the agricultural sector/areas (79%—81%) and not the cities. There are several reasons explaining this situation.

At first, up to the present the most important manifest hazard of CC in Athens is extreme temperatures and heat waves. However, people find it hard to imagine heat as a hazard and disaster in temperate regions (IFRCS, 2004). Even in cases that heat waves turn into disasters, these are silent. As Klinenberg (2002) puts it, "heat waves are slow, silent and invisible killers of silent and invisible people."

Secondly, Athenians are used to heat waves and high summer temperatures, and they have become resilient and adaptive to such extreme conditions at both levels of the human individual and responsible institutions. Indicative of people's responses are their habit to resort to nearby beaches on extremely hot days and summer weekends or take advantage of second house availability. Besides, the elderly, one of the most vulnerable groups to heat waves, are more or less integrated into the daily life of families, thus preventing loss of life due to isolation and insufficient assistance of the aged in case of a heat wave crisis. After all, public and private old-age-care facilities are equipped with air conditioners.

Thirdly, vertical social segregation in the central districts of Athens Municipality, i.e., the most exposed to heat waves in Greater Athens, functions as a factor ameliorating exposure. Higher social categories of people who live in upper-floor apartments, which are more exposed to extreme temperatures, can afford installation of air conditioners, while migrants and vulnerable groups living in lower-floor apartments experience lower levels of physical exposure. Oddly enough, vertical social segregation in central Athens alleviates inequalities at least with regard to exposure to extreme temperatures.

Fourthly, there is a lack of agencies to disseminate information on CC. The curricula of primary and secondary education do not include the necessary information and instructive messages on individuals' and collective responsibility for CC. Governance and public consultation procedures that are normally embedded in planning processes might function as a mechanism of consolidation of a culture of responsibility for CC mitigation and adaptation. However, there is a poor experience and tradition of public involvement in planning decision-making in Greece. Therefore, the plans, strategies, and programs mentioned in paragraph 16.2, those that include actions and measures of mitigation and adaptation to CC, have been elaborated and institutionalized without prior inclusive involvement of the civil society. Public consultation procedures do take place according to law, but these are of a limited range and actually activate only professionals and experts, not the general public.

Indeed, low levels of CC perception and awareness as well as the economic crisis are among the most important barriers to introduction and implementation of CC mitigation and adaptation actions/measures. Low public awareness affects how planning institutions think and therefore minimizes the degree of consideration of the CC factor in statutory planning documents. Although ROPA 2014–20 and MPA 2021 have been almost simultaneously elaborated planning documents, they show big differences regarding the degree of consideration of the CC factor. This is because the first bears the imprint of the EC Regional Policy while the second reflects domestic priorities. Furthermore, low public perception and awareness of CC may result in underspending of the funds of ROPA that are intended for CC mitigation and adaptation. Nevertheless, the economic crisis has an ambivalent effect; on the one hand it lowers CC perception as the adversities of the crisis become first-priority concerns of the general public, and on the other it reduces private vehicle traffic and forces households to practice energy saving.

In general terms, Athens does not lack resilience assets for adaptation to CC, especially at the level of individual households. Among these assets are social solidarity and a cohesive social structure based on the extended family unit; vertical social segregation; high rates of second home ownership, principally along the Attica coastline; quick, easy, and cheap access to the plentiful beaches of Attica; former experience of extreme summer temperatures; and knowledge about the appropriate self-protection responses. What Athens really lacks is the spatial planning and urban design component of mitigation and adaptation strategies and measures, and this lack results in an urban open space that is highly exposed and vulnerable to CC. This is due to a Greek culture hostile to spatial planning, totally outdated GUSPs, incapability of administration and government (at the central, regional, and local levels) to coordinate development programs with spatial plans, and a hard, ineffective, and time-consuming process of spatial planning institutionalization and implementation. As a result, public large-scale spatial interventions to serve CC mitigation in Athens are totally missing. What appears are only scattered local authority or private eco-architectural initiatives managing greening and adaptation of individual buildings or building complexes such as schools or museums. However, few of their users, if any, have perceived them as a response to CC.

References

Allen, J., Barlow, J., Leal, J., Maloutas, T., Padovani, L., 2004. Housing and Welfare in Southern Europe. Blackwell, London.
Bank of Greece, 2011. The Environmental, Economic and Social Impacts of Climate Change in Greece. Bank of Greece Publications, Athens.
Emmanuel, D., 2015. Social aspects of access to home ownership. In: Maloutas, T., Spyrellis, S. (Eds.), Athens Social Atlas. Harokopio University, French School of Athens, Hellenic Statistical Authority and National Centre for Social Research. Available at: http://www.athenssocialatlas.gr/en/article/access-to-home-ownership/.
Founda, D., 2011. Evolution of the air temperature in Athens and evidence of climatic change: a review. Advances in Building Energy Research 5 (1), 7–41.
Founda, D., Giannakopoulos, C., 2009. The exceptionally hot summer of 2007 in Athens, Greece – a typical summer in the future climate? Global and Planetary Change 67, 227–236.
Founda, D., Papadopoulos, K.H., Petrakis, M., Giannakopoulos, C., Good, P., 2004. Analysis of mean, maximum and minimum temperature in Athens from 1897 to 2001 with emphasis on the last decade: trends, warm events, and cold events. Global and Planetary Change 44, 27–38.
Giorgi, F., Lionello, P., 2008. Climate change projections for the Mediterranean region. Global and Planetary Change 63, 90–104.
Hellas, W.W.F., 2008. Solutions for Climate Change: A Vision for Sustainability for Greece of 2050. Scientific report, Athens (in Greek).
Klinenberg, E., 2002. Heat Wave: A Social Autopsy of Disaster in Chicago. The University of Chicago Press, Chicago.
Koutsoyiannis, D., 2002. On the covering of Kifissos river. Newspaper Machetiki of Moschato. June 8, Athens (in Greek).
Krommyda, V., Economou, D., 2016. Planning for climate change in the Attica basin. Spatial directions toward mitigation and adaptation in western athens. 9th Pan-Hellenic Conference HellasGIs, Athens 8/12/2016. Available at: https://www.researchgate.net/publication/311512051_SCHEDIASMOS_GIA_TEN_KLIMATIKE_ALLAGE_STO_LEKANOPEDIO_ATTIKES_CHORIKES_KATEUTHYNSEIS_PROLEPSES_KAI_PROSARMOGES_GIA_TE_DYTIKE_ATHENA.
Maloutas, T., Spyrellis, S., 2015. Vertical social segregation in Athenian apartment buildings. In: Maloutas, T., Spyrellis, S. (Eds.), Athens Social Atlas. Harokopio University, French School of Athens, Hellenic Statistical Authority and National Centre for Social Research. Available at: http://www.athenssocialatlas.gr/en/article/vertical-segregation/.
Ministry of Economy and Development, 2016. Partnership Agreement 2014–2020, Attica Operational Programme. Available at: https://www.espa.gr/en/Pages/staticOPAttica.aspx.
Ministry of Environment and Energy, 2016. National Strategy for Adaptation to Climate Change. Athens (in Greek). Available at: http://www.ypeka.gr/Default.aspx?tabid=303.
Municipality of Athens and the 100 Resilient Cities Programme (Rockefeller Foundation). Athens Resilience Strategy for 2030, 2017. Available at: http://www.100resilientcities.org/wp-content/uploads/2017/06/Athens_Resilience_Strategy_-_Reduced_PDF.compressed.pdf.
Nikolaidou, M., Hatzichristou, E., 1995. Registering and Assessment of Devastating Floods in Greece and Cyprus (Diploma thesis). Department of Water Resources, Hydraulic and Maritime Engineering, National Technical University of Athens, Athens (in Greek).

Pantazis, P., Psycharis, G., 2015. Residential segregation based on taxable income in the metropolitan area of Athens. In: Maloutas, T., Spyrellis, S. (Eds.), Athens Social Atlas. Harokopio University, French School of Athens, Hellenic Statistical Authority and National Centre for Social Research. Available at: http://www.athenssocialatlas.gr/en/article/income-groups/.

Pavleas, S., 2015. The "smart specialization strategy" in the region of Attica. In: Maloutas, T., Spyrellis, S. (Eds.), Athens Social Atlas. Harokopio University, French School of Athens, Hellenic Statistical Authority and National Centre for Social Research. Available at: http://www.athenssocialatlas.gr/en/article/smart-specialisation/.

Public Issue, 2008. Survey of the Greek Public Opinion on Climate Change. Available at: http://www.publicissue.gr/982/climate-change/.

Sapountzaki, K., Chalkias, C., 2014. Urban geographies of vulnerability and resilience in the economic crisis era: the case of Athens. A|Z Journal Special Issue "Cities at Risk" 11 (1), 59–75.

Special Unit for Climate Protection and Energy, Reusswig, F., Hirschl, B., Lass, W., 2014. The feasibility study "climate neutral Berlin 2050". Berlin: Senate Department for Urban Development and the Environment. Available at: https://www.pik-potsdam.de/members/lass/climate-neutral-berlin-20150-a-feasibility-study.

Tolika, K., Anagnostopoulou, C., Maheras, P., Vafiadis, M., 2008. Simulation of future changes in extreme rainfall and temperature conditions over the Greek area: a comparison of two statistical downscaling approaches. Global and Planetary Change. 63, 132–151.

Triantafyllopoulos, N., 2015. The building stock of central Athens. In: Maloutas, T., Spyrellis, S. (Eds.), Athens Social Atlas. Harokopio University, French School of Athens, Hellenic Statistical Authority and National Centre for Social Research. Available at: http://www.athenssocialatlas.gr/en/article/real-estate/.

Vlahou, O., 2010. Geographical Distribution, Variability and Impacts of Extreme Weather Events in Greece Based on Daily Press Reports and Surface Measurements (Diploma thesis), Geography Department, Harokopio University, Athens (in Greek).

Zoulia, I., Santamouris, M., Dimoudi, A., 2008. Monitoring the effect of urban green areas on the heat island in Athens. Environmental Monitoring and Assessment. https://doi.org/10.1007/s10661-008-0483-3.

CHAPTER

17

Sustainability of Climate Policy at Local Level: The Case of Gaziantep City

Osman Balaban, Bahar Gedikli
Middle East Technical University, Ankara, Turkey

17.1 INTRODUCTION

The international community has been responding to climate change for several decades, although at a slow pace. The United Nations Framework Convention on Climate Change (UNFCCC) was followed by other international agreements and meetings, such as the Kyoto Protocol and the Conference of the Parties (COP) meetings, which evolved around the aim of climate change mitigation and adaptation since 1992. In December 2015, the world's nations created a new agreement, the Paris Agreement, which will replace the Kyoto Protocol from 2020 onwards.

Throughout the evolution of climate policy, urban responses by city governments across the world have gained particular importance (Bulkeley et al., 2012). With their authorities over land-use planning, water and waste management, energy use, and transportation, cities can play significant roles in reducing greenhouse gas (GHG) emissions as well as in enhancing climate change resilience. Major local policies and actions for climate change include measures for controlling growth, encouraging nonmotorized transportation, reducing energy use in buildings, promoting the use of renewable energy, retrofitting urban buildings, and increasing the amount of green areas in and around cities (Balaban, 2012).

Nevertheless, the progress toward an effective climate policy is yet to be sufficient. One reason behind this poor progress is that climate-related actions and policies at the local level are mostly individual and piecemeal efforts. Limited authorities of city governments and hierarchical relations, technical and financial difficulties, lack of required expertise, and priorities other than the climate problem usually prevent cities from developing a systematic and strategic policy framework for climate change (Næss et al., 2005; Davies, 2005; Balaban and Senol Balaban, 2015).

This chapter presents a case study from Turkey in order to discuss proactive leadership as an important, yet insufficient factor for successful climate policy making at the city level. Gaziantep is a million-plus metropolitan city located in the Southeastern part of Turkey and has been the frontrunner of climate policy among Turkish cities for a decade. However, the city has lost its distinct position in recent years probably due to regional and local political fluctuations, leaving behind interesting lessons for political sustainability of climate policy.

17.2 THE CLIMATE POLICY IN TURKEY

Climate change mitigation and adaptation entered into the agenda of both central and local administrations in Turkey in the last decade after the initiation of negotiations for the European Union (EU) accession. Turkey became an official party to the UNFCCC in 2004 and the Kyoto Protocol in 2009, but has not ratified the Paris Agreement as yet. The major national policy document for climate change is the National Climate Change Strategy document that was prepared in 2009–10. This strategy document was followed by the National Climate Change Action Plan and the National Climate Change Adaptation Strategy and Action Plan introduced in 2011 and 2012, respectively. The Ministry of Environment and Urbanization is the coordinating agency of climate policy in Turkey. A specific department for climate change was established within the Ministry in 2009.

Despite the legal and institutional steps, the national government's commitment to climate change policy is not strong. This manifests itself in Turkey's weak intended contribution, which promises a reduction of 21% from future increase in total GHG emissions, and in the government's slow pace in ratifying the Paris Agreement. The Turkish government has signed the Paris Agreement on 22 April 2016 but it has not yet been ratified. The possibility of ratification in the near future seems to be low, as the government insists on having a special position to access climate finance, despite being listed as an annex-I country of the UNFCCC.

The weak commitment of the national government to climate policy also leads to limited progress in urban climate change governance in Turkey. Cities are not provided with clear guidance and strong support from the center. The goals and instructions for local authorities in national climate policy documents are either broad or nonbinding. Furthermore, climate change–related goals are barely mainstreamed into other related legal and policy frameworks. Thus, under such circumstances, involvement of local governments in climate governance is mostly shaped by proactive leaders or some motivated policy actors at the local level. Recent research indicated that many cities in Turkey have already introduced climate change policy in some way, but their policies and actions show great variety (Gedikli and Balaban, 2018).

At present, urban climate governance in Turkey can be characterized by efforts of some frontrunners. Gaziantep is one of the frontrunners of urban climate governance in Turkey. Gaziantep Metropolitan Municipality is the first city government in Turkey that prepared a climate change action plan and established a climate change bureau under the directorate of environmental issues to deal specifically with climate change. Furthermore, the city government presented its environmental policies and initiatives at the Rio+20 Conference in 2012. In the following discussion, we elaborate on the major aspects of the climate policy making in Gaziantep. The information provided is based mainly on the interviews conducted with key policy actors in Gaziantep Metropolitan Municipality as well as on recent academic work on the city.

17.3 THE CASE STUDY OF GAZIANTEP

17.3.1 A Brief Introduction About the City

Gaziantep is one of the principal cities of Turkey. As of 2015, the city of Gaziantep had a population of approximately 2 million people, and was listed as the eighth biggest city of Turkey by population. The city is located in Southeastern Anatolia, one of the underdeveloped regions of the country (Fig. 17.1). Despite the poor economic conditions of its wider region, Gaziantep has shown remarkable economic development based on industry and commerce, especially after 1980, and is often regarded as the "chief of Anatolian tigers" (Bayırbağ, 2010). On the other hand, as it is located in the vicinity of Syria, Gaziantep has been adversely impacted by the Syrian civil war. Many refugees have come to the city in the last 5 years and deepened the challenges faced by the city government, such as the need for jobs and shelter for newcomers, ensuring social cohesion among refugee populations.

The roots of climate policy making in Gaziantep date back to 2004, during the term of the former mayor of the city, Asım Güzelbey. He served as the metropolitan mayor for two terms until 2014. He is known for his visionary administration and innovative attempts for policy making and implementation in various fields of urban development. Our interviewees indicated that one of his objectives was to move the municipal administration beyond the conventional approach by involving contemporary policy fields and strengthening the international relationships of the municipality. In this respect, he is acknowledged for appointing the right people with required skills and capacity to the right positions. So, the interviewees noted the high in-house capacity and expertise as a major factor that made Gaziantep one of the most successful local governments in Turkey in the recent decade. Climate policy is one of the policy fields in which Gaziantep has shown significant progress and taken concrete actions.

17.3.2 Climate Governance in Gaziantep

The former mayor has appointed an active, enthusiastic, and knowledgeable expert as the head of the Directorate of Environmental Protection and Control, which is the environmental office of the municipality. The interviewees mentioned that the environmental manager of the city of Gaziantep was the mastermind of environmental policy making in the city between 2004 and 2014. The good match between the mayor and the environmental manager, as both were open-minded and innovative, was also highlighted as a driving force for environmental policies and actions taken.

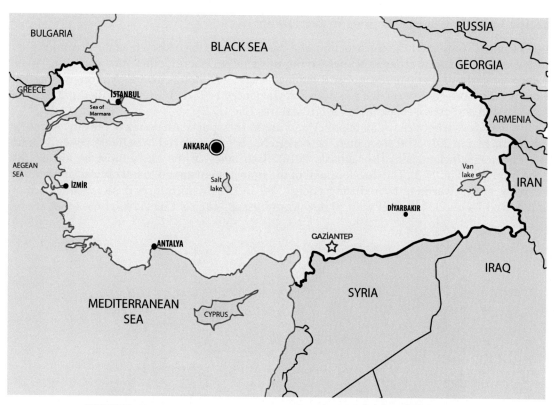

FIGURE 17.1 Gaziantep's location and provincial borders. *Authors' elaboration.*

One distinctive aspect of urban climate governance in Gaziantep was the establishment of a "climate change bureau" under the environmental office of the municipality. The bureau was established to deal with the climate problem in Gaziantep by means of mitigation and adaptation policies and actions. Meanwhile, an energy bureau was also established to develop policies and take actions for energy reduction and efficiency. However, the coordination between the energy bureau and climate change bureau was not well developed, as per our interviews.

The progress in climate governance was not limited to the institutional reforms mentioned herein. In 2010, the city of Gaziantep initiated the preparation of a climate change action plan in collaboration with the French Development Agency and Gaziantep University. The French side funded 60% of the total cost of the plan, whereas the remaining 40% was covered by the municipality. Gaziantep University provided technical support to the plan, especially to the preparation of the GHG inventory. The Climate Change Action Plan of Gaziantep was the first city-level climate change action plan in Turkey. The plan identified industry, service, transport, and residential sectors as the major GHG emitters in the city, and declared a per capita GHG reduction target of 15% by 2023, according to which particular strategies of the plan were prepared (Gaziantep Metropolitan Municipality, 2011). In its current form, the action plan of Gaziantep is basically a mitigation plan, in which adaptation is barely a concern in the actions specified.

Although the achievement of a climate change action plan was a significant and innovative attempt by the municipality, its implementation was not sufficiently monitored after its approval and enactment. An important factor for that was the lack of knowledge and expertise of the municipality staff for monitoring the implementation of the plan. The foreign counterparts provided support only in the preparation of the plan, but had no role in capacity building. At the time we conducted the interviews (December 2014), city officials in Gaziantep were in search of opportunities for training the staff to prepare and monitor GHG inventories. Such problems restrained the implementation of the climate change action plan, and no significant progress has been achieved in reducing GHG emissions. As several years passed after the preparation of the plan, the need for updating the plan with more recent data and a clear methodology has become urgent. If the plan is updated by using recent data and clear methodology, the related municipality staff should be trained about the calculation of sectoral energy consumption levels and associated emissions, and preparing inventories.

Nevertheless, there is no foreseen update of the climate change action plan. The new mayor took over the office with a different agenda and priorities, which led to termination of some actions directly related to climate policy. For instance, the climate change bureau was closed soon after the new mayor took over the office.

17.3.3 Specific Climate Change Actions in Gaziantep

The city government completed the construction of a passive house (also known as the eco-house) in one of the major urban parks in the city in order to showcase the most recent energy efficiency technologies as well as to increase the awareness about energy efficiency and renewable energy among citizens. Although it has become quite common in many cities in recent years, the eco-house in Gaziantep is Turkey's first certified passive house, which combines both passive and active sustainable design features (Fig. 17.2).

Parallels have been drawn between climate policy and urban transportation policy of the city. A tram system was developed and launched in 2010, and since then the system has been improved in network coverage and by replacing engines with more efficient ones (Kocakusak, 2016). Two lanes of the city's most important avenue were allocated to the tram line (Fig. 17.3). The development of the tram system aimed to not only address traffic problems but also to combat global warming by reducing energy use in the urban transport sector. Furthermore, the city government renewed its public bus fleet with 50 new Compressed Natural Gas (CNG) buses that are more energy

FIGURE 17.2 The eco-house in Gaziantep. *By the authors.*

FIGURE 17.3 The tram system in Gaziantep. *By the authors.*

efficient and less polluting. However, apart from transportation, climate policy had no significant influence over other fields of urban planning and development.

In line with the developments in climate policy between 2004 and 2014, the city government also strengthened its relations with international networks and counterparts. The former mayor participated in the Rio+20 meeting in 2012 in Rio de Janeiro and presented the city's environmental policies and actions. Furthermore, the city of Gaziantep has become a member of several transnational municipal networks since 2010. The city joined three networks, including Energy Cities in 2010 and ICLEI and Eurocities in 2012, and has become an associated city for EU-GUGLE-Sustainable Renovation Models for Smarter Cities Project (Bütün, 2016).

The governance reforms and particular actions discussed herein constituted the main aspects of climate policy making in Gaziantep as a frontrunner city in Turkey. The achievements in Gaziantep may not seem new or remarkable when compared to leading international examples. However, they should be acknowledged as significant steps, considering that the local climate policy development is in its infancy in Turkey. One important shortcoming of climate policy in Gaziantep case is that adaptation has barely been a concern in the actions taken so far. Although the term *adaptation* is mentioned several times in the climate change action plan, it is not translated into clear targets and concrete actions. None of the actions specified in the plan focus solely on climate change adaptation. Another shortcoming was indicated by our interviewees as the inadequate relations between environment and urban development departments. Both departments deal with climate policy within their fields of responsibility. The work of different departments that deal with climate change mitigation and adaptation, if elaborated together, can lead to more effective responses.

17.4 CONCLUDING DISCUSSION

Climate change is an emerging policy field at both national and local levels in Turkey. The national government has taken certain steps with preparation of national frameworks and programs, nevertheless these efforts were not effectively transmitted to local governments to address the climate problem at the local level. At present, cities are not provided with clear guidance and support from the center in their efforts to develop a local climate policy. Under these circumstances, urban climate governance in Turkey is limited to the efforts of some frontrunners. Gaziantep is a pioneering city of climate policy in Turkey due in large part to the particular actions taken between 2004 and 2014.

The major reason for Gaziantep to be the frontrunner of local climate policy was the proactive leadership of the former metropolitan mayor. He brought contemporary issues and policy fields like climate change into the municipal agenda and encouraged the city officials to push forward these issues. Proactive leadership of mayors or officials is important as a start-up initiative that generates and raises the awareness on climate change at a particular locality. However, the periods of the mayors terminate, and therefore such leadership may not ensure the continuity of the initial actions and the sustainability of the local climate policy. This was the case in Gaziantep. The new mayor has come with different concerns and priorities regarding the future development of the city. One of the actions of the new municipal administration was to close the climate change bureau and lessen the priority given to climate change. As climate change was not institutionalized and rooted in the municipal administration, the new mayor could easily decide to give priorities to other conventional and emerging issues.

What needs to be done to overcome the problems of individualistic policy settings is a matter for another paper. However, one thing should be emphasized here. During their terms, proactive leaders should prioritize awareness raising among citizens and help strengthen the NGOs or community groups that deal with climate change through participatory practices. Wider involvement of local community groups in policy making may safeguard the continuity and sustainability of climate change policy.

References

Balaban, O., 2012. Climate change and cities: a review on the impacts and policy responses. METU Journal of the Faculty of Architecture 29 (1), 21–44.

Balaban, O., Senol Balaban, M., 2015. Adaptation to climate change: barriers in the Turkish local context. TeMA-journal of Land Use, Mobility and Environment 8 (Special Issue ECCA 2015), 7–22.

Bayırbağ, M.K., 2010. Local entrepreneurialism and state rescaling in Turkey. Urban Studies 47 (2), 363–385.

Bulkeley, H., Castan Broto, V., Edwards, G., 2012. Bringing climate change to the city: towards low carbon urbanism. Local Environment: The International Journal of Justice and Sustainability 17 (5), 545–551.

Bütün, G.D., 2016. The Impact of Transnational Municipal Networks on Climate Policy-making: The Case Study of Gaziantep, Nilufer and Seferihisar Municipalities (Thesis submitted to the Graduate School of Social Sciences of Middle East Technical University. Ankara/Turkey).

Davies, A., 2005. Local action for climate change: transnational networks and the Irish experience. Local Environment: The International Journal of Justice and Sustainability 10 (1), 21–40.

Gaziantep Metropolitan Municipality, 2011. Climate Change Action Plan of Gaziantep. Available at: http://www.afd.fr/webdav/shared/PORTAILS/PAYS/TURQUIE/PAGE%20D%27ACCUEIL/Gaziantep-CCAP-TR-final-20111102.pdf.

Gedikli, B., Balaban, O., 2018. An evaluation of local policies and actions that address climate change in Turkish metropolitan cities. European Planning Studies 26 (3), 458–479.

Kocakusak, D., 2016. Combating Climate Change: Critical Evaluation of Climate Change Action Plans on Urban Scale (Thesis submitted to the Graduate School of Social Sciences of Middle East Technical University. Ankara/Turkey).

Næss, L.O., Bang, G., Eriksen, S.H., Vevatne, J., 2005. Institutional adaptation to climate change: flood responses at the municipal level in Norway. Global Environmental Change 15 (2), 125–138.

CHAPTER

18

Toward Integration: Managing the Divergence Between National Climate Change Interventions and Urban Planning in Ghana

Stephen Kofi Diko
University of Cincinnati, Cincinnati, OH, United States

18.1 BACKGROUND

The Kumasi metropolis is the second largest city in Ghana. It is also the capital of the Ashanti Region, one of 10 political regions in the country. The 2010 Population and Housing Census puts the population of the metropolis at 1,730,249. It constitutes 36.2% of the region's population of 4,780,380 and approximately 0.9% (214.3 sq. km) of the land area (Ghana Statistical Service, 2014). It is bordered to the north by two districts, Kwabre East and Afigya Kwabre districts; two districts to the west, Atwima Kwanwoma and Atwima Nwabiagya districts; two municipalities to the east, Asokore Mampong and Ejisu-Juaben municipalities; and one district to the south, Bosomtwe district (Fig. 18.1).[1]

Since the reform of decentralized decision-making and planning in Ghana in 1988, a new structure of local governance was initiated where districts, municipalities, and metropolises are delineated by population and governed by unified local government authorities called assemblies.

According to the Local Government Act of 1993 (Act 462), a metropolitan assembly covers a jurisdiction with over 250,000 people, a population of over 95,000 for a municipal assembly, and a district assembly for a jurisdiction with population between 75,000 and 95,000 (Institute of Local Government Studies, 2010; Diko and Akrofi, 2013). These authorities perform administrative and planning functions to promote development (Ahwoi, 2010; Institute of Local Government Studies, 2010). In the Kumasi metropolis, the Kumasi Metropolitan Assembly (KMA) is the development authority for socioeconomic and physical planning and development.

As part of this process, metropolitan, municipal, and district assemblies (MMDAs) are responsible for preparing 4-year medium-term development plans (MTDPs) to guide the development of their jurisdictions (Mensah, 2005). These plans are facilitated by Planning Coordinating Units (PCUs), who use guidelines prepared by the National Development Planning Commission (NDPC) to ensure that policies, programs, and projects, among other initiatives for implementation in these plans, align with Ghana's development agenda (NDPC, 2006, 2009, 2013). These guidelines are based on National Development Policy Frameworks (NDPFs), which are also prepared every 4 years. The NDPFs guide and set the agenda at all levels and sectors of national development (Fig. 18.2). Thus climate change interventions (CCIs) in Ghana's NDPFs relating to sustainability, adaptation, mitigation, and/or resilience are expected to translate into sectoral as well as district-, municipal-, and metropolitan-level interventions.[2] This

---

[1] In November 2017, the Government of Ghana created 38 new districts. This changed the total number of districts from 216 to 254. New municipalities were also carved out from some assemblies. Sub-metropolitan areas such as Oforikrom, Kwadaso, Old Tafo, Asokwa, and Suame, which hitherto were part of the Kumasi metropolis are now municipalities. See Citifmonline (2017): http://citifmonline.com/2017/11/18/list-of-new-districts-municipal-assemblies-infographic/ [Accessed on 17th February 2018].

[2] Climate Change Interventions (CCIs) are defined here as the various forms of development efforts that are used to guide climate actions and investments. It includes policy or development goals, objectives, strategies, programs of actions, and projects/activities/actions plans.

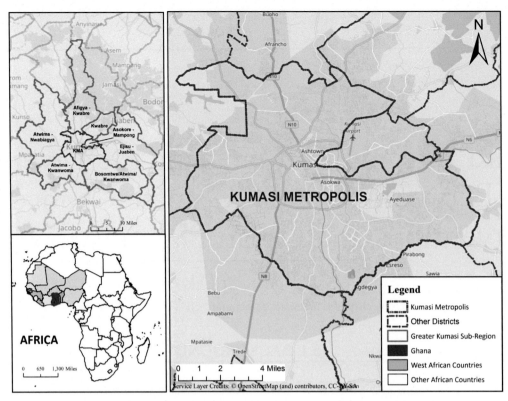

FIGURE 18.1 Location of the Kumasi metropolis. *Author's elaboration.*

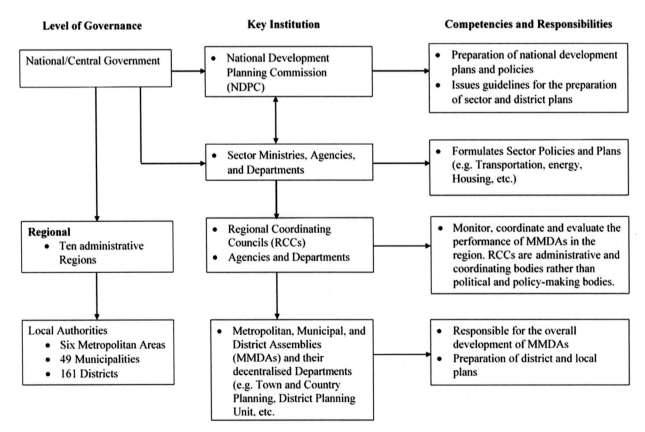

FIGURE 18.2 Structure of governance and development planning in Ghana. *Adopted with permission from Acheampong, R.A., Ibrahim, A., 2015. One nation, two planning systems? Spatial planning and multi-level policy integration in Ghana: mechanisms, challenges and the way forward. Urban Forum, 7.*

is evident in the NDPFs for the 2010–13 and 2014–17 periods. The former, for instance, aimed to "integrate/mainstream [the] impact of climate change into sectoral and district plans" in Ghana (NDPC, 2010, p. 145).

Furthermore, there are a number of specific national climate change policy frameworks that are separate from these NDPFs. The NDPC through its NDPFs takes into consideration all available national and sectoral policies and plans, including those on climate change. Therefore, it is expected that MMDAs align their development plans with NDPFs and these national climate change policy frameworks as part of their plan preparation processes. However, studies on climate change policy raise concern about a lack of integration between national climate change policies and local-level interventions (Deslatte and Swann, 2016; Ziervogel et al., 2016; Adu-Boateng, 2015; Lee and Painter, 2015; Shemdoe et al., 2015).

18.2 CLIMATE CHANGE ISSUES

This chapter takes cognizance of the lack of integration between national climate change policies and local intervention to examine whether there is a divergence or convergence of Ghana's national agenda on climate change—NDPFs and national climate change policy frameworks—and urban planning. The aim is to understand the framings of climate change issues, identify the focus of CCIs, and the level of integration of national CCIs in plans prepared by MMDAs. To do this, the chapter reviews four urban development plans for the Kumasi metropolis. These plans are three KMA MTDPs for the 2006–09, 2010–13, and 2014–17 periods, and the Greater Kumasi Comprehensive Plan (GKCP). The GKCP is a subregional master plan for the Kumasi metropolis and its adjoining five districts and two municipalities (Fig. 18.1). Together, these seven MMDAs constitute the Greater Kumasi subregion, which covers a total area of about 1100,391 sq. miles (2850 sq. km) and has a total population of 2,764,091. The master plan consists of a spatial development framework, structure plan, and implementation plan to guide the growth and development of the subregion (MESTI, 2013; Acheampong et al., 2016). The GKCP was prepared by the Physical Planning Department of the KMA with funding from the Japan International Cooperation Agency.

This chapter is organized into three sections (after this background discussion). The first section examines national planning documents to understand the framings of climate change issues. The second section examines the four urban development plans for the Kumasi metropolis to also understand the framings of climate change issues. In the last section, the areas of divergence or convergence in these two sets of documents are presented with suggestions on ways to integrate CCIs in urban planning within the Kumasi metropolis, and Ghana in general.

18.2.1 Climate Change Issues in National Development Policy Frameworks[3]

Beginning from 1996 up to 2014, the NDPC has facilitated and prepared five NDPFs in Ghana. The first was the *Ghana Vision 2020* prepared for the 1996–2000 period. Two poverty reduction papers were also prepared between 2003 and 2009: *Ghana Poverty Reduction Strategy* (GPRS I) from 2003 to 2005 and the *Growth and Poverty Reduction Strategy* (GPRS II) from 2006 to 2009. In the last 7 years, the NDPC has prepared two additional NDPFs: The *Ghana Shared Growth and Development Agenda* (GSGDA I) covers 2010 to 2013 and GSGDA II for the 2014 to 2017 period. For all the NDPFs, *Guidelines for the Preparation of Medium-Term Development Plans* were prepared for MMDAs to guide their plan preparation processes. The review of the NDPFs reveal that there were concerns for the "Environment" and "Environmental and Natural Resource Management" such as deforestation, coastal erosion, and drought. Yet, the term *climate change* only appeared sometime later. Specifically, the first appearance of *climate change* was in the GPRS II, which began in 2006 (Fig. 18.3).

In the Vision 2020, the NDPC identified some environmental challenges facing Ghana, including: "pollution, deforestation, soil and coastal erosion and inefficient waste management" (NDPC, 1995, p. vi). In response, the Commission identified preemptive measures to understand "the causes of environmental degradation [that] presents Ghana with an opportunity to avoid many of the errors committed by the industrialized countries in the past" (1995, p. 39). The aim was to reduce "ecological and environmental degradation" and promote "an efficient management system and environmentally sound development of all water resources in the country" (NDPC, 1995, p. 63). This continued in GPRS I, which also identified challenges with the "environment." The policy framework

---

[3] This section has been reproduced with some modification from a paper submitted to the *2016 Annual Reducing Urban Poverty Graduate Paper Competition*, Urban Sustainability Laboratory. Washington D. C.: Woodrow Wilson International Center for Scholars. This is also published by the agency in Owusu-Daaku and Diko (2017).

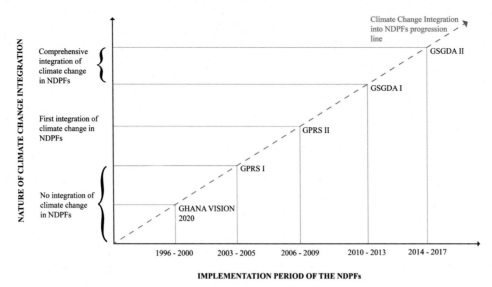

FIGURE 18.3 Progression of climate change integration in National Development Policy Frameworks in Ghana. *Author's elaboration.*

connected environmental factors to poverty, explaining that poverty "may be caused or exacerbated by ... exposure to shocks due to limited use of technology to stem the effects of droughts, floods, army worms, crop pests, crop diseases, and environmental degradation" (NDPC, 2003, p. 3). The object was to align national goals with international goals through "environmental sustainability and regeneration" (NDPC, 2003, p. 35) and a long-term objective "to maintain a sound environment and to prevent all forms of environmental degradation" (NDPC, 2003, p. 39). There was no allusion to climate change in GPRS I. In effect, climate change issues in NDPFs prior to GPRS II were not targeted, focused, or seen as binding factors that affect Ghana's development. By extension, while environmental issues were expected to form part of the issues MMDAs integrate into their MTDPs, the absence of climate change framings implied a lack of attention to climate change issues in these plans (Owusu-Daaku and Diko, 2017).

The first evidence of the term climate change appeared in GPRS II. Policies, goals, and strategies were framed using "climate variability" and "climate change." In problematizing climate change, GPRS II explicitly related the natural environment, environmental degradation, and natural resources to climate change. In one instance, the GPRS II problematized that the "adverse environmental factors such as climate variability and land/soil degradation continue to be challenges posed to the growth potential of the agricultural sector" in Ghana (NDPC, 2005, p. 15). The policy framework also framed poverty within climate change by emphasizing that climate change impacts especially "drought and desertification" has "vulnerability and exclusion" ramifications (NDPC, 2005, p. 115). GPRS II thus marked the onset of climate change framings in Ghana's national development as seen in the NDPFs.

In GSGDA I and II, appreciation and integration of climate change issues were stronger and pervasive. These NDPFs were prepared, taking cognizance of the need to confront climate change impacts in Ghana. In GSGDA I, for example, there was a call for "environmental sustainability as well as determine[ing] the impact pathways of climate change and the areas of national vulnerability for appropriate policy interventions" (NDPC, 2010, p. 5). GSGDA I and II conspicuously identified climate change as a key development issue and used "climate change" in its policy goals, objectives, and strategies.

As indicated earlier, the GPRS II period ushered in an era of climate change integration. Despite earlier intervention by nongovernment actors (see Würtenberger et al., 2011), it was the GPRS II period that consolidated efforts toward national action on climate change. During its implementation, the National Climate Change Committee (NCCE) was formed and together with the Ministry of Environment, Science, and Technology (MEST) initiated processes to integrate climate change issues in national development (Würtenberger et al., 2011; Owusu-Daaku and Diko, 2017). Consequently, a number of national sectoral climate change vulnerability and adaptation assessments were conducted and four national climate change policy frameworks were formulated. The national climate change–specific policy frameworks are: The *National Climate Change Policy Framework* (NCCPF) and the *National Climate Change Adaptation Strategy* (NCCAS) prepared in 2010, the *Ghana National Climate Change Policy* (NCCP) in 2013, and the *National Climate Change Policy Action Programme for Implementation* (NCCPAPI) in 2015 (Owusu-Daaku and Diko, 2017).

In these climate change documents, it was apparent that the government of Ghana identifies climate change as a key development issue needing urgent attention and integration into Ghana's development efforts. These documents outline in detail climatic zones, sectors, regions, and livelihoods at risk of climate change impacts, the nature of vulnerability and risks, and the interventions government aims to implement. The NCCPF, NCCAS, and NCCP provide a general focus and objectives of climate change policy whereas the NCCPAPI outlines specific actions to confront climate change impacts in Ghana for different groups, sectors, and regions at risk.

18.2.2 Climate Change and the Guidelines for the Preparation of Medium-Term Development Plans

Following these NDPFs, the climate change discourse that emerged sought to improve awareness of and to integrate climate change issues into Ghana's development at the local, regional, and sectoral levels. This was to be achieved using both district and sectoral guidelines for the preparation of MTDPs. This structure is influenced by the local governance and development planning processes in Ghana (Fig. 18.2). In the following discussion, attention is paid to only guidelines for preparing district MTDPs to form the basis for comparison and critical analysis of urban development plans for the Kumasi metropolis.

At the time of writing this chapter, guidelines for the preparation of MTDPs for Ghana Vision 2020 and GPRS I were not accessible. Hence, attention is dedicated to those after the GPRS I. Three guidelines for the preparation of MTDPs have been prepared by the NDPC after GPRS I. They outline the roles of different actors in the preparations of MTDPs[4] and the step-by-step actions involved in the process. Overall, while the processes are slightly different due to merging or rewording of steps for some of the guidelines, they are broadly similar. For all the three guidelines reviewed, there were steps for the following: conducting a performance review of previous plans; compiling a district profile; harmonizing community needs and aspirations to avoid conflicts and consolidate where appropriate; linking harmonized community needs and aspirations to national development priorities; prioritizing needs and aspirations; and formulating development goals, objectives, strategies, programs, programs of action and action plans, financial plans and budgets, monitoring and evaluation frameworks, as well as disseminating plans for the MTDPs.

The review reveals no mention of climate change issues in GPRS II guidelines; this was despite it being adopted in the GPRS II policy framework. Even environmental issues received little attention as the only connection made was with the Millennium Development Goal (MDG) 7, to "ensure environmental sustainability" by 2015. This was in part because the GPRS II guidelines that guided the MTDP preparation for the 2006–09 period explained that the "Guidelines have been developed to incorporate dimensions of development including the MDGs ... and environmental sustainability" (NDPC, 2006, p. i). This was different in GSGDA I and II guidelines as they emphasized that MTDPs pay "special consideration" to "environmental concerns" and "climate change issues" (NDPC, 2009, p. 14; NDPC, 2013, p. 16).

In the GSGDA I guidelines, climate change was listed as an issue that must be discussed in the district profile (see NDPC, 2009, p. 9), which was absent in GPRS II guidelines. In the prioritization of development needs and aspirations, MMDAs were expected to use "climate change issues" and "environmental concerns" as criteria in the decision process. These criteria were also expected to be integrated into the "formulation of development programmes" by MMDAs and they were to "indicate the mitigation measures to be undertaken to address the effect of their programmes on the environment using for example, the Strategic Environmental Assessment [SEA] tool, [and] climate change strategies" (NDPC, 2009, p. 19). The guidelines further provided a list of potential CCIs to guide MMDAs. This is not surprising because from the onset of the guidelines, the NDPC explained that the "Guidelines have been designed to serve three purposes [including]: ... ensure that DMTDPs integrate cross-cutting issues... [such as] SEA, Climate Change, Vulnerability...in development programmes and projects for sustainable development" (2009, p. 2).

All climate change discussions in the GSGDA I guidelines were present in GSGDA II guidelines with additional expectations from MMDAs. First, they were expected to go beyond simple profiling to use advance data collection approaches such as Climate Vulnerability Capacity Analysis tools to examine their climate change capacity.[5] Additionally, in conducting the performance review for the 2010–13 MTDP, MMDAs were expected to examine the extent to which "Implementation of cross-cutting issues such as ... environment, [and] climate change..." issues were integrated into the MTDP prepared and implemented during the GSGDA I period (NDPC, 2013, p. 9). They were also expected to consider the possibility of promoting and developing a green economy. Thus "environment, climate

---

[4] For details on the roles, see chapter two of each of the guidelines for preparing MTDPs for 2006–09, 2010–13, and 2014–17.

[5] For an example of this tool, refer to Climate Vulnerability and Capacity Analysis Handbook by Care International. http://www.careclimatechange.org/files/adaptation/CARE_CVCAHandbook.pdf.

change and green economy" became part of the criteria for prioritizing and formulating development programs in the GSGDA II guidelines (NDPC, 2013, p. 21).

There is a strong connection between climate change issues in the NDPFs and guidelines for preparing MTDPs, with the exception of the observation made for GPRS II. This is not surprising for two reasons. First, it is the same organization that facilitated and prepared both documents. Secondly, since MMDAs are expected to have MTDPs that align with NDPFs, the guidelines are obviously a reflection of this expectation. However, whether these expectations manifest in the prepared MTDPs is another issue. This is the focus of the sections that follow where plans prepared between 2006 and 2014 period have been examined to appreciate the extent to which climate change issues have been integrated into these plans.

18.2.3 Urban Climate Change Interventions in National Development Policy Frameworks

A number of CCIs are found in the GPRS II, GSGDA I and II, and the four national climate change policy frameworks. In these documents, CCIs can be put into three framings. Table 18.1 provides a summary of how the CCIs have been framed in the NDPFs. In all cases, only the "policy objectives" associated with climate change framings have been counted. The first framing relates to CCIs that directly capture the term *climate change* at the "key focus area," "issues" (problematizing), "policy objectives," and/or "strategy" level. The second framing relates to CCIs that capture climate change at both the issues and strategy level. The third framing relates to CCIs that directly capture climate change at either the issues or strategy level. Overall, there are only two policy objectives in GPRS II, 15 in GSGDA I, and 14 in GSGDA II that are associated with climate change framing.[6]

The policies and strategies in the three NDPFs with climate change framing are multidimensional, relating to all aspects of development. They include a need to create awareness on climate change issues, conduct research and build a knowledge base on climate change for Ghana, promote sustainable and alternative livelihoods for those at risk, implement effort to reduce carbon dioxide emissions, promote conservations and forestry, encourage the use of sustainable energy sources, adopt effective waste management systems, develop disaster response and early warning systems, ensure effective water resource management, among many others. For strategies, climate change is framed in 10 strategies in GPRS II, 25 in GSGDA I and 29 in GSGDA II. These CCIs are similar to the focus areas that the various national climate change policy frameworks promote.

The NCCP, for instance, delineates 5 policy areas and 10 strategic intervention areas (see Table 18.2). Compared to the NCCAS and NCCPF, the vision of NCCP—"to ensure a climate-resilient and climate-compatible economy while achieving sustainable development through equitable low-carbon economic growth for Ghana" (MESTI, 2012, p. ix)—is more encompassing. The NCCPF, on the other hand, has three focus areas captured by its objectives: low-carbon growth; effective adaptation to climate change; and social development. In contrast, the NCCAS focuses only on adaptation strategies. The goal of the NCCAS is "to enhance Ghana's current and future development to climate change impacts by strengthening its adaptive capacity and building resilience of the society and ecosystems" (UNEP, UNDP, 2010, p. 17). NCCAS has 10 priority areas on climate change adaptation dealing with early warning systems, sustainable livelihoods, land-use management, research and awareness creation, environmental sanitation, water resources, agriculture, health, and energy. NCCAS and NCCPF are both outcomes of the formative processes toward the preparation of NCCP—thus the similarities that emerge. Not surprisingly, the focus areas of both NCCAS and NCCPF are also similar to those that the NCCP delineates. All these documents are expected to shape and guide specific interventions at the national, sectoral, regional, and district levels.

TABLE 18.1 Climate Change Interventions in National Development Policy Frameworks

Type of CCI Framing	Vision 2020	GPRS I	GPRS II	GSGDA I	GSGDA II
Framing 1	0	0	1	3	4
Framing 2	0	0	0	9	7
Framing 3	0	0	1	3	3
Total	0	0	2	15	14

Author's elaboration.

---

[6] In cases where the policy objectives directly target climate change as a policy priority or focus, all strategies under the section were counted.

TABLE 18.2 List of Strategic Focus Areas of Ghana's National Climate Change Policy

Policy Themes	Strategic Focus Areas
A. Agriculture and Food Security	1. Develop Climate-Resilient Agriculture and Food Security Systems
B. Disaster Preparedness and Response	2. Build Climate-Resilient Infrastructure 3. Increase Resilience of Vulnerable Communities to Climate-Related Risks
C. Natural Resource Management	4. Increase Carbon Sinks 5. Improve Management and Resilience of Terrestrial, Aquatic, and Marine Ecosystems
D. Equitable Social Development	6. Address Impacts of Climate Change on Human Health 7. Minimize Impacts of Climate Change on Access to Water and Sanitation 8. Address Gender Issues in Climate Change 9. Address Climate Change and Migration
E. Energy, Industrial, and Infrastructural Development	10. Minimize Greenhouse Gas Emissions

Author's elaboration.

NCCPAPI resulted from these three national climate change policy frameworks and provides a detailed outline of the specific CCIs that need to be implemented to confront climate change impacts. A review of this document with emphasis on urban areas reveals that urban CCIs constitute 19 actions covering four program interventions areas, namely: (1) flood prevention activities; (2) conservation of trees through agroforestry, on-farm practices, and greening of urban areas; (3) improved drainage in urban areas; and (4) renewable energy development (See MESTI, 2015).

The expectation is that climate change issues and the CCIs outlined in these NDPFs will be integrated into development planning at all levels in Ghana. These are explicitly stated as strategies in GSGDA I and II. The strategy was to "integrate/mainstream impact of climate change into sectoral and district plans" (NDPC, 2010, p. 145) and also to "promote planning and integration of climate change and disaster risk reduction measures into all facets of national development planning" (NDPC, 2010, p. 179; NDPC, 2014, p. 212). To this end, the review turns to the Kumasi metropolis to examine the degree of integration of climate change issues into urban planning.

18.3 CLIMATE CHANGE AND URBAN PLANNING IN THE KUMASI METROPOLIS

In this section, the chapter focuses on climate change and CCIs in urban planning in the Kumasi metropolis by reviewing four urban development plans for the metropolis.

18.3.1 Climate Change and Urban Development Plans of the Kumasi Metropolis

In the 2006–09 and 2010–13 MTDPs, there was no mention of climate change or CCIs in the plan. The closest indication of a connection to climate change issues and CCIs can be found under discussions on "environmental sustainability" as result of the incorporation of the MDGs. Interventions, objectives, and projects in these plans have some implications for climate change but were not framed as such. They can be put into three areas, namely: (1) promoting environmental sanitation and a clean environment; (2) conducting environmental impact assessments and environmental audits; and (3) urban reforestation.

These interventions were expected to deal with development challenges such as decreasing arable land, declining green spaces, invasion and encroachment on water bodies, flooding (KMA, 2006, 2009), inadequate sanitation and drainage infrastructure (KMA, 2009). In the 2010–13 MTDP, there was a clear divergence from the GSGDA I guidelines, which explicitly called for "special considerations" to climate change issues in prioritization and programs of action (NDPC, 2009, p. 14).

In contrast, there was a slight paradigm shift in the 2014–17 MTDP. A section of the metropolitan profile in the plan was dedicated to climate change—an expectation from the GSGDA II guidelines. The plan also alluded to one national climate change policy framework, the NCCAS. Discussions on climate change issues were made from a global level and narrowed to the metropolitan level. For the Kumasi metropolis, the plan problematized climate change by articulating the causes and consequences of climate change. Development issues such as population growth, declining urban green spaces, solid waste disposal, and use of chlorofluorocarbon-emitting household appliances were framed as climate change challenges. The plan also identified CCIs that KMA can adopt to confront these climate change challenges.

For a plan that has a long-term focus (2012–30), the GKCP unfortunately followed similar patterns in the 2006–09 and 2010–13 MTDP for the Kumasi metropolis. Climate change appeared only four times in the whole of the GKCP. These emerged in environmental impact assessments (EIAs) for proposed projects on transportation, water supply, liquid waste, solid waste, and the electricity sector. Urban challenges, development goals, objectives, and strategies were not framed within climate change. Although the plan makes considerations for "environmental conservation and disaster management"—the third pillar of the GKCP (MESTI, 2013, p. 2-1)—these were not framed or connected to climate change. To establish legitimacy, the GKCP in some cases made references to other NDPFs, one of which was the GSGDA I guidelines, where the plan incorporated the need for a strategic environmental assessment (SEA). The preparation of the GKCP is not guided by the NDPC guidelines. Nonetheless, it is surprising that by using the GSGDA I guidelines to partly establish legitimacy, the plan failed to integrate the call for "special considerations" for climate change issues by the same guideline.

18.3.2 Climate Change Interventions in Urban Development Plans

In this section, the chapter concentrates on CCIs in urban development plans for the Kumasi metropolis. As observed earlier, it was only the 2014–17 MTDP of the Kumasi metropolis that explicitly framed development within climate change issues and explicitly identified CCIs. Subsequently, this section focuses on CCIs identified in this plan.

First, climate change is seen in one objective and strategy for the metropolis. This related to a challenge of "inadequate knowledge" about climate change issues (KMA, 2013, pp. 184, 215). In response, the KMA aimed to provide "training" to "equip community members with life skills to…climate change" (KMA, 2013, pp. 242, 254). However, a closer look at the objective and strategy shows that they had been adopted almost word for word from the GSGDA II policy framework. This is a challenge because it shows that climate change issues have not been adequately contextualized to the metropolis.

Additionally, the KMA sought to implement "mitigation and adaptation measures to climate change" in the Kumasi metropolis through "afforestation" and "dredging of silted rivers and streams." These were captured in the profile section of the 2014–17 MTDP as projects whose implementation began from 2013 and early 2014, respectively. For instance, the afforestation project is an initiative under "the Kumasi Urban Forestry Project, [which] is geared at planting and growing one million trees along the major driveways of Kumasi … to … help reduce the effects of climate change" (KMA, 2013, p. 144). This example illustrates a reframing of some development interventions from the preceding plan as CCIs by the Assembly, a realization that points to the co-benefits of CCIs or that CCIs have strong connections to other development interventions. Unfortunately, a further review of objectives and strategies show that this understanding did not manifest in new interventions in the 2014–17 MTDP. For instance, projects to "reverse forest and land degradation" were not framed as CCIs in the plan.

Since one objective was framed within the context of climate change, it was expected that the plan will delineate projects/activities with locations for implementation, output indicators, time schedule for implementation, indicative budget, sources of funding, and implementation agencies, as part of the MTDP's action plan section. This notwithstanding, there was no mention of this objective and its associated projects in this section.

Interestingly, there were a minimum of 96 projects that have climate change implications in the 2014–17 MTDP action plan section: 17 projects in 2014, 47 in 2015, 13 in 2016, and 19 in 2017. These projects relate to six objectives, namely: (1) to reverse forest and land degradation; (2) to manage waste, reduce pollution and noise; (3) to promote resilient urban infrastructure development, maintenance, and provision of basic services; (4) to develop recreational facilities and promote cultural heritage and nature conservation in urban areas; (5) to mitigate and reduce natural disasters and reduce risks and vulnerability; and (6) to create and sustain an efficient transport system that meet user needs. They can also be grouped into three main initiatives, namely: education and awareness, urban greenery and afforestation, and disaster prevention and management systems. These are clearly similar to the CCIs in the NDPFs and national climate change documents.

The review reveals, however, that while these interventions can provide some opportunity to confront the challenges of climate change impacts, they were not framed as CCIs. This is crucial in that the absence of a proper climate change context and framing can potentially affect the nature of the designed interventions, their capability and capacity to tackle climate change impacts, as well as how effective these interventions can be adopted as CCIs.

18.4 MANAGING THE NATIONAL AND URBAN PLANNING DIVERGENCE

This chapter provides evidence that the NDPC guidelines for the preparation of MTDPs will not be adequate to achieve climate change integration in urban planning in Ghana. This is despite the NDPC's aim to integrate climate change issues at all levels of national development and planning. Findings from this chapter affirm Adu-Boateng's (2015) observation of the climate change policy divergence at the national and local level in Ghana. Apparently, there is limited awareness and understanding of development co-benefits, tensions between national and local needs, and a political dispensation that does not support CCIs in Ghana's urban planning—hence the divergence (Adu-Boateng, 2015).

This divergence also hints that the NDPC has had little oversight over the preparation of MTDPs in Ghana. They have not been able to ensure that climate change expectations delineated in their own guidelines are conformed to. Indeed, the process of climate change integration is progressively taking hold in Ghana. However, it will be critical to redefine the processes of climate change integration into MTDPs and other urban development plans beyond the NDPC guidelines. This chapter has provided evidence that the NPDC guidelines have not been effective in helping to integrate climate change issues into MTDPs. One apparent reason is the nature of the guidelines, which do not fully present a climate change integration framework to guide MMDAs as to "how" to integrate climate change issues when preparing MTDPs.

To enhance climate change integration into MTDPs, the NDPC will need to ensure that "special considerations" to climate change expectations into urban plans are adhered to by MMDAs.

Again, there is a need for a paradigm shift toward climate co-benefit as many of the interventions that were formulated in KMA's urban development plans for the Kumasi metropolis can potentially help combat climate change impacts. Opportunities thus exist to optimize the climate co-benefits of these interventions. However, this will demand an in-depth appreciation of how climate co-benefit interventions operate (Bollen et al., 2009; Nemet et al., 2010; Jiang et al., 2013; Doll et al., 2013; de Oliveira, 2013). While this potential exists, there is a tendency for path dependency as urban planners can easily treat these development interventions simply as CCIs (Low and Astle, 2009; Matthews, 2013; Matthews et al., 2015). But, the effectiveness of such interventions as CCIs will depend on how they are adequately framed. Framing these interventions within climate change will thus help to identify the specific types of climate change impacts these interventions can tackle and whether they are the appropriate interventions to deal with such impacts.

Furthermore, each MMDA in Ghana will need to identify communities and people that are particularly vulnerable to climate change impacts and the forms of their vulnerability. This is critical for setting the needed targets and goals on climate change (Deslatte and Swann, 2016). Without proper diagnosis, urban planning authorities will not be able to identify synergies and overlaps amongst the different approaches to confront climate change impacts such as mitigation, adaptation, and resilience (Simon and Leck, 2015). Thus proper climate change vulnerability assessments and the identification of the approaches to manage them are vital entry points for climate change integration. As Ziervogel et al. (2016) study has shown, having a climate action plan is fundamental to begin processes of integration with local, regional, and national climate change policies.

Yet, studies across Africa (Elias and Omojola, 2015; Shemdoe et al., 2015; Ziervogel et al., 2016), and specifically Ghana (Adu-Boateng, 2015), indicate that the capacity of urban planning institutions to deal with climate change issues is often lacking. In this regard, urban planners will need to be trained and the capacity of planning institutions enhanced on the different climate change vulnerability assessment methodologies and climate change action planning.

For instance, in Ghana, the NDPC expects MMDAs to conduct detailed climate change vulnerability assessments to inform the CCIs they adopt in their MTDPs. The evidence from the review of urban development plans for the Kumasi metropolis reveals that this aim has not been attained. Thus training urban planners, in this case those at the various MMDAs, on different climate change vulnerability assessment methodologies, climate change action planning, climate change integration, and climate co-benefit will help develop their capacity. In fact, it will help facilitate processes and enhance efforts to integrate different CCIs into urban development plans, identify the CCIs that are most effective for specific climate change impacts, and identify those who can contribute to and benefit from their implementation (Deslatte and Swann, 2016; Ziervogel et al., 2016; Lee and Painter, 2015).

The roles of the NDPC, NCCE, and MEST are also paramount. They can plan and implement initiatives to increase awareness and understanding of the various national climate change policy frameworks such as NCCPF, NCCAS, NCCP, and NCCPAPI among MMDAs in Ghana. Together, such training will help MMDAs to have the capacity to integrate CCIs horizontally into MTDPs and vertically with national climate change policies (UN-Habitat, 2014; Shemdoe et al., 2015).

For the Kumasi metropolis, there is potential for partnerships on climate change with nongovernmental organizations and academic institutions to develop the needed capacity as well as develop a climate change action plan that can be integrated into urban development plans for the metropolis. The KMA can, for instance, partner with the Kwame Nkrumah University of Science and Technology to utilize their climate change experts to help conduct vulnerability assessments and capacity training. This will help facilitate an interdisciplinary learning process of shared knowledge about climate change for the metropolis, as in the case of Bergrivier Municipality, South Africa (Ziervogel et al., 2016).

This chapter has endeavored to present how climate change issues and CCIs are discussed in NDPFs in Ghana, including the national climate change policy frameworks. In this review, it is apparent that climate change issues have progressively been integrated in national development planning compared to urban planning. Despite the expectation through guidelines for MMDAs to prepare MTDPs that give "special consideration" to climate change issues, the framing of development problems and interventions does not take cognizance of climate change impacts in Ghana's urban areas. Clearly, there is some significant degree of divergence, as the Kumasi metropolis case has shown. Subsequently, the findings and suggestions—an approach that takes into account capacity building on climate change issues as well as climate change action planning that will form the basis to facilitate climate change integration into urban development plans—are imperative toward climate change integration in Ghana.

References

Acheampong, R.A., Agyemang, F.S.K., Abdul-Fatawu, M., 2017. GeoJournal 82, 823–840. https://doi.org/10.1007/s10708-016-9719-x.

Acheampong, R.A., Ibrahim, A., 2016. One Nation, Two Planning Systems? Spatial Planning and Multi-Level Policy Integration in Ghana: Mechanisms, Challenges and the Way Forward. Urban Forum 1–18. https://doi.org/10.1007/s12132-015-9269-1.

Adu-Boateng, A., 2015. Barriers to climate change policy responses for urban areas: a study of tamale metropolitan assembly Ghana. Current Opinion in Environmental Sustainability 13, 49–57. https://doi.org/10.1016/j.cosust.2015.02.001.

Ahwoi, K., 2010. Rethinking Decentralization and Local Government in Ghana- Proposals for Amendment. In: Const Rev Ser, vol. 6. The Institute of Economic Affairs, Accra. Available at: http://dspace.africaportal.org/jspui/bitstream/123456789/36072/1/crs-6.pdf?1.

Bollen, J., Guay, B., Jamet, S., Corfee-Morlot, J., 2009. Co-benefits of Climate Change Mitigation Policies: Literature Review and New Results. OECD Econ Dept Working Papers, 693. OECD Publishing, Paris. https://doi.org.proxy.libraries.uc.edu/10.1787/224388684356.

Deslatte, A., Swann, W.L., 2016. Is the price right? Gauging the marketplace for local sustainable policy tools. Journal of Urban Affairs 38, 581–596. https://doi.org/10.1111/juaf.12245.

Doll, C.N.H., Dreyfus, M., Ahmad, S., Balaban, O., 2013. Institutional framework for urban development with co-benefits: the Indian experience. Journal of Cleaner Production 58, 121–129.

Elias, P., Omojola, A., 2015. The challenges of climate change for Lagos Nigeria [case study]. Current Opinion in Environmental Sustainability 13, 74–78.

Ghana Statistical Service, 2014. Population and Housing Census. District Analytical Report. Kumasi Metropolitan Area. Ghana Statistical Service, Accra, p. 2010. Available at: http://www.statsghana.gov.gh/docfiles/2010_District_Report/Ashanti/KMA.pdf.

Institute of Local Government Studies, 2010. A Guide to District Assemblies in Ghana. Institute of Local Government Studies and Friedrich-Ebert-Stiftung Ghana, Accra. Available at: http://library.fes.de/pdf-files/bueros/ghana/10487.pdf.

Jiang, P., Chen, Y., Geng, Y., Dong, W., Xue, B., Xu, B., Li, W., 2013. Analysis of the co-benefits of climate change mitigation and air pollution reduction in China. Journal of Cleaner Production 58, 130–137.

Kumasi Metropolitan Assembly, 2006. Metropolitan Medium Term Development Plan, 2006–2009. KMA, Kumasi.

Kumasi Metropolitan Assembly, 2009. Metropolitan Medium Term Development Plan, 2010–2013. KMA, Kumasi.

Kumasi Metropolitan Assembly, 2013. Metropolitan Medium Term Development Plan, 2014–2017. KMA, Kumasi.

Lee, T., Painter, M., 2015. Comprehensive local climate policy: the role of urban governance. Urban Climate 14, 566–577. https://doi.org/10.1016/j.uclim.2015.09.003.

Low, N., Astle, R., 2009. Path dependence in urban transport: an institutional analysis of urban passenger transport in Melbourne, Australia, 1956–2006. Transport Policy 16, 47–58.

Matthews, T., Lo, A.Y., Byrne, J.A., 2015. Reconceptualizing Green infrastructure for climate change adaptation: barriers to adoption and drivers for uptake by spatial planners. Landscape and Urban Planning 138, 155–163.

Mensah, J.V., 2005. Problems of district medium-term development plan implementation in Ghana: the way forward. International Development Planning Review 27, 245–270.

Ministry of Environment, Science, Technology, Innovation, 2012. Ghana National Climate Change Policy. MESTI, Government of Ghana, Accra. Available at: http://www.pef.org.gh/documents/climate-change/national-climate-change-policy.pdf.

Ministry of Environment, Science, Technology and Innovation, 2013a. Town and country planning department (TCPD), Japan international cooperation agency (JICA). In: The Study on the Comprehensive Urban Development Plan for Greater Kumasi in the Republic of Ghana. Republic of Ghana, Accra. Draft final report. Main Text Volume 1–3. Spatial Development Framework (SDF) for Greater Kumasi Sub-Region.

REFERENCES

Ministry of Environment, Science, Technology, Innovation, 2015. National Climate Change Policy Action Programme for Implementation: 2015–2020. MESTI, Government of Ghana, Accra. Available at: https://www.weadapt.org/sites/weadapt.org/files/ghana_national_climate_change_master_plan_2015_2020.pdf.

National Development Planning Commission, 1995. Ghana – Vision 2020 (The First Step: 1996-2000). Presidential Report on Co-Ordinated Programme of Economic and Social Development Policies (Policies for The Preparation of 1996-2000 Development Plan). Government of Ghana, Accra. Available at: https://s3.amazonaws.com/ndpc-static/CACHES/NEWS/2015/07/27//Vision+2020-First+Step.pdf.

National Development Planning Commission, 2003. Ghana Poverty Reduction Strategy 2003–2005. An Agenda for Growth and Prosperity. Government of Ghana, Accra. Available at: https://s3.amazonaws.com/ndpc-static/pubication/GPRS+2003-2005_February2003.pdf.

National Development Planning Commission, 2005. Growth and Poverty Reduction Strategy (GPRS II), 2006–2009. NDPC, Accra. Available at: https://s3.amazonaws.com/ndpc-static/pubication/(GPRS+II)+2006-2009_November+2005.pdf.

National Development Planning Commission, 2009. Guidelines for the Preparation of District Medium-term Plans under the Medium-term Development Policy Framework 2010–2013. NDPC, Accra. Available at: https://s3.amazonaws.com/ndpc-static/CACHES/PUBLICATIONS/2016/04/16/DMTDP+Preparation+Guide+2010-2013.pdf.

National Development Planning Commission, 2013. Guidelines for the Preparation of District Medium-term Development Plan under the Ghana Shared Growth and Development Agenda II, 2014–2017. NDPC, Accra. Available at: https://s3.amazonaws.com/ndpc-static/CACHES/PUBLICATIONS/2016/01/21/DMTDPs+Guidelines_30-12-2013.pdf.

National Development Planning Commission, 2006. Guidelines for the Preparation of District Medium-term Development Plan under the Growth and Poverty Reduction Strategy 2006–2009. NDPC, Accra. Available at: https://s3.amazonaws.com/ndpc-static/CACHES/PUBLICATIONS/2016/04/16/DMTDP+Preparation+Guide+2006-2009.pdf.

National Development Planning Commission, 2010. Medium-term National Development Policy Framework: Ghana Shared Growth and Development Agenda (GSGDA I), 2010–2013. NDPC, Accra. Available at: https://s3.amazonaws.com/ndpc-static/pubication/(GSGDA)+2010-2013_December+2010.pdf.

National Development Planning Commission, 2014. Medium-term National Development Policy Framework: Ghana Shared Growth and Development Agenda (GSGDA II), 2014–2017. NDPC, Accra. Available at: https://s3.amazonaws.com/ndpc-static/pubication/GSGDA+II+2014-2017.pdf.

Nemet, G.F., Holloway, T., Meier, P., 2010. Implications of incorporating air-quality co-benefits into climate change policymaking. Environmental Research Letters 5, 1–9.

Owusu-Daaku, K.N., Diko, S.K., 2017. The Sea Defense project in the Ada east district and its implications for climate change policy implementation in Ghana's Peri-urban areas. In: Annual Reducing Urbn Pov Grad Pap Comp, Urban Sustainability Laboratory. Woodrow Wilson International Center for Scholars, Washington D.C., pp. 28–49.

De Oliveira, P., Jose, A., 2013. Learning how to align climate, environmental and development objectives in cities: lessons from the implementation of climate co-benefits initiatives in urban Asia. Journal of Cleaner Production 58, 7–14.

Shemdoe, R., Kassenga, G., Mbuligwe, S., 2015. Implementing climate change adaptation and mitigation interventions at the local government levels in Tanzania: where do we start. Current Opinion in Environmental Sustainability 13, 32–41. https://doi.org/10.1016/j.cosust.2015.01.002.

Simon, D., Leck, H., 2015. Understanding climate adaptation and transformation challenges in African cities. Current Opinion in Environmental Sustainability 13, 109–116. https://doi.org/10.1016/j.cosust.2015.03.003.

UN-HABITAT, 2014. State of African Cities Report 2014: Reimagining Sustainable Urban Transitions. UN-HABITAT, Nairobi.

United Nation Environmental Programme, United Nations Development Program, 2010. National climate change adaptation strategy. CC DARE: climate change and development – Adapting by reducing vulnerability. A Joint UNEP/UNDP Programme for Sub-saharan African. Available at: https://s3.amazonaws.com/ndpc-static/CACHES/PUBLICATIONS/2016/04/16/Ghana_national_climate_change_adaptation_strategy_nccas.pdf.

Würtenberger, L., Bunzeck, I.G., van Tilburg, X., 2011. Initiatives related to climate change in Ghana. Towards coordinating efforts. Clim Dev Know Net 1–36. Available at: http://cdkn.org/wp-content/uploads/2012/04/Ghana-initiatives-mapping-climate-change-May2011.pdf.

Ziervogel, G., Archer van Garderen, E., Price, P., 2016. Strengthening the knowledge–policy interface through co-production of a climate adaptation plan: Leveraging opportunities in Bergrivier Municipality South Africa. Environment and Urbanization 28, 455–474. https://doi.org/10.1177/0956247816647340.

CHAPTER 19

Spatial Planning for Climate Adaptation and Flood Risk: Development of the Sponge City Program in Guangzhou

Meng Meng[1], Marcin Dąbrowski[1], Faith Ka Shun Chan[2,3], Dominic Stead[1]

[1]Delft University of Technology, Delft, The Netherlands; [2]University of Nottingham Ningbo China, Ningbo, China; [3]University of Leeds, Leeds, United Kingdom

19.1 INTRODUCTION

In China, like elsewhere in Asia, pluvial flooding events are occurring increasingly often, wreaking havoc across many cities (Yu et al., 2015). According to an investigation by the Ministry of Housing and Urban-Rural Development (MoHURD) in 2010, in the period from 2008 to 2010, 231 out of 351 Chinese cities studied (62%) were affected by pluvial flooding. Among those, 74.6% experienced waterlogging of 0.5 m or more in depth and 90% experienced waterlogging of at least 0.15 m in depth. In 79% of the affected cities studied, the stagnant water lingered for at least 30 min before it could be discharged by the drainage system (Hou et al., 2012). As such, urban pluvial flooding (or surface water flooding) has become the "new normal" in most Chinese cities.

Since 2014, the MoHURD, an important government ministry at the national level, has been promoting a policy to improve the cities' resilience to pluvial flooding in a context of rapid urbanization and climate change—the Sponge City Program. Alongside this notion, the program supports the separation of the sewer and rainwater systems and the application of low-impact development measures as a means to raise the capacity of cities to cope with storm water (Construction Department of MoHURD, 2014). The sponge city metaphor was formalized in the national document *Technical Guideline for the Construction of Sponge Cities: Rainwater System Based on Low Impact Development* launched in November 2014, combining the ambitions in terms of resilience to flood risk with a pursuit of a more sustainable way to build an attractive and livable urban environment (Ministry of Housing & Urban-rural Development, 2014).

Achieving the ambition and implementing the program locally, through spatial interventions, remains a challenge. This chapter sheds more light on how the local spatial planning system responds to a national policy in resolving the flood risk through a new policy-framing process. By doing so, it adds to the literature on effectiveness and implementation in urban climate adaptation policies and on the integration of flood resilience concerns in spatial planning.

Guangzhou, a delta city in China, is the fourth main industrial and commercial hub in China (alongside Beijing, Shanghai, and Shenzhen) and is expanding rapidly. The city is highly exposed to flooding and has been ranked as one of the most vulnerable cities that will be exposed to flooding by the 2070s considering the projected global sea-level rise, the intensity and frequency of storms, and expansion of the city's assets and populations (Hallegatte et al., 2013). Guangzhou's plight illustrates a widespread problem affecting many Chinese and Asian cities generally—the rapid expansion of the urban fabric, which worsens the vulnerability to the negative consequences of the changing climate.

In this chapter, information is mainly drawn from the analysis of a set of policy documents related to spatial planning and sponge city development. This is complemented by insights from a series of interviews with regional and local policy makers, researchers, private and public planning institutions, and civil engineers. Transcript materials from six interviews are presented in this chapter to help illustrate the evolution of the nexus between spatial planning and flood risk management and the implementation of the Sponge City Program in Guangzhou.

The chapter starts with a description of the socioeconomic and institutional features of Guangzhou and its exposure to flood in the context of climate change. Then, the chapter reviews the *Guangzhou Sponge City Plan (SCP)* and analyzes how this new tool has added to the transition in (1) problem framing of flood risk in spatial planning, in (2) shifting toward new climate adaptation measures (3), enforcement tools, and (4) governance practices. It is followed by a critical analysis of the cognitive, technical, institutional, and financial barriers and obstacles to the implementation of the *SCP*.

19.2 THE PROFILE OF GUANGZHOU: RAPID URBANIZATION AND EXPOSURE TO CLIMATE CHANGE

19.2.1 Demographic, Social, Economic, and Political Features in the Trend of Rapid Urbanization

Guangzhou (also known as Canton), a metropolis located in China's Pearl River Delta (PRD), is 120 km northwest of Hong Kong and 145 km northwest of Macau. Its total area of 7434 km^2 has a diversity of topography, ranging from a dense urban center area beside the Pearl River to a mainly agricultural and rural area in the north and east. The elevation generally increases from southwest to northeast, with mountains forming to the north of the city (Fig. 19.1). The southern part of the city is situated in a floodplain, with large swathes of land that have been reclaimed from the sea and onto which the city is currently expanding.

As one of China's first-tier cities, Guangzhou occupies a central position in the PRD both in administrative and socioeconomic terms, being the capital city of Guangdong province and the PRD region's major economic and commerce hub. After the launch of Deng Xiaoping's economic reform and open-door policy in 1978, this city was one of the pioneers in the shift toward open market economy. It underwent huge growth in gross domestic product from approximately 4 billion RMB in 1978 to 1960 billion RMB in 2016 (Guangzhou Statistics Bureau, 2016). Correspondingly, the population increased from 3 million to 13 million in this period, making it the third most populous city in mainland China, behind Shanghai (24.19 million) and Beijing (21.72 million) (Shanghai Government, 2017; Beijing Statistical Bureau, 2017).

19.2.2 Climate Change and Flood Vulnerability in Guangzhou

Guangzhou is one of the most striking examples of how urban expansion can exacerbate vulnerability to flooding in the context of the changing climate, which brings increased intensive rainstorms, frequent typhoons, and sea-level rise. The city is frequently affected by waterlogging. According to Guangzhou's *SCP*, there are more than 100 points in the city's central districts where waterlogging occurs (Dong et al., 2015, p. 257; see also Fig. 19.2 (left)). These are related to the extent of impervious paving in the densely built-up areas. An average of 87% impervious surface in the

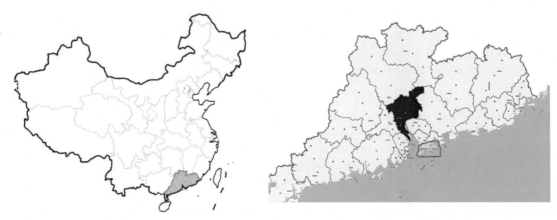

FIGURE 19.1 Map of the Guangdong Province in China (*left*) and the location of Guangzhou in Guangdong Province (*right*). *Authors' elaboration.*

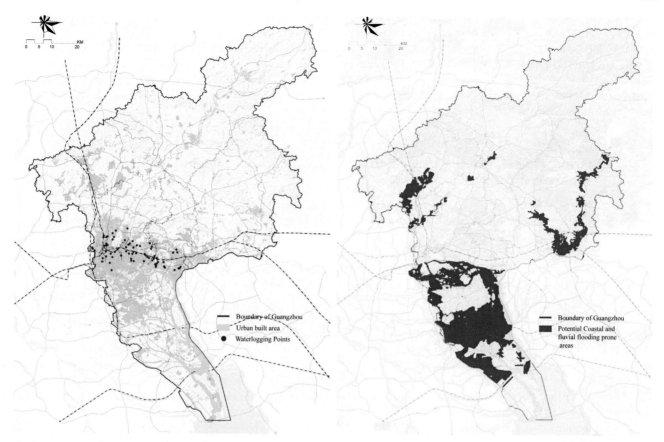

FIGURE 19.2 The waterlogging points in Guangzhou (*left*) and the areas prone to potential coastal flooding and fluvial flooding (*right*). *Authors' elaboration (based on Guangzhou Sponge City Plan, 2016–30).*

city center significantly hinders the infiltration of rainwater and contributes to the increase of the surface runoff at the source (Li et al., 2015; Guangzhou Water Affairs Bureau, 2015). Moreover, the urban drainage system is increasingly unable to deal with extreme rainstorms, which are occurring more frequently than before (Chan et al., 2014; Wu, 2010).

At present, sea-level rise is not of high concern for policy makers in Guangzhou, but it is poised to become a much more important issue if the current pattern of urban sprawl in southern areas continues and no dramatic improvement in the drainage and water storage system takes place. While being downplayed by the Guangzhou Water Affairs Bureau until recently (see Guangzhou Water White Paper from 2013), the *SCP* for Guangzhou does recognize the flood issues and indicates that an area of 970 km^2, mainly located in the southern districts (e.g., Nansha), is particularly vulnerable (Guangzhou Government, 2017, see also Fig. 19.2 (right)). Southern districts are also threatened by unexpected extreme weather. For example, the recent Typhoon Hato caused coastal flooding in southern Guangzhou (Nansha District) as well as Macau and Hong Kong (USA Today, 2017) when more than 8000 people were moved to emergency shelters (China National Radio, 2017).

19.3 SPONGE CITY: SHIFTING THE SPATIAL PLANNING FRAME

The formulation of national technical guidelines for the construction of sponge cities was led by MoHURD in 2014. This document identifies that a nationwide urban flood mitigation program should complement the traditional drainage system and excessive discharge system while relying on three nature-based approaches: (1) ecosystem preservation; (2) ecosystem restoration; and (3) low-impact development.

In response to the national program, a special section of the *Guangdong Provincial 5-Year Plan* (2016–20) was launched in 2016. In this document, four key strategies are identified: (1) restoration of the green-blue infrastructure; (2) reinforcement of the fluvial flooding defense and upgrade of the storm water discharge infrastructures; (3)

purification of polluted water environments; and (4) enhancement of the capacity in water supply and reuse (Guangdong Provincial Housing and Construction Department, 2016).

The *Guangzhou SCP (2016−30)* was published in 2017 and was jointly prepared by the Urban Planning Bureau, as lead agency, in collaboration with the Urban Water Affairs Bureau and a series of public and private planning institutions. This document relates flood risk and climate change to spatial issues, frames flood resilience solutions as multifunctional interventions, visualizes the runoff coefficient regulatory map, and emphasizes the leading role of the spatial planning system.

19.3.1 Previous Flood Concerns: A Section in the Latest Spatial Planning Documents With Limited Attention

In the Chinese context, flood affairs were previously regarded as the realm of flood risk management rather than spatial planning in most cases. Master plans,[1] which are the most important spatial policy document concerning urban development (Yu, 2014), thus, expend limited efforts on this topic. In recent masterplans for Guangzhou (2000−10 and 2010−20), flood issues were only discussed under the heading "flooding prevention and rainfall discharge," a small section in the master plan, weak relation to climate change and urban development in the planning discourse. Here, the previous elaboration in the latest two master plans are briefly discussed.

In the Guangzhou Master Plan 2000−10, flooding-related issues were not systematically described (see Table 19.1). There was no discussion of the causes of and factors affecting flooding. Available options identified were primarily engineering based: reinforcing the dikes and sluices; raising the ground level in low-lying areas when necessary; retrofitting the underground discharge system; dredging open canals; or suppressing the occupation of existing waterways and constructing artificial lakes. Moreover, these options were merely mentioned in passing and scattered across the document. This lack of detail makes the operationalization of these options difficult in practice.

The subsequent *Guangzhou Master Plan* (for 2010−20) resulted in change regarding problem-setting and policy options (see Table 19.1). Here, severe flood risk is associated with, for instance, the outdated underground discharge system, the erosion of open waterways and canal systems, and the substandard dykes, levees, and riverbanks in the case of coastal and fluvial flooding (Guangzhou Government, 2016, pp. 442−445). The document also claims that reducing floods can be integrated into the development of new urban areas by, for example, improving water quality and enhancing the ecological environment. Nonetheless, the power of this document is weakened by limited scientific knowledge on the impacts of climate change (especially for flood risk) in urban areas and an emphasis on engineering-based (structural) solutions. Most of all, the document provides no guidance on the role of planners in tackling flood risk.

19.3.2 New Understanding of Flood Risk as a Spatial Issue Related to Climate Change

The attention on the severe loss brought by flood events contributes to a separate and special spatial policy—*Guangzhou SCP*. The participation of spatial planners in the formulation process represents a major shift in the conceptualization of flooding as a significant spatial issue according to various interviewees, including one from Guangzhou's Land Resources and Urban Planning Committee:[2]

> Guangzhou SCP is a sort of master plan for water rather than economic development. It aims to alleviate the flooding problems exacerbated by urbanization and, simultaneously, construct an attractive and livable urban environment.

Moreover, the *Guangzhou SCP* brought a new recognition of the causes of flooding. On the one hand, it attributes flood risk to the negative consequences of urbanization such as excessive paving, limited open space for water retention and runoff, and the lagging construction of drainage system. On the other hand, the role of the climate change in

---

[1] A master plan is normally produced by the local spatial planning authorities on behalf of municipal governments for periods of 10 years. In the case of Guangzhou, the master plan is regarded as a reference for the local spatial planning system to plan further spatial interventions.

[2] In 2014, one branch of Guangzhou Land Resources and Housing Management Bureau was combined with Guangzhou Urban Planning Bureau to create the current Guangzhou Land Resources and Urban Planning Committee.

TABLE 19.1 Planning Policy Documents and Their Attention to Flood Prevention and Rainfall Discharge

Year	Key Official Institutions Involved	Policy Activities	Areas of Attention	Key Reflection of Flooding
2005	Urban Planning Bureau	Guangzhou Master Plan 2000–2010	A new comprehensive plan for land use and economic development; The discussion of flood affairs is under the subtitle "flooding prevention and rainfall discharge"	• No discussions of causes • No clear solutions • No definition of the responsibility of spatial planning system
2016	Urban Planning Bureau	Guangzhou Master Plan 2010–2020		• Failures of rapid urbanization • A proposed interaction between adaptive measures land use and environmental improvement • No definition of the responsibility of spatial planning system
2017	Guangzhou Land Resources and Urban Planning Committee (leader), Urban Water Affairs Bureau	Guangzhou Sponge City Plan 2016–2030	To make the idea precise, clear, and short, change the content in this column to *A new thematic plan for reducing flood risk in terms of climate change and rapid urbanization* A new comprehensive plan for resolving flooding regarding the climate change; Seek the way to mitigate the flood risk attributed to the negative implication of climate change and rapid urbanization	• Links between rapid urbanization and climate change • Mainstreaming framing regarding ecosystem preservation and restoration, low-impact development, hydrological infrastructure construction, water purification, water supply, and waterfront recreation • Structural and nonstructural adaptive measures proposed • Regulatory tools • Spatial planning system given a leading position in developing collaborative relationships

Based on Guangzhou Government (Ed.), 2005. Guangzhou Master Plan (2001–2010) (Draft for Approval). Guangzhou Government (in Chinese), Guangzhou Government (Ed.), 2016. Guangzhou Master Plan (2010–2020) (Draft for Approval). Guangzhou Government (in Chinese), and Guangzhou Government (Ed.), 2017. Guangzhou Sponge City Plan (2016–2030), Guangzhou Government (in Chinese).

the increasingly frequent flooding events is emphasized clearly, perhaps for the first time in the local spatial policy. As the document states,

> Due to climate change, there will be an increase in the frequency of storm events. Because of this, the occurrence of waterlogging will be much higher than ever… Climate change might lead to the increasing occurrence of typhoons and extreme tides, which pose a threat to the coastal areas. *Guangzhou Government, 2017, pp. 42, 43*

19.3.3 Toward Mainstreaming and Multifunctional Interventions

With the introduction of the *SCP*, flooding concerns have become a key issue, not merely relevant to flood risk management authorities but also for those dealing with urban development. This is also reflected in the new repertoire of key solutions put forward to implement the plan with the consideration of the ecology, safety, environment, and social identity (see Table 19.2). Initially, the preservation of green-blue network provides the basis for Guangzhou, acting as a sponge, to absorb the excessive rainfall (see Fig. 19.3 (left)). In addition, a comprehensive flood-resilience system based on structural (engineered) and nonstructural measures ("soft" solutions in the urban space) is suggested through a combination of efforts to reinforce flood defense infrastructure, construction of pumps, and upgrade of discharge system (structural), consolidation of the banks of canals (structural and nonstructural) and enhancing water retention and detention areas through wetlands, parks, and green-blue corridors (nonstructural). In parallel, another three solutions are proposed that aim to pursue a high quality of water, an increase in water reuse efficiency, sufficient water supply, and friendly waterfront space. By doing so, flooding mitigation measures are mainstreamed into the local agendas of urban development. An interviewee from Guangzhou Municipal Engineering Design and Research Institute explained:

> In practice, a project might have two or more options. For instance, the "softened" banks and widening of the canals (a pattern of nonstructural measures) along with a dredging project might act as a part of green-blue corridors which help drain the excess water… while providing an attractive place for recreation.

TABLE 19.2 Mainstreaming Flood Mitigation Into Local Agendas in Guangzhou

Consideration	Solutions
Ecological Concerns	Preserving the green-blue network, including mountains, forests, farmlands, wetlands, lakes, open waterways.
Safety Concerns	Enforcing and upgrading the structural infrastructure such as dykes, pumps, river banks, and drainage systems; optimizing the nonstructural infrastructure such as the increase of water retention areas (based on low-impact development)
Ecological and Environmental Concerns	Purifying the polluted water, including building of water treatment industries and ecopurification systems
Environmental Concerns	Improving water supplement and water recycling system
Social identity Concerns	Reconstructing the connection between water and citizens by facilitating the access to waterfront areas and arranging waterfront recreation

Based on Guangzhou Government (Ed.), 2017. Guangzhou Sponge City Plan (2016–2030), Guangzhou Government (in Chinese).

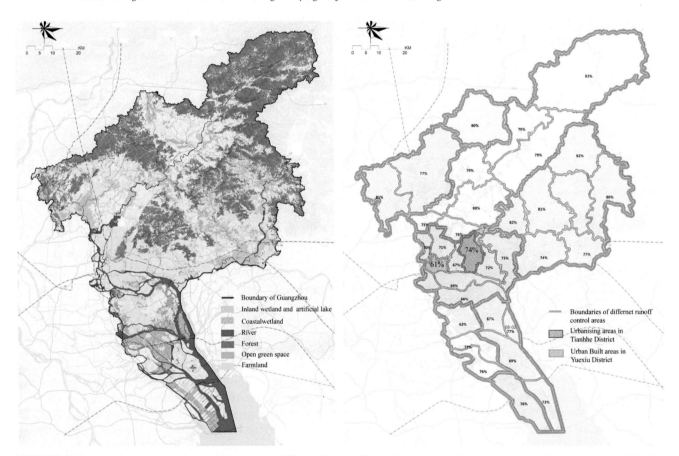

FIGURE 19.3 The preservation of green-blue network (*left*) and the runoff control map (*right*, the values presented in the map equal 1 minus the runoff coefficient). *Authors (based on Guangzhou Sponge City Plan, 2016–2030).*

19.3.4 New Regulatory Tools to Regulate the Competition for Land Between High-density Development and Water Retention and Detention

For spatial planners, the *SCP* document introduces the runoff coefficient to regulate the development of every piece of land in the city. It is an index relating the amount of runoff to the amount of precipitation received and is usually used in the low-impact development techniques (Wang et al., 2016). For example, the runoff coefficient of open green space is around 0.15, which means only 15% of precipitation flows into the water discharge system.

Green space is, therefore, a buffer that can reduce the volume of excessive water entering the discharge system. By contrast, asphalt pavement has a high runoff coefficient (around 0.8) and only a weak capacity for water infiltration. As a result, its large scale of usage would leave the pressure of excess water to the urban discharge system.

Based on this notion, the runoff coefficient is used to decide on the location and size of permeable areas such as wetlands, parks, and gardens, rather than high density of concrete forests. Fig. 19.3 (right), the runoff control coefficient map, shows how much precipitation is not allowed into the drainage system and is proposed to be stored by sponge infrastructure. The values equal 1 minus the runoff coefficient. Normally, a looser standard is set for the areas with high percentage of paving and difficult to be transformed while a tougher standard is for the opposite conditions. For instance, for a built area located in Yuexiu district at the old city center, the proposed runoff coefficient is 0.39 with 0.61 as the runoff control coefficient. It means 61% precipitation could not be discharged into the drainage system. By comparison, for an urbanizing area with abundant undeveloped land in Tianhe District, the proposed runoff coefficient is 0. 26.%; 74% should be stored by the sponge infrastructure (see Fig. 19.3 (right)).

19.3.5 A Rise of Status of Spatial Planning in a New Flood Governance

The *Guangzhou SCP* redefines the distribution of responsibilities in the spatial planning system. Guangzhou Land Resources and Urban Planning Committee (a combination of previous Guangzhou Land Resources Bureau and Guangzhou Planning Bureau) is highlighted in *SCP* as one of the key actors in flood mitigation. It is:

> responsible for policy making of comprehensive plans and detailed plans …, coordinating the interests between different administrators …, regulating the rules, building codes for construction… and inspecting the implementation of adaptation measures… *Guangzhou Government (2017), p. 120*

As such, the local spatial planning system formally and officially steps into the arena in resolving the flood risk. In parallel, it appeals for horizontal cooperation with another 13 bureaus such as Guangzhou Water Affairs Bureau, with rich experience in flood-mitigation projects; Guangzhou Finance Bureau, responsible for the funding support; and Housing and Construction Committee, working on the inspect of the project construction. Therefore, the roles of the planners are experiencing a change and new kinds of collaboration arrangements across policy silos because of the *SCP*.

19.4 CRITICAL ANALYSIS OF THE MUNICIPAL INTERPRETATION OF THE SPONGE CITY PROGRAM IN GUANGZHOU

In spite of the shifts in the content of Guangzhou's *SCP* and, arguably, progress in integrating efforts to ensure flood resilience in the context of climate change into spatial planning, there are at least four key obstacles for implementation of this plan: (1) recognition of the coastal flooding and socioeconomic trends; (2) attention to the changing and dynamic climate; (3) horizontal collaboration between spatial planning and flood risk management; and (4) costs of long-term development.

19.4.1 Potential Criticality in the Recognition of the Coastal Flooding and Socioeconomic Trends

The Sponge City Program mainly addresses urban surface water flood issues rather than coastal flood scenarios. This problem-framing preference could be found in the national and provincial documents, such as the *Technical Guideline for the Construction of Sponge Cities* and *Guangdong Provincial Sponge City Program*. Claimed sponge solutions are, thus, mainly used for absorbing the storm water, for instance, bioswales, rain gardens, artificial wetlands, ponds, sustainable drainage systems, etc. (Qiu, 2015; Xia et al., 2017). There are no spatial interventions taking coastal flooding into the consideration. In contrast, the *Guangzhou SCP* attempts to build a comprehensive flood resilience system with flood defense, retention, and discharge. Nevertheless, the enforcement of the flood defense, namely dykes, is still regarded as the main solution in coastal areas. Consequently, people might be under the impression that coastal flood-prone areas are totally safe behind the dykes and flood walls. However, when the extreme weather happens, it will be difficult for the communities to handle the emergency (IPCC, 2007). Thus, the city still has to find alternative solutions to address coastal flood risk and climate issues that the Sponge City Program is unable to offer.

Moreover, flooding hazards are always associated with the density of population and value of economic assets such as property and infrastructure (Schanze, 2007). *Guangzhou SCP* mainly concentrates on the physical aspects of vulnerability to flooding while ignoring scenarios for social and economic change. Such consideration could indeed draw attention to the future severely flood-prone areas yet promising for economic development, for instance, waterfront areas with high quality of natural amenities while close to tides. By doing so, potential conflicts could be discussed and a balance between land use for economic development and for water management could be found.

19.4.2 The Mismatch Between Dynamic Changes in Flood Risk and a Static Regulatory Index

Future flood risk due to climate change has not attracted much attention in policy documents. Neither has the issue of climate change adaptation. In *Technical Guideline for the Construction of Sponge Cities* and *Guangdong Provincial Sponge City Program*, there are no parts that explicitly relate the increased risks to climatic factors. Even in *Guangzhou SCP*, climate change and climate adaptation are merely used to raise the attention of politicians, planners, and engineers. There are no further explorations of how this can affect and threaten the city over time. It could be explained by the fact that climate experts become "the missing piece" in the policy-making process of Sponge City Program at multilevels. By this, the track of changing climate information was left aside, which could have provided feedback to test and optimize the spatial plans.

In an implementation, the runoff coefficient is introduced as a static regulatory index to control the density and underlying surface of every piece of land. Its virtue is that it provides a new tool for practitioners to use. However, the current runoff coefficient index will fail if rainfall intensity increases due to changing climate (which is expected for the Guangdong province and Pearl River Delta). Areas prone to coastal flooding, like the Southern Nansha district, will also be more and more vulnerable because of more frequent extreme weather and higher tides. At this time, the city is unable to deal with these future scenarios and the *SCP* glosses over this issue.

19.4.3 Potential Challenges in the Horizontal Collaborative Implementation Around the Corner

The significance of horizontal cooperation between spatial planning with other professions is recognized in Sponge City Program. At the national level, this program is led by MoHURD and supported by the Ministry of Finance and Ministry of Water Resources. This joint venture has increased the legitimacy of professional collaboration between spatial planners and hydrological and financial experts in practice. This collaborative relationship is formalized in the *Technical Guideline for the Construction of Sponge Cities* by the contention that "In practice, spatial planning is proposed to take effect with the support from other fields such as finance, greening, transportation, drainage, architecture, and hydrology." Similar arguments could be found in the *Guangdong Provincial Sponge City Program* and *Guangzhou SCP* as well.

In spite of the above, horizontal cooperation between spatial planning and other professions is challenging in local implementation. One of the reasons is the lack of knowledge on water management among local spatial planners. According to the plan, spatial planners are expected to coordinate actions on preserving of green infrastructure and design of water retention areas. The former is relatively familiar and manageable for them but the latter is more complex and requires expert hydrological knowledge. For the spatial planners, how to calculate the runoff coefficient and interpret it at the submunicipal level could be a problem. As one spatial planner involved in the formulation of *Guangzhou SCP* mentioned,

> I don't know how to calculate the runoff coefficient index, let alone to apply the index as post-assessment criteria to inspect the proposed projects. The mission of calculation is left to Guangzhou Water Affairs Planning Design & Survey Research Institute. The engineers are the real captains. We just put their outcomes into the document.

As a result, spatial planners do not have a leading position in cooperation with engineers in the field of flood risk management.

In addition, the discussion of coastal flooding takes up a small portion of the Sponge City Program compared with the pluvial and fluvial flooding. As a result, the Pearl River Committee, a regional authority responsible for the reinforcement and maintenance of dykes along Guangzhou southern coastlines, is excluded from the policy-making process at the provincial and local levels, let alone the definition of their position in the sponge city document. This exclusion may become an obstacle for joint working, resulting in the neglect of coastal flooding in spatial plans process.

19.4.4 Potential Conflicts Between Limited Subsidy and Long-term Investment in the Top-down Affairs

As mentioned before, the implementation of Sponge City Program is a top-down process. National authorities propose the political vision and generic principles and encourage the provincial and local authorities to operationalize these principles in their jurisdictions. Subnational authorities have the power to undertake some modifications to accommodate their specific context, as long as it corresponds to the national policy *Technical Guideline for the Construction of Sponge Cities*.

Apart from political pressure, an important driving force in promoting this program locally is the financial incentive from the central government. These sentiments are shared in interviews from Guangzhou Municipal Engineering Design and Research Institute, Guangzhou Urban Planning, Design and Survey Research Institute, and Turen Urban Planning Company, who are involved in the formulation of *Guangzhou SCP*. By 2016, 30 cities had received funding from the national government for pilot projects on sponge city development (Economy and Construction Department of the Ministry of Finance (2015); Pengpai News (Shanghai) 2016). The cities receive 3-year continuous funding of between 400 and 600 RMB million per year (Economy and Construction Department of the Ministry of Finance, 2014). Clearly, sponge-city infrastructures are expensive (Xia et al., 2017) and the construction does not bring economic benefits in the short term. Thus, new funding mechanisms may be needed. A participant in the policy-making process makes the case:

> one goal of the formulation of Guangzhou Sponge City Plan is to pursue the subsidy. It is not a problem since the financial incentives attract the attention of local authorities to take actions. The problem is if they achieved the proposed subsidy and ran out the money, what would be the next step to push forward this program and avoid becoming a temporal political movement?

19.5 CONCLUSIONS

This chapter focuses on the Chinese experience at the local level and reveals how spatial planning is involved in climate adaptation initiatives under a new metaphor—the sponge city. It does so by tracing its innovation in the local spatial planning system and its distinction and cohesion with the national and regional documents. This research contributes to the literature on the transition in the field of spatial planning in the framing of adaptation in local planning documents and the interpretation of adaptation planning in a multilevel governance context.

Through a top-down process, this national program has triggered some rethinking of flood risk in the field of spatial planning. In the case of Guangzhou, the study indicates the *Guangzhou Sponge City Plan 2016–30* makes a transformation in spatial planning. Specifically, climate change, as a new factor, has begun to attract the attention of local policy makers and affect the policy formulation in the spatial planning system. In response to its negative impacts, integrated flood adaptation measures are proposed by using the runoff coefficient to regulate the development of every piece of urban land and promoting a mixture of structural and nonstructural interventions. To ensure its implementation, flood adaptive initiatives are integrated with other urban issues such as water purification, ecological improvement, and increase in social well-being. Importantly, spatial planners are officially expected to enter the arena of climate adaptation in the face of the flood risk.

As a result, climatic factors have been introduced into problem setting, establishing mainstreaming adaptation solutions in local agendas, formulating regular tools for operation, and defining a clear position of spatial planning responsibilities in flood governance. Guangzhou provides a useful reference case for other cities that are seeking to translate national sponge city policies to the city scale. Nevertheless, various obstacles still need to be overcome for Guangzhou's *SCP* to be implemented. First, a lack of attention to coastal flooding might severely limit potential options for flood resilience. Second, a rigid and static regulatory tool, such as the runoff index introduced by the *SCP*, might not be well suited considering that the dynamic and uncertain climate change impacts may play out differently from the current expectations. Third, horizontal cooperation between spatial planning and other professions might result in a deadlock on account of the lack of hydrological knowledge among the planners and limited experience in working across sectoral boundaries. Fourth, the cost might be a concern for a long-term operation of *SCP* unless a sustainable way is found to balance the costs and benefits.

References

Beijing Statistical Bureau, 2017. Statistical Communique of Shanghai on Economy and Social Development in 2016. Available at: http://www.bjstats.gov.cn/tjsj/tjgb/ndgb/201702/t20170227_369467.html.

Chan, F.K.S., et al., 2014. After sandy: rethinking flood risk management in Asian coastal megacities. Natural Hazards Review 15 (2), 101–103.

China National Radio, 2017. The Recovery of Nansha District after the Hato Typhoon. Available at: http://news.cnr.cn/native/city/20170824/t20170824_523917230.shtml.

Construction Department of Ministry of Housing & Urban-rural Development, 2014. Working Paper for the Construction Department of Ministry of Housing & Urban-rural Development in 2014. Available at: http://www.mohurd.gov.cn/wjfb/201402/t20140214_217084.html.

Dong, H., et al. (Eds.), 2015. Annual Report on Urban Construction and Management of Guangzhou in China. Scientific report, Social Science Academic Press (in Chinese).

Economy and Construction Department of the Ministry of Finance, 2014. Financial Support from the Central Government on the Pilot Sponge City Construction. Available at: http://jjs.mof.gov.cn/zhengwuxinxi/tongzhigonggao/201501/t20150115_1180280.html.

Economy and Construction Department of the Ministry of Finance, 2015. The List of Appointed Pilot Sponge Cities. Available at: http://jjs.mof.gov.cn/zhengwuxinxi/tongzhigonggao/201504/t20150402_1211835.html.

Guangdong Provincial Housing and Construction Department (Ed.), 2016. Guangdong Provincial Sponge City Programme: A Special Section in the 13th 5 Year Plan (2016–2020. Guangdong Provincial Housing and Construction Department (in Chinese).

Guangzhou Government (Ed.), 2005. Guangzhou Master Plan (2001–2010) (Draft for Approval). Guangzhou Government (in Chinese).

Guangzhou Government (Ed.), 2016. Guangzhou Master Plan (2010–2020) (Draft for Approval). Guangzhou Government (in Chinese).

Guangzhou Government (Ed.), 2017. Guangzhou Sponge City Plan (2016–2030). Guangzhou Government (in Chinese).

Guangzhou Statistics Bureau (Ed.), 2016. Guangzhou Statistical Information Manual. Scientific Report, Guangzhou Statistics Bureau (in Chinese). Available at: http://www.stats.gov.cn/english/.

Guangzhou Water Affairs Bureau (Ed.), 2015. Guangzhou City Centre Area Rainwater Discharge System Plan. Guangzhou Water Affairs Bureau (in Chinese).

Hallegatte, S., et al., 2013. Future flood losses in major coastal cities. Nature Climate Change 3 (9), 802–806. https://doi.org/10.1038/nclimate1979.

Hou, Y., et al., 2012. The analysis of current urban storm water problem and the discussion of countermeasures. In: China Water & Wastewater the 9th Annual Conference Publication.

IPCC, 2007. Climate Change 2007-impacts, Adaptation and Vulnerability: Working Group II Contribution to the Fourth Assessment Report of the IPCC. Cambridge University.

Li, B., Zhao, Y., Fu, Y., 2015. Spatio-temporal characteristics of urban storm waterlogging in Guangzhou and the impact of urban growth (in Chinese). Journal of Geo-information Science 17 (4), 445–450.

Ministry of Housing & Urban-rural Development (Ed.), 2014. Technical Guideline for the Construction of Sponge Cities: Rainwater System Based on Low Impact Development (Trail). Ministry of Housing & Urban-rural Development (in Chinese).

Pengpai News (Shanghai), 2016. The Second Batch of 14 Pilot Sponge Cities: Each Would Receive 1.2 Billion to 1.8 Billion Subsidy (In Chinese). Available at: http://money.163.com/16/0425/12/BLGFA73F00253B0H.html.

Qiu, B., 2015. The connotation, approach and perspective of Sponge city (LID) (in Chinese). Water and Wastewater Engineering 4 (3), 1–7.

Schanze, J., 2007. Flood risk management – a basic framework. In: Schanze, J., Zeman, E., Marsalek, J. (Eds.), Flood Risk Management: Hazards, Vulnerability and Mitigation Measures. Springer Science & Business Media, pp. 1–20.

Shanghai Government, 2017. Statistical Communique of Shanghai on Economy and Social Development in 2016. Scientific Report, Shanghai Government (in Chinese). Available at: http://www.shanghai.gov.cn/nw2/nw2314/nw2318/nw26434/u21aw1210720.html.

USA Today, 2017. Typhoon Hato Batters Hong Kong and Macau. Available at: https://www.usatoday.com/picture-gallery/news/world/2017/08/23/typhoon-hato-batters-hong-kong-and-macau/104874414/.

Wang, W., et al., 2016. Discussion on the rainwater runoff control (in Chinese). Water and Wastewater Engineering 42 (10), 61–69.

Wu, T., 2010. Flooded streets in the city centre area of Guangzhou: reasons and Resolutions (in Chinese). Guangdong Water Resources and Hydropower 9, 19–21.

Xia, J., et al., 2017. Opportunities and challenges of the Sponge City construction related to urban water issues in China. Science China Earth Sciences 60 (4), 652–658.

Yu, K., et al., 2015. "Sponge city": the theory and practice. City Planning Review 39 (6).

Yu, L., 2014. Chinese City and Regional Planning Systems. Ashgate.

CHAPTER 20

Climate Change and Australian Local Governments: Adaptation Between Strategic Planning and Challenges in Newcastle, New South Wales

Giuseppe Forino, Jason von Meding, Graham Brewer
University of Newcastle, Callaghan, NSW, Australia

20.1 INTRODUCTION

Newcastle is the largest city of New South Wales (NSW) after Sydney and is located in the Hunter Valley region, a climate change hotspot (Evans, 2008). Newcastle is also Australia's largest coal port and one of the largest in the world. Since the 18th century, the local economy of both the Hunter Valley and Newcastle has been largely dependent on coal, energy, and timber industries (Bulkeley, 2000; Evans, 2008).

Therefore, this area significantly contributes to greenhouse gas emissions and ultimately to climate change. In this way, an analysis of how Newcastle copes with climate change can provide interesting insights on the ongoing debate about climate change adaptation (CCA) in Australia. Climate policy in Newcastle started in the 1990s, when the Newcastle City Council (NCC) began a process of energy efficiency and mitigation by replacing inefficient heating and lighting (Bulkeley, 2000). Approximately one decade later, CCA discussion also started simultaneously with the worldwide increase of CCA debate.

This chapter aims to provide an overview of CCA measures and recommendations within the strategic planning by NCC. Toward this goal, the chapter reports initial findings from a research project (2014–17) on governance of CCA and disaster risk reduction integration in some local governments areas (LGAs) of the Hunter Valley. A content analysis of publicly available documents on the NCC's website (by February 2016) was performed (see more in-depth results in Forino et al., 2017).

To provide directions in Newcastle for important aspects within the functioning of a city such as land-use and development, environmental protection, economic development, social and community well-being, and so on. Collecting the whole body of documents by NCC is helpful in providing an understanding on how climate change is considered and embedded within the everyday urban practices, rather than being part of isolated initiatives and directions.

Furthermore, four interviews with staff officers of NCC (between May and September 2016) provided additional primary data (see Table 20.1), while specific literature on Australia and CCA provided complementary data. Further development of the presented analysis can be found in other publications by the authors (Forino et al., 2017; 2018a; 2018b).

Results are presented under the form of a narrative able to describe CCA challenges and potential for implementation in Newcastle. The chapter does not aim at investigating the whole body of strategic planning by NCC including CCA, but it can describe the trend undertaken in Newcastle in recent years.

TABLE 20.1 Interviewees' Codes and Their Main Topics Covered Into the Newcastle City Council

Int. 1	Water Management
Int. 2	Bushfire mitigation and environmental protection
Int. 3	Community health and welfare
Int. 4	Environment and sustainability

20.2 CLIMATE CHANGE ADAPTATION BETWEEN LAND USE AND DEVELOPMENT

The city of Newcastle has large development along the coast. Therefore, climate change can contribute to natural hazards risk in the area by increasing the frequency and intensity of coastal hazards associated with sea-level rise (erosion, flooding, inundations), as well as extreme storm events (e.g., The City of Newcastle, 2010c, 2012a,b, 2015a) and bushfires (Newcastle Bush Fire Management Committee, 2012). According to the interviewee Int. 1, CCA measures are primarily required for the material of assets and infrastructure (e.g., roads, drain systems, pavements, footpath), and related life cycle, mainly in order to adapt to sea-level rise and flood risk. Such measures are based on current information about the predictions on climate change and related impacts and include seawalls, land-use and planning control, raising buildings, retreat areas, relocation, and zoning. Plans such as the *Newcastle Local Environmental Plan* (NSW, 2012) and the *Development Control Plan* (The City of Newcastle, 2012c), for example, set regulations for development on coastal areas and along the Hunter river floodplain. Accordingly, recommendations for development consent are provided in terms of compatibility to coastal hazards risks and to flood risk on the Hunter river floodplain. The *City-wide Floodplain Risk Management Study and Plan* (The City of Newcastle, 2012a), similarly, recommends house raising in areas at flood risk, particularly for properties that are frequently flooded and on which future development is expected. However, high associated costs, limited state government funding, and the required cooperation of property owners can contribute to delay the full implementation by many years. Likewise, two plans for beach reserves, the *Merewether Beach Reserves Public Domain Plan* (The City of Newcastle, 2010b) and the *South Stockton Reserves Public Domain Plan* (The City of Newcastle, 2012d), recommend adaptive measures for potentially long-term effects of climate change (e.g., bike and walk pathways) for the material and design of asset and infrastructure. The *Development Control Plan* also recommends promoting an effective bushfire risk management into the development consents as well as in land use (The City of Newcastle, 2012c).

In addition, NCC also considered CCA for coping with heat waves. For the councils in the Hunter Valley, heat waves have been acknowledged as an issue for the well-being and health of local communities (HCCREMS, 2014). In this way, strategic planning recommended designing and implementing public shade and weather protection to support walking in heat conditions (The City of Newcastle, 2010a) or to adapt to heat in proximity of bus stops, pedestrian-oriented locations, and car parks (The City of Newcastle, 2012c). Some documents also support CCA to heat waves through trees and vegetation. For example, the *Local Planning Strategy* (The City of Newcastle, 2015b) recommends urban forest programs through tree management, compensatory planting, and landscaping; likewise, the *Street Tree Masterplan* (The City of Newcastle, 2011) recommends the use of alternative species adaptable to new extreme weather conditions and temperatures.

20.3 CLIMATE CHANGE ADAPTATION AND THE TARGET ON LOCAL COMMUNITIES

NCC is also giving attention to issues that climate change poses to local communities and to how communities can enact CCA. For example, the *Community Strategic Plan* (The City of Newcastle, 2013a) provides adaptive trajectories for communities through the Objective "Environment and climate change risks and impacts are understood and managed." Such Objective becomes a milestone for other documents such as the *Newcastle Environmental Management Strategy* implementation (The City of Newcastle, 2013c). While it urges enacting CCA in light of expected climate change—related events (e.g., bushfires and seasonal rainfalls), it also recognizes that the implementation of CCA in Newcastle is difficult. In fact, CCA still has high costs, but these costs are unevenly distributed among taxpayers as benefits mainly occur on the coasts where CCA is urgent. Additionally, local communities do not yet fully accept the occurrence of climate change; therefore, their engagement in CCA should be strengthened. Finally, the implementation of CCA is still time-consuming, as it relies on time-consuming activities such as attracting funding and determining appropriate triggers for implementation (The City of Newcastle, 2013c). For example, the *Social Strategy 2016–19* (The City of Newcastle, 2015c) refers to the aforementioned *Community Strategic Plan* (The City of Newcastle, 2013a) in the Focus Area 5—Community Strategy through the initiative "build community readiness by engaging the community in risk management processes including the development and implementation of action plans." In this regard, a community-led experience of CCA is undertaken by the volunteers of The City of Newcastle Landcare, a partnership established between NCC and Landcare Network Newcastle to involve local communities in environmental management, including CCA. According to the interviewee Int. 2, these volunteers in Newcastle work on adaptive measures such as rehabilitation, soft landscaping, and revegetation of natural areas in order to control soil erosion, e.g., of dune systems, and provide natural shade on coastal areas and beaches. While with this partnership NCC recognizes the usefulness of adaptive efforts by local communities, however, Int. 2 also argues that volunteers cannot perform major adaptation works, e.g., on dune systems. Therefore, this partnership should be considered as a support to the strategic planning and initiatives by NCC. While NCC is not the only council in the Hunter Valley to partner with Landcare groups, this partnership is significant for CCA in Australia as the Landcare Network is also a partner of the federal government for enacting CCA measures. In fact, under the *National Climate Resilience and Adaptation Strategy 2015* (Australian Government, 2015), the federal government funded $450 million within the National Landcare Program to undertake natural resource management activities, including local sustainable agriculture priorities.

20.4 CHALLENGES FOR CLIMATE CHANGE ADAPTATION

The previous discussion has shown the commitment by NCC in undertaking CCA measures; however, at least three challenges exist and hinder effective implementation. First, the actual commitment by NCC on CCA and generally on climate change is still problematic. As interviewee Int. 3 claims, in fact, beyond the contents of the strategic planning there is a poor understanding by NCC of the climate change impacts on local communities. This is due to that climate change represents a sensitive issue not just or politics at the federal or state levels, but also at the local level, as in Newcastle at the federal or state level but also at the local level (e.g., in Newcastle and within NCC). Similarly, another interviewee (Int. 4) says that within NCC the opinion about climate change varies based on the political flag of the electoral mandate. Int. 3 strikes a similar chord arguing that in the last decade climate change was included into the political agenda of NCC, also thanks to the establishment of a team devoted to environmental issues, including sustainability and climate change. However, around 5 years ago NCC went through a restructuring of the organization to save money on staff, and this team was dismantled. From then, climate change issues had no more a specific strategic direction coming from NCC.

Second, the lack of power by councils challenges CCA. Local governments are increasingly recognized worldwide as important actors for governing climate change—related issues through CCA (Measham et al., 2011; Mukheibir et al., 2013; Serrao-Neumann et al., 2014: McGuirk et al., 2015). However, their effectiveness in enacting CCA mainly varies based on size, location, assets, portfolios, legislative mandates, budgets, and stakeholders' inclusion (Measham et al., 2011; Heazle et al., 2013; Fallon and Sullivan, 2014; Macintosh et al., 2015; Serrao-Neumann et al., 2014, 2015; McGuirk et al., 2015). In addition, the consideration of CCA as a "local responsibility" (Nalau et al., 2015) actually hides the lack of necessary support and leadership by higher government levels to local governments in terms of adequale resource, skills and knowledge for implementing CCA (Measham et al., 2011; Nalau et al., 2015). For example, Int. 4 advocates that the state government seems basically to provide indications to the councils, but then the councils become primarily responsible for decisions and are often perceived by public opinion as being ineffective or making unpopular choices. Therefore, according again to Int. 4, the state government should be the primary stakeholder as it is in a better position to find and allocate budgets and resources. Int. 4 sheds light on the case of coastal erosion in Newcastle, pointing out that it is meaningless that each council implements CCA in an isolate way. Rather a uniform approach among neighboring areas would allow to more effectively respond to the existing challenges, also with the support by higher level of governments.

Third, as the Newcastle Employment Lands Strategy also claims, while coal and coal-associated industries are essential for the local economy, for many people coal represents an energy source that has the potential to adversely influence climate change (The City of Newcastle, 2013b). The Newcastle port has three coal export terminals, and in October 2015 the Planning Assessment Commission approved a fourth terminal. This is expected to increase coal capacity of the port by around 70 million tons per year, with projected costs of $5 billion and an expected creation of 1500 construction jobs and 80 permanent coal jobs in NSW when completed (NSW Planning Assessment Commission, 2015). This approval was strongly criticized by social and environmental movements and groups, leading to a street protest in Newcastle on May 2016, where thousands of people called for moving from coal economy towards a more sustainable economy in the whole Hunter Valley. Such approval implies that in the whole Hunter Valley, including Newcastle, local economy will continue to rely on coal and therefore to feed the fossil fuel market, which ultimately contributes to climate change. This is confirmed by a recent decision by the NSW government, which gives consent for the expansion of extractive industries of oil, coal, and coal seam gas in the whole Hunter Valley (NSW, 2015). Also in this case, the federal and the state governments are responsible for making decisions about local economic development, which is based on pollution and high energy consumption, rather than looking for alternatives, which can lead to a postcarbon society and challenge the dominant neoliberal paradigm (Evans, 2008). Therefore, addressing the root causes of climate change would represent the only effective way to enact CCA in Newcastle.

20.5 CONCLUSION

This chapter has briefly discussed main CCA measures in strategic planning by NCC and the related challenges. According to findings, strategic planning by NCC recommends working on the life cycle of materials of assets and infrastructure, as well as development control in areas at risk on the coasts, in flood plains, or in areas exposed to bushfire risk. Strategic planning also recommends tree planting and shade and weather protection for coping with heat wave risk. Finally, it provides ways to include local communities in understanding and taking adaptive actions for climate change—related risks. For example, local communities are involved in CCA measures through the Landcare Network, which has a long, established partnership with NCC.

Nevertheless, challenges emerge for CCA. Climate change is still a political issue in NCC, therefore the commitment by the organization mainly relies on the political color of the electoral mandate. Also, a lack of power by councils and the related lack of leadership by higher levels of government in providing adequate resources and budgets to councils do not allow NCC to have clear directions and guidelines and to provide an integrate response to issues in common with other local governments. Finally, the expansion of coal industries and related infrastructure severely undermines any possibility of CCA measures implemented by NCC being effective. In terms of urban metaphors (smart, resilience, transition), these are not considered yet as part of strategic planning by NCC. *Resilience* is certainly a word and a concept that is recurrent within some of the strategic planning documents produced in Newcastle, but it cannot be considered an urban metaphor able to provide a comprehensive understanding, implementation, and action in terms of CCA efforts by NCC. While opportunities certainly exist for including these metaphors within strategic planning, efforts are necessary to increase awareness about, and to eventually drive action on, the real challenges for effective CCA in Newcastle, related to fossil fuels and highly polluting local economy.

Acknowledgments

The authors thank the staff officers of the Newcastle City Council who participated in the research.

References

Australian Government, 2015. National Climate Resilience and Adaptation Strategy 2015 (Canberra).
Bulkeley, H., 2000. Down to earth: local government and greenhouse policy in Australia. Australian Geographer 31 (3), 289–308.
Evans, G., 2008. Transformation from "carbon valley" to a "Post-Carbon society" in a climate change hot spot: the coalfields of the Hunter Valley, New South Wales, Australia. Ecology and Society 13 (1), 39.
Fallon, D.S.M., Sullivan, C.A., 2014. Are we there yet? NSW local governments' progress on climate change. Australian Geographer 45 (2), 221–238..
Forino, G., von Meding, J., Brewer, G., van Niekerk, D., 2017. Climate change adaptation and disaster risk reduction integration: strategies, policies, and plans in three Australian local governments. International Journal of Disaster Risk Reduction, 24, 100–108.
Forino, G., Von Meding, J., Brewer, G.J., 2018a. Governing the integration of climate change adaptation into disaster risk reduction: insights from two Australian local governments. In: Forino, G., Bonati, S., Calandra, L.M. (Eds.), Governance of Risk. Hazards and Disasters, pp. 127–142.
HCCREMS, 2014. Heatwave Planning Template for Lake Macquarie and the Central Coast. N. Hunter Councils.

REFERENCES

Heazle, M., Tangney, P., Burton, P., Howes, M., Grant-Smith, D., Reis, K., Bosomworth, K., 2013. Mainstreaming climate change adaptation: an incremental approach to disaster risk management in Australia. Environmental Science & Policy 33, 162–170.

Macintosh, A., Foerster, A., McDonald, J., 2015. Policy design, spatial planning and climate change adaptation: a case study from Australia. Journal of Environmental Planning and Management 58 (8), 1432–1453.

McGuirk, P., Dowling, R., Brennan, C., Bulkeley, H., 2015. Urban carbon governance experiments: the role of Australian local governments. Geographical Research 53 (1), 39–52.

Measham, T.G., Preston, B.L., Smith, T.F., Brooke, C., Gorddard, R., Withycombe, G., Morrison, C., 2011. Adapting to climate change through local municipal planning: barriers and challenges. Mitigation and Adaptation Strategies for Global Change 16 (8), 889–909.

Mukheibir, P., Kuruppu, N., Gero, A., Herriman, J., 2013. Overcoming cross-scale challenges to climate change adaptation for local government: a focus on Australia. Climatic Change 121 (2), 271–283.

Nalau, J., Preston, B.L., Maloney, M., 2015. Is adaptation a local responsibility? Environmental Science & Policy 48, 89–98.

Newcastle Bush Fire Management Committee, 2012. Bush Fire Risk Management Plan.

NSW, 2012. Newcastle Local Environmental Plan.

NSW Planning Assessment Commission, 2015. NSW Planning Assessment Commission Determination Report Port Waratah Coal Services Terminal 4, Newcastle LGA. Available at: http://www.pwcs.com.au/media/1773/t4_determination_report_20160805.pdf.

NSW, 2015. Draft Hunter Regional Plan.

Routledge, Forino, G., von Meding, J., Brewer, G., 2018b. Challenges and opportunities for Australian local governments in governing climate change adaptation and disaster risk reduction integration. International Journal of Disaster Resilience in the Built Environment. in press.

Serrao-Neumann, S., Crick, F., Harman, B., Schuch, G., Choy, D.L., 2015. Maximising synergies between disaster risk reduction and climate change adaptation: potential enablers for improved planning outcomes. Environmental Science & Policy 50, 46–61.

Serrao-Neumann, S., Harman, B., Leitch, A., Low Choy, D., 2014. Public engagement and climate adaptation: insights from three local governments in Australia. Journal of Environmental Planning and Management 58 (7), 1196–1216.

The City of Newcastle, 2010a. Hunter Street Revitalisation. Final Strategic Framework.

The City of Newcastle, 2010b. Merewether Beach Reserves Public Domain Plan.

The City of Newcastle, 2010c. Newcastle Coastal Revitalisation Strategy – Master Plan Report.

The City of Newcastle, 2011. Newcastle Street Tree Masterplan.

The City of Newcastle, 2012a. Newcastle City-wide Floodplain Risk Management Study and Plan. Final Report.

The City of Newcastle, 2012b. Newcastle Cycling Strategy and Action Plan.

The City of Newcastle, 2012c. Newcastle Development Control Plan.

The City of Newcastle, 2012d. South Stockton Reserves Public Domain Plan.

The City of Newcastle, 2013a. Newcastle 2030. Our Vision for a Smart, Liveable and Sustainable City. Newcastle Community Strategic Plan (Revised 2013).

The City of Newcastle, 2013b. Newcastle Employment Lands Strategy.

The City of Newcastle, 2013c. Newcastle Environmental Management Strategy.

The City of Newcastle, 2015a. Coastal Plan of Management.

The City of Newcastle, 2015b. Local Planning Strategy.

The City of Newcastle, 2015c. Social Strategy 2016–2019 (Draft September 2015).

SECTION IV

CITIES DEALING WITH CLIMATE CHANGE: TRANSITION INITIATIVES

CHAPTER 21

Transition Initiatives: Three Exploration Paths

Angela Colucci

Co.O.Pe.Ra.Te. ldt, Pavia, Italy

21.1 TRANSITION INITIATIVES: THREE EXPLORATION PATHS

In the literature debate, transition initiatives are investigated through a differentiated lens (Chapter 3) in relation to innovative and creative proposals and solutions (grassroots innovations, social innovation, etc.), in relation to community-led governance processes (coproduction and codesign, participative urban interventions, etc.), in relation to social aspects (commons and public services, urban and public polices, etc.) and others.

A transition initiative is a process driven (and acted) by multiple (local) actors and in which activities aim to improve (modify) services, approaches, routines, practices, and/or infrastructures existing within the city region boundaries (Frantzeskaki and Kabisch, 2016).

Under the transition initiatives umbrella concept there is a large range of local community-led initiatives that could be clustered in differentiated and flexible geographies in relation to the governance of processes, to the main focus (issue or lever for mobilization activation) and to the tools used and developed. This chapter focuses on the "practices" and "initiatives" exploration referring to the third chapter (Chapter 3) that approaches "transition metaphor" from a (more) theoretical point of view highlighting communalities and peculiarities emerging from the literature debate. Transition initiatives are in general acting on "commons": interventions activated are oriented to the improvement of public spaces, common environments, common goods (and services) through community-led actions generating positive benefits on environmental pressure phenomena (related to climate change), and in community strengthening. A commonality that emerged from the literature debate is that environmental issues are explicitly integrated in the goals in a large number of transition initiatives in terms of climate change mitigation and adaptation strategies or in terms of resilience capacities strengthening—it was assumed also a criteria in selecting the transition initiatives practices.

The three transition exploration paths are: the Transition Towns Network, some selected transition initiatives acting on urban commons (most localized in US and European urban contexts), and the experience of the Italian Observatory of Resilience Practices. The Transition Network is a consolidated experience (with more than 10 years of activity) able to manage a global network of worldwide local initiatives. From the Transition Network geography will be explored guidelines and tools for supporting the local transition initiatives activation, consolidation, and stabilization. The transition initiatives on commons geography includes selected experiences working on urban commons improvement and oriented to urban decay regeneration through social innovation and citizens-led initiatives that are explicitly oriented to urban adaptation facing climate change phenomena (water management, urban greening, and other issues) and community resilience strengthening.

The Italian Observatory of Resilience Practices (ORP) is a project aimed at local practices networking and support. Resilience practices include acting with an inclusive, participatory, and multi-issues approach. From the ORP experience we present the most relevant barriers and opportunities merging from the dialogues activated between the Observatory and Practices.

21.2 TRANSITION NETWORK

In 2005, Rob Hopkins developed a proposal for the Energy Descent Action Plan (EDAP) for the town of Kinsale, exploring "how the town of Kinsale might make the Transition to a lower-energy future" (Hopkins, 2008, p. 123). From this experience, that engaged citizens and local actors, emerged a larger vision of "how Kinsale would reduce its fossil fuel dependence while at the same time building and strengthening the town's resilience to the challenges that climate change and peak oil would present." The Transition Towns approach and ideas find its origins in this participatory process: Transition Initiatives' aim is "to act as catalysts for a community to explore and come up with its own answers" (Hopkins, 2008, p. 134) facing the challenges presented by peak oil and climate change.

Since 2006, when the first Transition Initiative of Totnes (England) was officially launched, the Transition Towns movement has grown fast in the United Kingdom and at a global scale; currently there are about 850 local Transition Initiatives. The popularity and the success of Transition movement is based in the capacities in providing tools (large range of applied and practical tools) to citizens groups to take positive visions and tangible actions facing peak oil, climate change (in supporting and strengthening the networking and local and global scale and in valorizing citizens' skill and energies [social resources]) (Hopkins, 2010).

21.2.1 Transition Town Network Evolution

Initially the first guidelines and tools developed in supporting the Local Transition Initiatives by Transition Network were based on the "Twelve steps of transition" that are presented in the Transition Handbook (Hopkins, 2008).

The 12 steps (summarized and commented on in Table 21.1) include both some principles of action and some operative suggestions for transition initiatives activation and management. For instance, steps like "Creating a leading group," "Formation process and awareness raising," or "Big initial event" refer to actions (a sort of to-do list) and

TABLE 21.1 The Twelve Steps for Transition

	The 12 Steps for Transition
1	Creating a Leading Group
2	Formation process and awareness raising
3	Laying the foundations (local initiatives have to recognize and activate alliance and synergies with other actors, both institutions and associations)
4	Big initial event (Events play relevant role in engaging the local community. In Transition Initiatives, events have different aims: celebration of volunteers' efforts, success of local initiatives, involvement of local community, etc.)
5	Create subgroups (The subgroup is a useful model for teamwork. Examples of topic areas are: food, waste, energy, education, youth, economics, transport, and water.)
6	Open Space (Open Space technology)
7	Developing practical and visible actions (every initiative needs, from the earliest steps, to create something concrete and tangible in the community)
8	Facilitating the largest "redevelopment" (valorization of skills commonly used in the past that nowadays are "hidden" and providing workshops to community)
9	Building relationships with the local administration
10	Honoring the community elders
11	Let things go how they go
12	Creating an Energy Descent Plan (the project to be undertaken in order to become free from the threat of peak oil and climate change. Each action plan for energy degrowth will have to expose "a project for a future of degrowth, resilience and relocation", and also must list "a series of concrete steps to achieve it" (Hopkins, 2008). The Energy Descent Plan should be the main project to which all Transition Initiatives should aspire, because it is aimed at creating a project for the transition from today's dependence on oil to a resilient society)

Authors' elaboration based on Hopkins, R., 2008. The Transition Handbook. From Oil Dependency to Local Resilience. Green Books Ltd., Devon (UK).

provide practical suggestions for actions implementation. Steps like "Honoring the community elders," "Let things go how they go," and others refer to general values that have to be assumed as cross-cutting principles in the implementation and realization of all actions and initiatives.

Since the beginning of Transition Network it is possible to underline the crucial role of learning-and-awareness tools, the codesign (engagement of the whole community in defining the transition path) and coproduction tools and the communication tools both in disseminating the good local practices through the global network and in celebrating the results and the goals achieved through internal and local events (tangible results and visible events of activation and celebration) (Hopkins, 2008, 2011).

In the film *In Transition 1.0 From Oil Dependency to Local Resilience* (the first film produced by Transition Network, In Transition 1.0; 2009) Hopkins underlines the relevance for Transition Initiatives of building a recognizable identity and visibility and the relevance in activating dialogues and alliances with local authorities and other existing associations, groups, and local movements. In particular, the Local Transition Initiative has to present itself to local authorities and to local actors as a credible and influential interlocutor. This aspect of recognizability of transition initiative in local context is also demonstrated by the attention played by several tools targeted to local initiative placement in the local context through celebration events devoted to presenting tangible results and list of actions to be activated.

In recent years the Transition Network evolved and developed an improved approach with the launch of Transition 2.0 (presented in several documents published on the Transition Network website) and summarized and presented in the film *In Transition 2.0* (a Green Line Production published by Transition Towns on November 25, 2013).

The main aspects of this new transition approach are based on the principle of "Head, Heart, and Hands" where doing transition successfully is about finding a balance between "The Head: we act on the basis of the best information and evidence available and apply our collective intelligence to find better ways of living; The Heart: we work with compassion, valuing and paying attention to the emotional, psychological, relational and social aspects of the work we do; The Hands: we turn our vision and ideas into a tangible reality, initiating practical projects and starting to build a new, healthy economy in the place we live" (Transition Network, 2016).

Instead of 12 steps the renovated approach proposes a more flexible framework of "essential elements" to reach/gain the success of the transition initiative. The essential elements are not configured as "steps" (linear or temporal path) but as principles that are assumed and implemented as values in doing transition initiatives and are similar to a declaration of assumption of principles shared and assumed as common ground of all peoples and communities involved in transition initiatives:

We respect resource limits and create resilience — The urgent need to reduce carbon dioxide emissions, greatly reduce our reliance on fossil fuels and make wise use of precious resources is at the forefront of everything we do.

We promote inclusivity and social justice — The most disadvantaged and powerless people in our societies are likely to be worst affected by rising fuel and food prices, resource shortages and extreme weather events. We want to increase the chances of all groups in society to live well, healthily and with sustainable livelihoods.

We adopt subsidiarity (self-organization and decision making at the appropriate level) — The intention of the Transition model is not to centralize or control decision making, but rather to work with everyone so that it is practiced at the most appropriate, practical and empowering level.

We pay attention to balance — In responding to urgent, global challenges, individuals and groups can end up feeling stressed, closed or driven rather than open, connected and creative. We create space for reflection, celebration and rest to balance the times when we're busily getting things done. We explore different ways of working, which engage our heads, hands and hearts and enable us to develop collaborative and trusting relationships.

We are part of an experimental, learning network — Transition is a real-life, real-time global social experiment. Being part of a network means we can create change more quickly and more effectively, drawing on each other's experiences and insights. We want to acknowledge and learn from failure as well as success — if we're going to be bold and find new ways of living and working, we won't always get it right first time. We will be open about our processes and will actively seek and respond positively to feedback.

We freely share ideas and power — Transition is a grassroots movement, where ideas can be taken up rapidly, widely and effectively because each community takes ownership of the process themselves. Transition looks different in different places and we want to encourage rather than unhelpfully constrain that diversity.

We collaborate and look for synergies — The Transition approach is to work together as a community, unleashing our collective genius to have a greater impact together than we can as individuals. We will look for opportunities to build creative and powerful partnerships across and beyond the Transition movement and develop a collaborative culture, finding links between projects, creating open decision-making processes and designing events and activities that help people make connections.

We foster positive visioning and creativity — Our primary focus is not on being against things, but on developing and promoting positive possibilities. We believe in using creative ways to engage and involve people, encouraging them to imagine the future they want to inhabit. The generation of new stories is central to this visioning work, as is having fun and celebrating success. *Transition Network (2016)*

21.2.2 Transition Town Initiatives: Geographies and Distribution

The Transition Network is coordinated at global level by the Transition Network located in Totness. Two levels of organization were developed in order to support local initiatives: a geographical organization based on national and regional hubs and a thematic organization based on thematic working groups developing tools and guidelines (guidelines, training, etc.). Thematic groups are related, for instance, to food production and permaculture, re-economy, art and creativity, food transformation and delivery, energy, and others. The national and regional hubs (that could be also preexisting organizations who support transition in their part of the world) play a coordination role in engaging institutions, national organizations, and other social movements and in catalyzing local transition initiatives. Hubs organize at national- and international-level working groups, communication strategies and events, and national and international meetings. In 2017, 25 National and Regional Hubs (Table 21.2) were mapped on the Transition Network website (Transition Network website, accessed December 1, 2017).

Transition Initiatives are distinguished in relation to three implementation or stabilization level of transition: the Initial stage (where a group of people, the Leading Group, decide to start an Initiative of Transition); Mulling (where the Leading Group activates a dialogue with Transition Network Ltd., the Initiatives at mulling step are included in transition initiatives map); Official Transition Initiative (where the Leading Group defines the declaration documents of initiative guideline and obtains the status of Official Initiative).

There are some 836 Transition Initiatives that have been mapped, with some few differences between the map and database of Transition Network Ltd. Website and the database of single national and regional hubs (Table 21.2).

TABLE 21.2 National and Regional Hubs and the Number of Local Transition Initiatives in Relation to Geographical Continents

Continent	National/Regional Hub	Number of Local Transition Initiatives
Europa	Belgium, Croatia, Denmark, France, Germany, Hungary, Ireland, Italy, Latvia (Permaculture Association), Luxembourg, Netherlands, Norway, Portugal, Romania, Scotland, Slovenia, Spain, Sweden	431 (255 in UK)
Oceania (Australia and New Zealand)	No Hubs	62
Asia	Israel, Japan	14 (1 in Israel)
Africa	No Hubs	4
North America	Transition US (United States)	317
South America	Argentina, Brazil, Chile, Mexico	8
Total	25	836

Author's elaboration based on the Transition Network database (https://transitionnetwork.org accessed November 15, 2017).

21.3 TRANSITION INITIATIVES ACTING ON URBAN COMMONS AND PUBLIC LIFE

The second exploration path is related to the codesign regeneration and transformation process in urban contexts engaging a large range of population where the engagement is not only related to the design (problem setting and problem solution identification phases) but it also influences the implementation, realization, and management phases where communities and citizens groups directly act on the public spaces and public life improvement and revitalization.

The US pavilion at Biennale di Venezia 2012 (title was common ground) presented the first results of a "collection" and mapping of spontaneous interventions promoted and managed by local communities on the "urban commons" (where on the public spaces). "Spontaneous Interventions: Design Actions for the Common Good" celebrated and presented initiatives of citizens able to transform critical urgencies of urban context into new opportunities to bring back to the public life. The selected practices included both more structured and more "direct" interventions but underlining the role of "action" of citizens that act personally "finding creative sources of funding, using every tool at hand to network and form tribes, mobilizing for the sake of shared passions, and simply making things happen—these are the *modi operandi* of a new class of citizen activists who are changing the shape of cities today" (Ho, 2012). This spontaneous interventions exhibition (which also took place in Chicago in 2013 and in Governors Island, NY, in 2014/2015) and a related website are sources of inspiration and show new approaches to urban public life design and transformation able to valorize the synergies between stewardship of places by local community and structured vision for future urban scenarios dealing with urban climate issues (mitigation and adaptation). A large range of initiatives refer to mitigation and adaptation issues in implementing actions, for instance, as depavement of parking lots, urban public spaces greening, shared (informal) mobility services or collective resources management, citizens awareness (on climate change issues and on more sustainable individual and community behaviors).

The capacity of community-led projects and interventions to drive more systemic change in urban complex systems is an interesting assumption presented in the book *Handmade Urbanism* (Rosa and Weiland, 2013)—strong partnerships create a "productive environment" where creativity and vitality spill over into surrounding urban public space, creating what the book's editors refer to as a "ripple effect." Proposing the "Lighter, Quicker, Cheaper-style Approach," the authors present several interventions in five cities (Mumbai, São Paulo, Istanbul, Mexico City, and Cape Town) acting in urban informal or critical settlement and where the process of regeneration is based on community engagement and large partnerships and the benefits overcame the specific initial goal or issue activating a more large urban systemic change that includes the local community stewardship for urban places and activate processes toward more sustainable condition and scenarios. The practices presented show in some cases success in finding an equilibrium and an integrated solution between community-led and institutional governance process in urban regeneration.

Spontaneous interventions, transition initiatives, and grassroots initiatives—all these initiatives are acting on public spaces and public life and the diffusion and growth of these processes for urban commons renovation and improvement highlights the urgency of urban design and urban planning "toolbox" renovation in order to recognize, valorize, and learn from these initiatives (Banerjee, 2001). The urgency in reactivating the disciplinary debate toward renovated approaches in public life design is also suggested by the role that the public space can play in urban climate solutions and in reaching urban sustainability goals.

The public space is a strategic key asset to reach the goals of urban agenda (UN-Habitat) toward sustainable urban development. In 2011, UN-Habitat's Governing Council gave the program a clear opportunity and direction through Resolution 23/4 (UN-Habitat, 2017) to consolidate our agency-wide work on public space. UN-Habitat's member states have mandated the agency to develop an approach that promotes the role of public space in meeting the challenges of our rapidly urbanizing world, to coordinate various global partners and experts on public space, and to directly assist cities in their initiatives on public space. Since then, UN-Habitat has actively promoted public space as an important component for prosperity in cities:

> "37. We commit ourselves to promoting safe, inclusive, accessible, green and quality public spaces, including streets, sidewalks and cycling lanes, squares, waterfront areas, gardens and parks, that are multifunctional areas for social interaction and inclusion, human health and well-being, economic exchange and cultural expression and dialogue among a wide diversity of people and cultures, and that are designed and managed to ensure human development and build peaceful, inclusive and participatory societies, as well as to promote living together, connectivity and social inclusion" **UN-Habitat (2017) A/RES/71/256**

The public space (and public life) became crucial not only in relation to the achievements related to social vitalization and inclusion but also a strategic infrastructure to achieve more environmental aims connected both to health and safety of populations and communities and to support the life of cities through urban ecosystem services and integrating principles and solutions able to improve the management of water, energy, food flows, and the metabolism of cities (decreasing the impact and negative output from the urban system) (Colucci, 2016a).

From a resilience perspective, to talk about public spaces and common goods brings up a wide range of issues that are rarely considered during design and policy-making processes. Numerous community-led initiatives are acting on urban environment issues and on resources (also approached in terms of cultural and awareness values and technical capability of community awareness and behaviors). One example of urban transition practices of sharing expertise and skills to maximize resource management is the Repair Café Network—the meeting of people sharing their skills to help each other repair out-of-order appliances and devices (a worldwide network), promoting local initiatives promoting reuse and recycling, as well as toward a strengthening of the sense of community and the temporary or long-term regeneration of not-used urban spaces where Repair Cafés are placed.

The Depave project is a growing initiative that began in 2007 in order to transform desolate or underused paved surfaces (often not-used parking lots) into lively green spaces. Aiming to contrast the excessive expansion of impervious surfaces in urban contexts, the interventions had a huge impact also on the social dynamics of the neighborhoods, involving inhabitants in the stewardship of urban place and environment. Starting in Portland (Oregon) the Depave initiative was replicated in other US cities (e.g., Seattle, Tacoma). The Depave initiative gives several positive benefits locally on urban greening improvement (and urban biodiversity), on drainage management, on urban heat islands mitigation, and on the re-use of recycled materials (e.g., the removed pavements). The Depave initiatives are clearly oriented to the climate mitigation and adaptation issues, demonstrated also by the production of an annual report demonstrating the "benefit" to urban local climate reached through community-led initiatives.

A large range of urban community-led initiatives deal with food. Initiatives based on local food chains, urban food production, and food security (culture of healthy and quality food) have grown worldwide. In the metropolitan context, these initiatives could play a relevant role in improving urban greening and generating positive impacts in local urban climate and water drainage regulation (Colucci, 2016b). For instance, the initiatives of reconversion of rooftop grey surfaces into green places to be used for different food productions (from community and educational gardening to farming) positively impact urban climate polices (microclimate regulation, water drainage regulation), citizens' awareness of healthy diet (projects on the school rooftop), and urban community and public life enhancement (gardens and social market become nodes of urban public life network). In New York City (in recent years also with the institutional support) several positive and successful initiatives were developed, for instance: Eagle Street Farm (2009) and Brooklyn Grange (2010), two rooftop farms on tertiary buildings; and Greenhouses Vinegar Factory and Forest House (pioneers since 1993) on the top of residential buildings. These initiatives contribute to resilience and adaptation measures in urban contexts considering the mitigation of heat islands, urban microclimate regulation, water drainage regulation (rainwater re-use, storage, etc.), and urban biodiversity and contribute also indirectly in the reduction of food provisioning distances and transportation impacts that became "short" and local (community and local market of urban agriculture). Initiatives on urban farming often integrate the cultivation dimension with the cultural one stimulating the rediscovery of foods and traditional techniques of food production and promoting healthy life stile and behaviors.

The initiatives on urban commons often develop, as an indirect benefit, the improvement of community stewardship of public space recognized as common good. In Italy, for example, since 2012, the project Zappata Romana developed an online community map of green and open spaces existing in the municipality of Rome. The project is at the same time a powerful tool in boosting community stewardship of rural and open land, in coordinating single initiatives of regeneration and management of spaces by communities, and an incredible instrument against the illegal appropriation (and settlement) of marginal/periurban open spaces. The project, launched as a community mapping, became also a process of codesign transforming isolated pockets of open space in a strategic and systemic network of open and green spaces bringing environmental, economic, and social advantages into the periurban context of the metropolitan area of Rome.

21.4 RESILIENCE PRACTICES OBSERVATORY

The Resilience Practices Observatory (RPO) officially started in April 2015 (Dezio et al., 2016) and is a project coordinated by the Department of Architecture and Urban Studies (DASTU) of the Politecnico di Milano, the association REsilienceLAB, Department of Science, Planning and Territorial Policies (DIST) of the Politecnico di Torino, the University of Molise and the Fondazione Lombardia per l'Ambiente and supported by Fondazione Cariplo.

The Observatory has as its overall strategic objective the capacity building of territorial subjects to enhance resilience, considering all territorial entities, institutions, and communities involved in a long-term transition project, toward a stronger, aware and adaptive society, and the subjects themselves, which animate the RPO. This is why the RPO project design has been based on four main axes (Fig. 21.1).

1. Mapping resilience initiatives at national level. Here the concept of "mapping" refers to the understanding of characteristics that differentiate initiatives and to the understanding of the territorial areas mainly involved in the resilience practices. This proved to be an important step toward better understanding of the potentials of the multidimensional nature of the resilience concept "in action," when applied to urban and territorial practices.
2. Develop a set of tools and design criteria to support the dissemination of resilience practices. The goal is to build and offer new tool sets and project design criteria to support the expansion and reinforcement of resilience practices and the building of solutions.
3. Produce developments in the cultural and methodological paths focusing on capacity building and innovation in knowledge approaches. The strategy relies on promoting scientific advancement about resilience and transition themes, based on applied research to territory and community. This to produce advances in scientific research in terms of conceptual and methodological innovation.
4. Promoting the involvement of a huge actors' network. The definition of conceptual tools, both interpretative and for planning, aims at involving, in a process of mutual magnetization, the promoters and the community through the implementation of innovative and shared routes. This approach is based on a multisector and interdisciplinary perspective.

The RPO, as mentioned before, does not assume one specific definition of resilience but prefers to recognize many different definitions in order to include and host the larger possible range of resilience practices. The RPO assumes that the resilience of complex systems is the sum of capabilities of the system (social ecosystem) to activate responses to phenomena of stress and shock, positively adapting to new conditions and maintaining its identity and its functionality over time. The aim of RPO is to give and produce conceptual frames supporting clear orientations in the complexity, leaving (and highlighting) possible aspects of synergies. There is not a univocal "definition of resilience practice" but a "concept of resilience practice," where process and actions have to be able to strengthen the capabilities and the resilience properties of the system itself.

Since 2015, the Observatory has engaged about 101 practices articulated in:

- 41 practices connected to the CARIPLO foundation call on "resilient communities" (grant call targeted to NGO bodies that financed 41 projects in four annual editions from 2014 to 2016) mostly located in Lombardy;
- 47 practices engaged during the RPO networking activities (Resilience Practices Forum);
- 16 applied research projects (in which research and academic bodies play a coordination role).

The RPO organized two editions of the Forum of Resilience Practices (RPF) and several activities of knowledge exchange and knowledge coproduction. In particular, the RPFs were launched as innovative instruments in order to

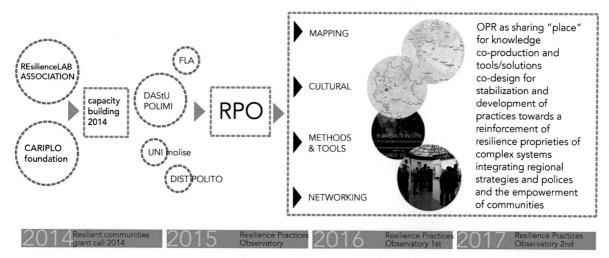

FIGURE 21.1 Resilience practices observatory: A synthetic timeline and the main four axes. *Authors' elaboration.*

engage and activate interdisciplinary and multiactor dialogues. During the two editions of the RPF (January 29, 2016; February 24, 2017) there were 86 practices or initiatives activated by NGOs or local institutions and 16 research projects applied to local contexts coordinated by universities or research centers directly involving the local communities.

The RPF has then been configured as a path that connects different worlds: the first one is the world of the practices, or of the actors that implement the practices, and of the actors' promoters, which are public and private bodies that support and promote the initiatives and the resilience processes. The second one is the world of research and innovation, represented by universities, practitioners, international networks, and national research centers.

Based on the practices engaged (86 Resilience Practices mapped for the RPF of 2017), the main issues (or thematic levers) are:

- Urban and territorial regeneration and local development (33% of practices). A large range of practices focuses on strategic vision for the redevelopment of depressed areas (internal areas but also peripheries and peri-urban areas in metropolitan contexts);
- Food chains and natural resources cycles including biodiversity and natural habitats improvement (43% of practices). In most of the mapped practices, the food topic is strictly connected to the natural and ecosystem preservation issues (ecosystem services, natural areas improvement, agroecology, etc.);
- Climate change mitigation and adaptation and territorial risks (24% of practices). Only a few practices focus on climate change adaptation as primary lever for practices promotion and those using this reading key refer to regional or territorial context and are usually connected to other existing processes already activated (Mayor Adapt, energy polices, etc.). Some practices on sustainable river basin management (characterized by large partnerships including institutions and national associations) include polices and measures for risk prevention. Few of these focus on communities' risk preparedness and prevention in terms of self-help preparation and of intervention for the reduction of building vulnerability.

During the RPF (2017), a first characterization of main categories of instruments and solutions adopted was presented. From the 86 Resilience Practices emerged five "categories" of instruments or tolls adopted to reach their strategic goals:

- Capacity building instruments (adopted as main instruments by 21 practices/24% of practices);
- Mapping/monitoring or supporting methods for decision-making instruments (adopted as main instruments by 18 practices/21% of practices);
- Local improvement interventions instruments (adopted as main instruments by 19 practices/22% of practices);
- Codesign and participation instruments (adopted as main instruments by 16 practices/19% of practices);
- Instruments promoting circular economies (or new economies) (adopted as main instruments by 12 practices/14% of practices).

RPO developed several activities of coproduction, dialogues, and workshops. Common aspects of barriers and of opportunities emerged from these activities could be summarized in relation to the governance and processes management (difficulties in implementing) and in relation to the economic dimension that is related to the practices stabilization and consolidation (long-period consolidation and upscaling).

Several critical aspects emerged in relation to the governance of resilience practices processes. In particular, the main barriers highlighted are: difficulties in managing (or designing) the governance process (often it is possible to underline a lack in defining explicit strategies in managing the governance process and in defining actors maps); a lack of virtuous contaminations and alliances between sectors and actors (often community-led initiatives do not involve stakeholders that could be able to support the practice in terms of implementation/management competencies, innovation in solution design and so on) and difficulties in promoting large partnerships and alliances (a low confidence in the feasibility and effectiveness of public-private cooperation emerged often as feeling perceived by the actors of practices). The processes promoted by public/institutional, private, and civil society (and/or addressed to different sectors) often do not converge.

To guarantee long-term visions and stabilization of Resilience Practices it is necessary to increase the level of awareness of how to deal with the economic and financial components that have to be integrated in the projects. It is necessary to increase the quality and the innovation contents of the economic and financial-based tools available. There are increasing opportunities in better integrating projects on urban territorial resilience with methodological approaches and action goals looking at their economic dimensions. This is particularly when referring to the more innovative tools and practices available in the so-called new economies, based on collaboration, sharing, and sustainability.

21.5 CONCLUDING REMARKS

It is possible to highlight a few aspects emerging from a comparison between three exploration paths presented that are able to support critical remarks useful in the fourth section's cases presentations.

In relation to the focus on urban climate issues, it is clear that Transition Network initiatives are more oriented to a mitigation action and that urban commons initiatives focus on both mitigation and adaptation actions. In the Practices of RPO the initiatives are focusing on both adaptation and mitigation but, at the same time, it clearly emerges that community-led initiatives are less able to produce and figure out strategies and visions having climate issues as a primary thematic lever for action and that climate issue policies and actions are proposed in institution-led initiatives (or in initiatives where the partnerships include institutional actors). In relation to the organizational topic, the Transition Network is characterized by a clear identity and relative closure in comparison to other geographies (Table 21.3).

The critical remarks refer to three main aspects: the role of Networking (and local-global scales); the partnerships of single initiatives and related impact on urban climate issues; and the aspect of monitoring and evaluation (both in relation to evaluation of initiative success degree and in relation to positive impacts/benefits on urban polices dealing with climate change mitigation and adaptation). The relevant role of "Networking" (structured or informal kind of networking and association among single initiatives) in supporting local and single transition initiatives emerges as a common aspect. The Transition Network and other less-structured (and more flexible) kinds of networking among local initiatives (DEPAVE, for instance) demonstrate the relevance of an organizational structure able to support single initiatives providing capacities building, training, and cultural instruments (recognition and identity) to local and isolated community-led interventions that are often fragile if not supported by strong partnerships. The networking organizational level also plays a relevant role in defining long-term and more complex strategic visions related, for example, to urban climate issues, low-carbon or environmental and social sustainability, and in providing conditions (knowledge, skills, organizational, recognizability, etc.) for upscaling and consolidation of community-led initiatives.

From literature on the Transition Network (Feola and Nunes, 2013; Barnes, 2016) and from the RPO experience it emerges that large and inclusive partnerships are one of the main factors for the success of transition initiatives: "Transition initiatives that elect to practice confrontational politics with local institutional governance structures could more rapidly and comprehensively advance community resiliency, economic localization, and low-carbon development" (Barnes, 2016, p. 23). RPO experience demonstrates how resilience practices able to become a consolidated process (with significant positive transition toward sustainable scenario of local urban complex systems) are based on partnerships including institutions, research and/or expert actors, private/economic actors, and communities/NGOs organizations.

In sociotechnological transition or in transition management literature (Chapter 3), the monitoring of transition process is crucial in reorienting polices and measures and in supporting fertile environments for innovation niches.

TABLE 21.3 Some Comparative Notes Among Transition Network, Urban Commons, and (Italian) Resilience Practices Observatory Geographies

	Urban Climate	Organization	Tools	Issues/Thematic Levers
Transition Town Network	Mitigation	Closed	Training, community building/engagement, coproduction/codesign, thematic guidelines/technical support tools	Fixed "list" at global level; locally based
Urban Commons (networking level)	Mitigation adaptation	Fluid	Capacity building, Coproduction/codesign, networking, thematic guidelines/technical support tools	Open and rich (greening, food, commons, public life, etc.)
Resilience Practice Observatory	Adaptation (institutional and research-led practices)	Fluid	Capacity building, Coproduction/codesign, networking	Open and locally based (community led initiatives more oriented to food/natural resources and urban renovation; institutional-led initiatives oriented to climate change and environmental issues)

Authors' elaboration.

At the same time, structured monitoring tools are rarely developed and implemented in grassroots innovations and proposed by different networks of local transition initiatives. Monitoring tools could be powerful instruments in documenting positive contributions to the community (also as a gateway to increased resources, both in terms of motivated volunteers and external supports, including grants) and in providing tangible data to demonstrate how the transition initiatives are making a difference in local and global sustainability. The monitoring also could be a powerful tool in supporting trajectory adjustments in actions during the process. The monitoring and assessment is not one of the core tools developed and proposed by Transition Network and by other transition networks (in general). Monitoring and assessment tools are developed in relation to specific issues (for example, in Transition Network to specific energy descent plan targets or to local currency implementation; in DEPAVE network to surfaces depaved, and other specific targeted goals). The monitoring (and assessment) of the process of transition itself is not proposed/developed in community-led initiatives.

In Transition Network two aspects are underlined (in the "7 essentials"): celebration and reflection. These two aspects could refer to the assessment sphere (celebrate results reached and reflect results) but in the Transition Network approach these two are oriented to the community activation aspects and to the internal/external recognition of community efforts (celebration of activities and initiatives that have been done) than to the assessment of the level of strategic goals achievement and of positive impacts (benefits) on urban systems deriving from local initiative actions implementation. In general, the actors of Transition Initiatives consider as relevant factors to evaluate or demonstrate the success of local Transition Initiative (Feola et al. 2013, 23; Feola et al. 2017) the level of networking and/or the level of community building/engagement more than quantitative targets achievement.

References

Barnes, P., 2016. Transition initiatives and confrontational politics: Guidelines, opportunities, and practices. Western Political Science Association 2016 Annual Meeting. http://wpsa.research.pdx.edu/papers/docs/Transition%20initiatives%20and%20confrontational%20politics%20(Barnes).pdf (accessed on 1st of December 2017).

Banerjee, T., 2001. The future of public space: beyond invented streets and reinvented places. Journal of the American Planning Association 67 (1), 9–24.

Colucci, A., 2016a. Peri-urban/peri-rural areas: identities, values and strategies. In: Colucci, A., Magoni, M., Menoni, S. (Eds.), Peri-urban Areas and Food-energy-water Nexus. Sustainability and Resilience Strategies in the Age of Climate Change. Springer-Verlag GmbH, Heidelberg.

Colucci, A., 2016b. La resilienza dei sistemi socio-ecologici. In: Colucci, A., Cottino, P. (Eds.), Resilienza tra Territorio e comunità. Approcci, strategie, temi e casi. Fondazione CARIPLO, Milano.

Dezio, C., Colucci, A., Magoni, M., Pesaro, G., Redaelli, R., 2016. Observatory of resilience practices: strategies and perspectives. In: Proceedings of the 1st AMSR Congress and 23rd APDR Congress. Sustainability of Territories in the Context of Global Changes, pp. 226–231.

Feola, G., Butt, A., 2017. The diffusion of grassroots innovations for sustainability in Italy and Great Britain: an exploratory spatial data analysis. The Geographical Journal 183 (1), 16–33. https://doi.org/10.1111/geoj.1215.

Feola, G., Nunes, R., 2013. Failure and Success of Transition Initiatives: A Study of the International Replication of the Transition Movement. Research Note 4, Walker Institute for Climate System Research, University of Reading. www.walker-institute.ac.uk/publications/research_notes/WalkerInResNote4.pdf.

Frantzeskaki, N., Kabisch, N., 2016. Advancing urban environmental governance: understanding theories, practices and processes shaping urban sustainability and resilience. Environmental Science & Policy 62 (2016), 90–98.

Ho, C.L., 2012. Spontaneous Interventions: design actions for the common good. Architect-Northbrook 20. http://www.spontaneousinterventions.org/reading#essays.

Hopkins, R., 2008. The Transition Handbook. From Oil Dependency to Local Resilience. Green Books Ltd., Devon (UK).

Hopkins, R.J., 2010. Localisation and Resilience at the Local Level: The Case of Transition Town Totnes (Ph.D. thesis). School of Geography, Earth and Environmental Science, Faculty of Science and Technology, University of Plymouth.

Hopkins, R., 2011. The Transition Companion: Making Your Community More Resilient in Uncertain Times. Green Publishing, Chelsea.

Rosa, M.L., Weiland, U.E. (Eds.), 2013. Handmade Urbanism: From Community Initiatives to Participatory Models. Jovis Verlag.

Transition Network, 2016. The Essential Guide to Doing Transition. Transition Network, Totnes.

UN-Habitat, 2017. United Nations A/RES/71/256 General Assembly Distr.: General 25 January 2017 Seventy-first Session Agenda Item 20-16-23021 [Resolution adopted by the General Assembly on 23 December 2016/without reference to a Main Committee (A/71/L.23)].

Films

In Transition 1.0 (2009) Emma Goude (Director), Transition Network (published on Vimeo platform by Transition Network. https://vimeo.com/8029815).

In Transition 2.0 (2012) Emma Goude (Director), Transition Network and Green Lane Films (Production) (published on Youtube platform by Transition Network. https://www.youtube.com/watch?v=FFQFBmq7X84).

Websites

Brooklyn Grange farms rooftops. https://www.brooklyngrangefarm.com/.

DEPAVE volunteer-driven organization web site. http://depave.org/.

Eagle Street Rooftop Farm. http://rooftopfarms.org/.
New York City Rooftop Greenhouses and Forest House (in Carrot City web site). http://www.ryerson.ca/carrotcity/index.html.
Observatory of Resilience Practices web site (in Italian). http://www.osservatorioresilienza.it/.
Repair Café Foundation. https://repaircafe.org/en.
REsilienceLAB association web site (in English and including a Resilience Practices Observatory presentation in English). http://www.resiliencelab.eu/.
Spontaneous Interventions: Design Actions for the Common Good. http://www.spontaneousinterventions.org/.
Transition Network Web site. https://transitionnetwork.org.
Zappata Romana (in Italian). http://www.zappataromana.net/.

CHAPTER

22

Transition Towns Network in the United Kingdom: The Case of Totnes

Alessia Canzian
Independent Researcher, Italy

22.1 TOTNES TRANSITION TOWN

The experience of Totnes Transition Initiative is discussed and presented in several Transition Towns Network publications and documents and it is used as an experimental case study in books on the Transition Network edited or coordinated by Hopkins (Hopkins, 2008, 2010, 2011; Brangwyn and Hopkins, 2008). This chapter, without any ambition to present a comprehensive history and critical discourses on this experience, aims to summarize and recap main activities realized during the Totnes Transition Initiative process in order to highlight these main issues and spheres of attentions and interest developed in a large range of citizens-led initiatives.

Totnes is a town in the county of Devon, in the southwest of England, which became the first Transition Town experiment and the first local initiative activated by the Transition Town movement that is a community-based approach that works with ecological sustainability and resilience (Connors and McDonald, 2010).

The process of local community engagement began in October 25 with a screening of the film *The End of Suburbia*, one of the first films on the subject of peak oil, and with a series of public talks (Hopkins, 2008, 2010). The event "Official Unleashing of TTT" that officially launched the initiative of Transition Town Totnes (TTT) occurred on September 26, 2006. This event was the Transition organization's formal launch and it was articulated in several meetings and events: film projections, training and conferences, brainstorming, and lunch events. Common characteristics of all events were the knowledge sharing on the peak oil, local economies, environmental aspects for the awareness of citizens, the communication based on film or participatory session (in which citizens could propose and explain their points of view and proposals), events for community building (lunch, cafés, and so on) (Hopkins, 2008, 2010).

Since then TTT has grown to have 10 working groups exploring different aspects of Transition. The process uses creative engagement tools, such as Open Space, World Café, Fishbowl, as well as running training workshops and having other "Great Reskilling" events, including a popular gardening course.

Totnes economy is interconnected with the local producers. From books in bookstores to beer that is drunk in pubs, everything is produced locally. To support this type of economy, a Local Entrepreneur Forum was created; it is an annual meeting for local entrepreneurs. The Forum was born thanks to the Project REconomy of Transition Town Totnes. REconomy Centre is a coworking and meeting place for local ethical enterprises, for people creating new livelihoods, and for community groups that are working for a strong, resilient, and equitable economy. The annual forum is a kind of business incubator that supports ideas and projects presented by local entrepreneurs. The support can be either financial or practical. The financial support is an amount of money. The practical support is some tools for the dissemination of ideas, for example, advertising campaigns

TABLE 22.1 Totnes Work Groups and Related Projects

Transition Groups	Transition Town Totnes (TTT) Projects
Arts	TTT Arts Network; TTT Film Festival; TTT Film Club
Food Group	Food-Link Project; Food Hub; Food in Community; Grown in Totnes; Incredible Edible
Incredible Edible	Borough Park and Totnes Station; Bridgetown fruit and nut trees; Follaton Arboretum; Steamer Quay and Longmarsh; Town Cemetery, Follaton
REconomy Project	Local Entrepreneur Forum; REconomy Centre; Totnes Local Economic Blueprint; Totnes Pound
Health and Well-being	Caring Town Totnes; Inner Transition; Earth Stories Project; Mentoring and Well-being Support; Transition Support Group; Keeping Totnes warm; Play Group; Transition Streets
Skill-share Project	No specific projects
Building, Housing and Energy	Draught busting; Energy Descent Action Plan (EDAP); Good Energy partnership; Keeping Totnes warm; Eco and Community Homes Fair 2017; Transition Homes CLT; Transition Streets
Waste and Resources	Refill Totnes
Transport	Cycling Group; DoctorBike; Totnes e-bike scheme

Authors' elaboration based on Totnes Transition Initiative website (https://www.transitiontowntotnes.org/).

or creation of a website. The total funds collected in Totnes have passed the 70 thousand pounds (source https://www.transitiontowntotnes.org/).

Totnes is organized into working groups divided according to key sectors of the Transition model outlined by Hopkins. These key sectors are: food and farming; medicine and health; education; economy; transport; and energy.

Each group has the task of organizing discussion, planning, and activities—always having climate change and peak oil as the focal point. Each group is divided into subgroups that have the role of coordinating a specific project (Table 22.1).

In general, the projects activated in Totnes have the purpose of decreasing the emissions of substances from fossil fuels and reducing the use of petroleum-based products.

The Health and Medicine group has organized public meetings to discuss the implication of energy descent on local health, for example, reducing access to plastic and other petroleum-based products.

Another group was focused on housing policies, working with the local planning authority and discussing localized control over planning and construction to boost the development and use of local materials in the house-building sector.

Another project aims to transform Totnes into "the Nut Capital of England" (Hopkins, 2008, p. 28). The project was introduced to encourage the planting of fruit trees for the realization of a resilient community.

These are only some of the 20 local projects active in Totnes. All the projects are developed by community members who assist either the practical, the communication, or the fund-raising aspects from the TTT Office.

The Totnes Transition Town aims to develop practical actions in order to reduce carbon emissions and dependence on fossil fuels. As with all the Transition Towns, the first goal is to strengthen the resilience of the communities that make up the local territory to prepare people for the shocks of peak oil and go beyond climate change and economic crisis.

In 2010, Totnes Transition Town Initiative published the Totnes Energy Descent Plan that is a result of a coproduction and codesign process involving a large partnership of actors and is based on a community engagement process (Table 22.2).

TABLE 22.2 Summarization of Totnes Energy Descent Plan

Extract from Totnes Descent Plan (Transition Town Totnes (2010), Transition in Action Totnes and District, 2030 An Energy Descent Action Plan. Totnes: Transition Town Totnes); website descated https://www.transitiontowntotnes.org/groups/building-and-housing/energy-descent-action-plan/ (accessed on May 15, 2018).

Introduction/structure
[Extracts from pp. 4–12]

An Energy Descent Action Plan is a guide to reducing our dependence on fossil fuels and reducing our carbon footprint over the next 20 years, during which we expect many changes associated with declining oil supplies and some of the impacts of climate change to become more apparent. In this EDAP we have built a picture of this future scenario based on visions of a better future.

- PART ONE, "Where we start from," sets out the assumptions that underpin this report. It sets out the 3 key assumptions, namely the imminent peaking in world oil production, climate change and the economic crisis. It also introduces the concepts of resilience and localisation. It closes with a look at Totnes and District, drawing together some of the key information about the area.
- PART TWO, "Creating A New Story," looks at why, as a culture, the stories we have about the future aren't up to the job, and why we need new ones. This Plan is, in effect, a story about how the community could make the transition away from its oil dependency.
- PART THREE, "A Timeline to 2030," looks at a range of subject areas. For each it sets out the challenges that Transition presents to them, and how things might progress if we carry on as usual and do nothing to start embracing the changes already underway.

The Town Totnes "Energy Descent Pathways" project was funded by Esmee Fairbairn Foundation and Artists Planet Earth. It is a process that had not previously been used, and a number of tools and approaches were developed during the process, which unfolded through a series of steps.
- Step One: Developing a Framework
- Step Two: Key Tools
- Step Three: Engage the Community
- Step Four: The Public Launch
- Step Five: Public Workshops (Two series of public workshops were hosted. The participants identified assumptions about the key drivers and changes they anticipated influencing the future.)
- Step Six: Back-casting on Strategic Themes
- Step Seven: Drafting the EDAP and Consultation (Strategic themes were identified from the back-casting worksheets and a general process for evolving idea and building the pathways was developed.)
- Step Eight: Implementing the EDAP timelines

Joined up Thinking
[Extracts from pp. 49–65]

Key Challenges identified
- Peak Oil; power down from 9 barrels pp/y in 29 to (possibly) 1 barrel per person by 2030
- Climate Change: keeping the lid on global temperature rise, getting carbon below 350 ppm Carbon sequestration: removing some carbon from the atmosphere
- Stabilising Population growth to around 7 billion and reducing thereafter
- Increasing renewable energy supplies to meet 50% of current energy demand
- Reducing consumption and waste to zero
- Repairing biodiversity
- Maintaining adequate clean water supplies with less energy inputs
- Society making an inner transition and taking responsibility

Indicators - Key Characteristics of a Resilient Community
1. Leadership is diversified and representative of age, gender, and cultural composition of the community.
2. Elected community leadership is visionary, shares power and builds consensus.
3. Community members are involved in significant community decisions.
4. The community feels a sense of pride.
5. People feel optimistic about the future of the community.
6. There is a spirit of mutual assistance and co-operation in the community.
7. People feel a sense of attachment to their community.
8. The community is self-reliant and looks to itself and its own resources to address major issues.
9. There is a strong belief in and support for education at all levels.
10. There are a variety of Community, Enterprise and Development (CED) organisations in the community such that the key CED functions are well served.
11. Organisations in the community have developed partnerships and collaborative working relationships.
12. Employment in the community is diversified beyond a single large employer.
13. Major businesses in the community are locally owned.
14. The community has a strategy for increasing independent local ownership.

Continued

TABLE 22.2 Summarization of Totnes Energy Descent Plan—cont'd

15. There is openness to alternative ways of earning a living and economic activity.
16. The community looks outside itself to seek and secure resources (skills, expertise, finance) that will address areas of identified weakness.
17. The community is aware of its competitive position in the broader economy.
18. Citizens are involved in the creation and implementation of the community vision and goals and have a CED Plan that guides its development.
19. There is on-going action towards achieving the goals in the CED plan.
20. There is regular evaluation of progress towards the community's strategic goals.

Working with Nature - Food Security and Food Production and Farming
[Extracts from pp. 67–103]

Food Production and Farming - Principles/Key Challenges:
- It will need to be well on the way to the 80% cut in carbon emissions by 2050 (as stated by UK Government policy).
- The concept of resilience, the ability at all levels to withstand shock, must be key, embodied in the ability of settlements and their food supply system, to adapt rapidly to rising energy costs and climate change.
- The need for improved access to nutritious and affordable food.
- The need for far more diversity than at present, in terms of species, ecosystems, produce, occupations, etc. to support food production systems.
- The need to increase the capacity of our soils to act as a carbon sink requires us to adopt more perennial, farming systems supporting grass and tree systems, as well as making good soil management and the building of organic matter in soils a priority.
- Stronger links to local markets than at present, supplying local markets by preference where possible.
- A much reduced dependence on fertilisers and other agrochemicals (ideally enabled by a shift to organic practices).
- Accompanying this will also be the need for a large increase in the amount of food produced from back gardens, allotments and other more "urban" food sources.
- The use of genetically modified crops has no place in a more sustainable agriculture.

Food Production and Farming - Resilience Indicators
- From the research that led to this Plan, we have identified a number of key indicators by which we can be sure that we are moving forward. These include:
- The percentage of the population with basic food production skills.
- The percentage of the population who feel confident in cooking with fresh produce.
- The percentage of food consumed locally which has been also grown locally.
- The number of people who feel they have access to good advice, skills and retraining in basic food production.
- The percentage of land (agricultural and urban) under utilisation for food production.
- Rates of obesity and chronic heart disease.
- The average body mass index.

Working with Nature - Health And Well-being
[Extracts from pp. 104–114]

Health and Well-being - Principles/Key Challenges:
- Growth in the number of people taking responsibility for their own health.
- People motivated to be healthy.
- People eat less junk food, sugar is rationed.
- Better health on less food.
- Reduced need to travel to healthcare.
- More localized provision with technical assistance/webcams (CAT—Computer assisted technology).
- New developments in health care and resources.
- Research more patient centered and connected.
- More integrated health care.
- A blur between regular and complementary medicine.

Health and Well-being - Resilience Indicators
- From the research that led to this Plan, we have identified a number of key indicators by which we can be sure that we are moving forward. These include:
- Depression trends/rates
- Obesity rates in children and adults
- Frequency of visits to the doctor
- The proportion of babies exclusively breastfed for 6 months or more
- Acres of land used to cultivate medicinal herbs
- Average age of dying
- Number of hours spent walking
- Number of meals per capita eaten alone by over 65s

TABLE 22.2 Summarization of Totnes Energy Descent Plan—cont'd

Working with Nature - Water Matters and Supporting Biodiversity
[Extracts from pp. 115 119]

Water Matters and Supporting Biodiversity - Resilience Indicators
- Hectares of deciduous woodland managed for nature conservation.
- Monitoring of Red Shanked Carder bumblebee population.40
- The total km of hedgerows.
- Number of mating pairs of otters (Operation Otter at Dartington/Devon Wildlife Trust).
- Numbers of Skylarks in the district.
- Monitoring of key bat species.
- % of households with bird tables and bat boxes.
- Cleanliness of main waterways in the area.
- Number of people actively involved in nature conservation.

Creative Energy Systems - Energy Security and T&D Renewable Energy Budget
[Extracts from pp. 120–167]

Energy Security - Principles/Key Challenges:
- Clear understanding and awareness of how energy is used
- Major reduction in use of energy at all levels of society
- Severe reduction in use of products with high embodied energy
- High priority for investment of time and finance in energy efficiency
- Recognition of the vulnerability of conventional energy supplies
- Development and provision of renewable energy supplies must be prioritised and brought on stream
- Rapid reduction in use of fossil fuel based energy
- Equitable approach to the sharing of all resources

Energy Security - Resilience Indicators
- % of houses with insulation to Passivhaus standards
- % of energy produced from local renewable sources to meet local (estimated) demand
- % of buildings with solar hot water collectors
- Number of people who feel well informed about energy issues
- Number of people concerned about energy security/climate change
- Reaching the Government target of reducing carbon emissions by at least 20% by 2020 (80% by 2050)

Creative Energy Systems - Transportation
[Extracts from pp. 168–187]

Transportation - Resilience Indicators
- % of people who walk for 10 min at least daily
- % of children who cycle or walk to school
- % of people who cycle or walk to work
- N° of people with access to a local bus
- Distance driven each year
- Overall split of journeys between walking, cycling, public transport and car

Creative Energy Systems - Building and Housing The Proposed Transition Zero Carbon Homes Code
[Extracts from pp. 188–208]

Zero Carbon Homes - Principles/Key Challenges:
- Meet the current highest standard for sustainable buildings (i.e., Passivhaus/exceeds level 6)
- Be designed so as to maximise natural lighting and solar space heating
- Eliminate toxic or highly-engineered materials and energy-intensive processes
- Be independent of fossil-fuel based heating systems
- Be designed for adaptability and dismantling: so as to allow the building to be subsequently adapted for a range of other uses
- Where appropriate, integrate working and living
- Ensure outdoor spaces are south facing with the minimum of overshadowing, so as to maximise the potential of the property/development to grow food
- Maximise grey water recycling and rain water capture
- Be built to address needs not speculation
- Adhere to good spatial planning to benefit communal interaction and shared open space
- Maximum use of locally produced materials: (defined as clay, straw, hemp, lime, timber, reed, stone)
- Maximum use of used and recycled building materials, particularly those on site
- The inclusion of water-permeable surfaces rather than hard paving, etc.

Continued

TABLE 22.2 Summarization of Totnes Energy Descent Plan—cont'd

Zero Carbon Homes - Resilience Indicators
- Percentage of houses that have been retrofitted to maximum possible standard.
- Number of second homes that have been let though the 'Homes for All' scheme.
- Number of houses with solar hot water panels installed.
- Number of builders that have undertaken the 'Construction in Transition' training course, which introduces them to a range of natural building materials and techniques.
- Heat emitted from buildings — as measured by an infrared scan from the sky.
- Trends in fuel poverty.
- Average amount of energy produced by buildings in Totnes and District.

Resourcing Localisation - Economics and Livelihoods
[Extracts from pp. 209—227]

Economics and Livelihoods - Principles/ Key Challenges:
- Checking the balance - Principals of Natural Capital/Holistic Economics:
- Community, Friends, Family (nurture our soul — valuing the heart)
- Biodiversity (provide sustenance for food and air, recreation, relaxation. Gaia,/valuing the head and the heart)
- Education and Skills (potential for our future and understanding of the world — valuing the hands and the head)
- Mineral Resources (inc. petroleum oil) (current economy based on this principal only — distorted with credit)

Economics and Livelihoods - Resilience Indicators
- The percentage of economic leakage out of the community
- The percentage the local community spend on locally procured business, goods and services.
- Percentage of major employers in the community that are locally owned
- Niche markets (in which unique opportunities exist) that take advantage of community strengths.
- The relative value by percentage of community owned major assets for the economic and social benefit of the community.
- The number of Totnes Pounds in circulation.
- Degree to which people perceive an openness to alternative forms of earning a living

Resourcing Localisation - Consumption and Waste
[Extracts from pp. 228—230]

Consumption and Waste - Principles/Key Challenges:
- Waste minimising resource management, infrastructure and policies
- Comprehensive range of recycling services and facilities
- Reducing consumption patterns — awareness and information; meeting needs not wants

Consumption and Waste — Resilience Indicators
- Overall waste volumes
- % of agricultural and sewerage waste to anaerobic digestion
- Reduction in packaging on goods

Nurturing Transition - Arts, Culture, Media and Innovation
[Extracts from pp. 231—243]

Arts, Culture, Media and Innovation - Principles/Key Challenges:
- Creative thinking and methods will need to be more widely shared and placed at the centre of how we think about our lives, our education system, and how we plan for our future.
- Visual and performance art is fully resourced both at the national level and the local level through education, adult education, art, drama, creative writing and music schools and investment in public art works to nurture society's creative talent.
- Visual and performance art should be used to influence our understanding of how the community can respond to peak oil and climate change.
- TV, radio, the Internet and printed media use and support, through their stories, documentaries and programmes, the visionary story of transition, helping people understand and enter the new paradigm with a positive approach.
- Printed media will need to consolidate to reduce the excess of paper and its high energy dependence. New ways of recording and dispersing the writings of journalists, creative writers etc.; perhaps through libraries, electronic billboards or more use of the Internet (if it can continue to exist).

Arts, Culture, Media and Innovation — Resilience Indicators
- The number of public art works commissioned each year
- The amount of funding allocated to art initiatives with a Transition theme
- The number of new business start-ups that are about making everyday household objects, at affordable prices, yet which incorporate art
- The % of local society engaged in transition projects and activities

Nurturing Transition - Inner Transition
[Extracts from pp. 243—255]

TABLE 22.2 Summarization of Totnes Energy Descent Plan—cont'd

Nurturing Transition - Principles/Key Challenges:
- To work with people's reactions to deep or rapid change such as decreasing energy supply, job loss, migration; and dealing with the unknown.
- To heal and support those (many of us) who have been conditioned and hurt by the current materialist and isolated lifestyle.
- Promoting understanding of what builds — and harms — healthy, strong, lasting relationships.
- To embody qualities that deepen our connection and help us stay present, to hold a timeless wisdom whilst bringing in practical projects which have a timescale.
- To acknowledge our interdependence with all other humans and living beings that allows us to develop ways of acting that embody Justice, Equality and Care for the Earth.
- To sustain ourselves emotionally, and develop ways to support ourselves and others when we are overstretched, or overwhelmed.
- To learn to work in groups so that we can cooperate more effectively.
- To learn to value and respect all members of the community, young and old, and those whose background or beliefs are very different from our own.
- To build a safe and supporting environment for all especially children and vulnerable people.

Nurturing Transition - Resilience Indicators
- In general I am satisfied with my life. (Footnote: This is the question used in the World Values Survey and in other international surveys measuring happiness, also used in Rob's survey with 94% agreeing or agreeing strongly.)
- Questions from the "Your recent feelings" section of the Happy Planet Index could be used to assess personal well-being (see reference below).
- I feel confident that in the future my needs and those of my loved ones will be met (Agree strongly to disagree strongly).
- On the whole I feel safe in my community.
- Connections with other people, nature and spiritual life.
- I feel included and welcome in my community.
- I know most/all of my neighbours.
- How often do you spend time outside in natural or green spaces?
- Do you consider yourself to be a spiritual person?
- Availability of Support.
- I can find support that is appropriate when I need it (from family, friends, community services or other organisations).

Nurturing Transition - Education, Awareness and Skills for Transition
[Extracts from pp. 256–274]

Education, Awareness and Skills for Transition/Principles/Key Challenges:
- A broader provision of many educational and personal development opportunities for all sectors of local society, promoting awareness and understanding about the challenges of peak oil and climate change and the role of transition.
- An ethos of sustainability has to be central to all our educational institutions in terms of what they teach as well as what they do as institutions.
- Our local schools could work more closely together and with the local community, in order to: share ideas, programmes and projects, and develop new ones.
- More agricultural and horticultural training and research.
- More apprenticeships and practical training for the diverse range of power-down skills.
- More time for reflection, at all ages.

Education, Awareness and Skills for Transition/Resilience Indicators (education)
- Continue to develop Transition Tales as a project for KS3
- Children become the main teachers and leaders of the education project
- As part of ongoing research into how people best learn, ask those who attend TTT events over 6–12 months to fill in a very small slip of paper which asks them to reflect on what helps them to learn best: e.g., I learn best through reading/listening/talking/group work/individually (on the shelf ones do exist).
- Establish the concept of "Transition Schools."
- Make use of the current opportunities such as the "new" work with parents: spot all such opportunities for making the most of the available resources.
- Actively link school and community projects to the story of Transition, e.g., intergeneration work.

Education, Awareness and Skills for Transition/Resilience Indicators (training and skill)
- Percentage of population who have trained in specific transition skills; academic, practical, personal development.
- Percentage of people whom, when asked, state that they feel confident in a range of skills (see above).
- Percentage of adults registered in postsecondary education.
- Percentage of children who walk or cycle to school.
- Percentage of students who reach 16 with a firm understanding of climate change and other environmental issues, as well as being familiar with practical solutions.

Continued

TABLE 22.2 Summarization of Totnes Energy Descent Plan—cont'd

Empowering People - Local Governance
[Extracts from pp. 275–279]

Local Governance in Transition Principles/Key Challenges:
[reference: Report of the City of Portland, US Peak Oil Task Force 27] Act Big, Act Now

1. Reduce total oil and natural gas consumption by 50% over the next 25 years. Leadership builds the public will, community spirit and institutional capacity needed to implement the ambitious changes. Leadership is needed to build partnerships to address these issues at a regional and state-wide level.
2. Inform citizens about peak oil and foster community and community-based solutions.
3. Engage business, government and community leaders to initiate planning and policy change. Urban Design addresses the challenge at a community scale.
4. Support land use patterns that reduce transportation needs, promote walkability and provide easy access to services and transport options.
5. Design infrastructure to promote transportation options and facilitate efficient movement of freight and prevent infrastructure investments that would not be prudent given fuel shortages and higher prices. Expanded efficiency and conservation programmes shape the many choices made by individual households and businesses.
6. Encourage energy-efficient and renewable transportation choices.
7. Expand building energy-efficient programmes and incentives for all new and existing structures. Sustainable economic development fosters the growth of businesses that can supply energy-efficient solutions and provide employment and wealth creation in a new economic context.
8. Preserve farmland and expand local food production and processing.
9. Identify and promote sustainable business opportunities. Social and economic support systems will be needed to help (Devonians, Totnesians, etc.) dislocated by the effects of fuel price increases.
10. Redesign the safety net and protect vulnerable and marginalised populations. Emergency plans should be in place to respond to sudden price increases or supply interruptions.
11. Prepare emergency plans for sudden and severe shortages.

Empowering People - Community Matters
[Extracts from pp. 280–288]

Strategic Themes: Vibrant Local Communities; Sharing Tasks and food; Friendships and Family; Transition Lifestyle
Youth Issues/Resilience Indicators
- Rates of smoking, substance abuse and alcohol consumption by mothers during pregnancy
- Breastfeeding rates at 6–8 weeks after birth
- Size of the poverty gap
- The number of children and young people killed or seriously injured on the roads
- Number of families with children under 18 where a parent is home outside of school hours and during school holidays.

Authors' elaboration (Extract from Totnes Descent Plan, Transition Town Totnes, 2010).

References

Brangwyn, B., Hopkins, R., 2008. Transition Initiatives Primer—Becoming a Transition Town, City, District, Village, Community or Even Island. Transition Network, Totnes. https://community-wealth.org/content/transition-initiatives-primer-becoming-transition-town-city-district-village-community-or.

Connors, P., McDonald, P., 2010. Transitioning communities: community participation and the transition town movement. Community Development Journal 46 (4), 558–572. https://doi.org/10.1093/cdj/bsq014.

Hopkins, R., 2008. The Transition Handbook: From Oil Dependence to Local Resilience. Green Books, Totnes.

Hopkins, R.J., 2010. Localisation and Resilience at the Local Level: The Case of Transition Town Totnes (Ph.D Thesis School of Geography, Earth and Environmental Science, Faculty of Science and Technology, University of Plymouth).

Hopkins, R., 2011. The Transition Companion. Green Books, Totnes.

Transition Town Totnes, 2010. Transition in Action Totnes and District 2030 an Energy Descent Action Plan. Transition Town Totnes, Totnes.

Website

Italia che Cambia website: http://www.italiachecambia.org/2016/10/viaggio-transizione-totnes-high-street/.
REconomy Centre (Totnes): https://reconomycentre.org/home/lef/.
Transition Town Totnes: https://www.transitiontowntotnes.org/.

CHAPTER 23

Transition Towns Network in Italy: The Case of Monteveglio

Alessia Canzian

Independent Researcher, Italy

23.1 MONTEVEGLIO TRANSITION TOWN

Monteveglio was the first Italian village to be officially recognized by the International Transition Network (Transition Towns Network website).

A key aspect of the Transition movement is achieving practical and tangible projects. Therefore, in this chapter, we will describe a number of practical projects that have been implemented in Monteveglio that have being expanded to many other parts of Italy.

However, first it is important to show what made Monteveglio Transition Town more innovative than other realities.

The transition idea arrived in Italy in 2007 when Cristiano Bottone, as promoter of the movement in Italy, decided to create the first Transition Town Group Guide to Monteveglio.

In 2009, there were municipal elections and within the group a debated was opened on whether or not to work directly with the municipality. Some members of the group decided to run for the municipal elections. The group decided that those who would be candidates for the elections would have to leave the Transition Town Group Guide to avoid an overlapping of roles. It was only a formal break because in fact people continued to work together. The people who were nominated for the elections were elected and brought the tools learned during the transition trainings in the world of politics. For example, the entire electoral program was drawn up with meetings in Coffee World mode. This fusion between the political world and the world of transition gave birth to the famous Resolution of Monteveglio (Comune di Monteveglio, 2009), a municipal resolution that has inspired great interest internationally (Transition Culture website, 2009).

The Municipal Deliberation 54/2009 was intended to provide the procedures for the implementation of environmental policies for the implementation of an Energy Descent Action Plan (EDAP) in order to make Monteveglio a postcarbon city.

The Municipality, therefore, declared to share with the Transition Movement:

- Context: the depletion of energy resources and the limits of development;
- Method: bottom-up;
- Goals: make communities more resilient;
- Optimism: given all the problems of our society we have the opportunity to build a better society for the future.

Simultaneously with the work of the Municipality, the Transition Town Guide Group has continued to follow the Hopkins model. All the transition initiatives follow the same logic—giving the information in order to give the choice to people to begin a social change.

Each project that was born in the context of the transition is based on the systemic vision, which interconnects the different aspects of reality to each other. Therefore, each project is closely related to the others. Following the trend of the natural system, the projects were born, evolved, changed, and transformed themselves.

The project *Strade in Transizione* (Roads in Transition) is intended to help families to reduce domestic dependence on fossil fuels, reducing carbon monoxide emissions and energy costs. The project aims to provide practical tools for change—small things that anyone can implement in their own home—in order to reduce costs and harmful emissions. The change that the project proposes is at domestic level, based mainly on the learning of new daily habits. Citizens who decide to participate in the project are divided according to their neighborhood. Later, the groups are accompanied by a facilitating support, in different thematic meetings concerning issues of energy, water, food, waste, and mobility. This program was offered to the residents of Monteveglio through CURSA funding (University Consortium for Socioeconomic Research and the Environment) (Monteveglio città in transitione Blog (a)).

The project *Regalami un Albero* (Give Me a Tree) plans to offer to Monteveglio citizens the opportunity to give a new tree to their territory. The intent of the project is to give a tree in all social situations that involve the exchange of gifts. In addition, the gift not only includes the tree but also the laying, fertilization, and care of it. It has been estimated that if in Monteveglio each child received for their birthday at least one tree, every year 600 new trees would be planted. The project involves not only the association "Monteveglio City in Transition" but also the association *Streccapogn*, the City Council, many classes of the town primary school, and the parks and biodiversity management organization of Eastern Emilia.

The cost for a tree is 25 euros and all the money collected is reinvested in the same project. For each tree donated a certificate is issued by the Town Council that holds the public register of all trees donated in Monteveglio. One can access the registry of certificates via the project website (Monteveglio città in transitione Blog (b)).

It should be specified that those who receive a tree as a gift do not become the owner of it, but that she/he simply makes its existence possible, suggesting to the community the difference between contributing to commons creation and private property. The aim of this project is to create resilience within the territory, creating natural capital for future generations.

The project *Alimentazione Sostenibile* (Sustainable Food Supply) includes a deep awareness of the transformation of the eating habits of citizens. The project aims to inform the community about agriculture and food supply history. In our society, industrialized agriculture and intensive farming strongly impact the planet's resources. Moving toward a sustainable future involves a profound transformation of the food sector and of our society's dietary habits, which are provoking harmful and devastating externalities to human health and to the environment. Through a meeting, lasting about an hour, the project is trying to involve people in the community to be interested in this issue. The module is updated to make it more comprehensive and comprehensible through feedback and observations of those who are participating at the meetings. The logic is always that of the Transition—informing as clearly and accurately as possible and then letting everyone decide on how to act with respect to the information received. The final objective is to develop a handbook that can provide practical and applicable directions considering the daily decisions of food consumption in order to start a new sustainable food path. The handbook is modified and updated through the suggestions and indications of those participating in the meeting (Monteveglio città in transitione Blog (c)).

The project *Firma Energetica* (Energetic Signature) was sponsored by the Municipality of Monteveglio in collaboration with the association Monteveglio City in Transition. The intent of the project is to detect trends in energy consumption of the families who participated in it. Families who were in the project have reported weekly the data about their electricity and gas consumption for regular monitoring of their energy consumptions. The ultimate goal is to show how good practices can reduce energy waste and costs (Monteveglio città in transitione Blog (d)).

The project *Mercatino del Riuso* (Reuse Market) aims to encourage the reuse of objects, rather than their disposal as waste, reducing the environmental impact of dumps. The reuse of objects can also be a good economic saving method, as well as a way to implicitly counteract the consumer society (Monteveglio città in transitione Blog (e)).

The project *Piedibus* (Walking and Bus) involves the testing of emission-free mobility, through walking. It consists of a group of children who go to school by foot, accompanied by two adults, a "driver" in front and a "controller" who closes the group. This "alternative bus" leaves from a terminal and follows a path marked by bus stops, exactly like a bus. Also the bus stops and the time of leaving are organized like a normal bus. Each child wears a reflective vest and on the way children learn useful things on road safety. This project is already a reality in many parts of the world and is having a lot of success in Italy (Monteveglio città in transitione Blog (f)).

The project *Banca della Memoria* (Bank of Memories) is a project that involves collecting a video database of testimonials from people older than 60 years of age. The intent is to create a virtual place where the community elders donate their knowledge and their life experiences without modern technology. On the Monteveglio in Transition website it is possible to watch videos about some residents of the community talking about their past lives. The videos are also loaded on a YouTube channel (Monteveglio città in transitione Blog (g)). This project shows how the Transition Initiatives wish to reach all levels of the community. Through the Bank of Memories, older people of the community feel involved and therefore included in the Transition process, and conversely, young people can benefit from learning traditional skills.

The project *Gruppo d'Acquisto Fotovoltaico e Solare Termico* (Photovoltaic and Solar Thermal Fair Trade Group) is the result of a first meeting that took place in the spring of 2009 between about 30 people interested in producing renewable energy for their own household consumption. At the end of the meeting the participants suggested the idea of creating a buying group for solar thermal systems. Many, however, were also interested in photovoltaics, so the group analyzed both technologies. The Fair Trade Group has been working to select some ethical providers who also offered installation with excellent value for the money. At its birth, in 2010, 11 families joined the group; a supplier was identified who would provide both technologies. However, the decision of the supplier was not an easy task, since the estimates of the different providers often revealed themselves to be confused and uncertain. Some people who took part in the purchase group creation process revealed that this last aspect was the main reason for renunciation of the conversion towards the use renewable energy. The group dynamic, conversely, could be the solution to this problem, showing the collaborative logic involving the project. Eventually, suppliers who aligned themselves with the collaborative logic of the group and who shared the directives of GAS (Anagram of fair trade group in Italian) were chosen as official suppliers of the project (Monteveglio città in transitione Blog (h)).

From these practical examples of transition it is understood that the relocation of resources and the economy can take initiation from civil society. However, the process cannot run only based on the individual actions of citizens, but they need also the support of the institutions.

23.2 CRITICAL ANALYSIS OF THE INITIATIVE

For a real reorganization of both the way of living and the whole economy, it is necessary to involve all parts of the system. We have to change the entire legislative scheme. We have to change the internal relations within a nation and the international relations. Then, if one really wants to get the Transition, one will have to gradually involve himself/herself in all the parts of this new social organization.

"A municipality may not become a city in transition even if it wants," says Cristiano Bottone, "a city becomes concretely in transition only when all citizens decide to live permanent in transition" (Canzian and Bottone, 2014).

The example of Monteveglio was a very interesting experiment because for the first time a direct collaboration with the City Council was launched. In 2009, an agreement was signed between the administration of the municipality of Monteveglio and the transition movement in Monteveglio. This collaboration helped the movement to work with the council and plan together all the different projects that were initiated.

It should be remembered that what happened in Monteveglio was a very interesting experiment because it was the first time that an institutional relationship took place in the Transition Initiatives. In addition, this collaboration was born in 2009, a time when even talking about certain issues was quite far removed from the political context.

The case of Monteveglio was also used as an example at the global level, because it was the first time that an official political body was involved.

However, since January 1, 2014, the municipality of Monteveglio no longer exists, as it became the town of Valsamoggia. The city was established through the merger of the municipalities of Bazzano, Castello di Serravalle, Crespellano, Monteveglio, and Savigno.

Valsamoggia is the second largest municipality of the area and the fifth in the number of inhabitants, 30,606, and is part of the Metropolitan City of Bologna.

Currently the Transition Town Guide Group is thinking about how to bring the experience of Monteveglio from a small to large scale. The intent is to be able to repeat the same practices and proposals in a micro context of a macro context without, however, losing quality.

References

Canzian, A., Bottone, C., March 2014. Interview with Cristiano Bottone, Conducted by the Author (Turin).
Comune di Monteveglio, 2009. Deliberazione della Giunta Comunale N. 92 del 26/11/2009. Available at: https://transitionitalia.files.wordpress.com/2009/12/delg0092-09.pdf.
Monteveglio città in transitione Blog (a), Strade in transizione project available (in Italian) at: https://montegliotransizione.wordpress.com/strade-in-transizione/.

Monteveglio città in transitione Blog (b), regala un albero project available (in Italian) at: https://montevegliotransizione.wordpress.com/2012/04/05/regala-un-albero/.

Monteveglio città in transitione Blog (c), Alimentazione sostenibile project available (in Italian) at: http://montevegliotransizione.wordpress.com/progetto-alimentazione-sostenibile/.

Monteveglio città in transitione Blog (d), Firma Energetica project available (in Italian) at: http://montevegliotransizione.wordpress.com/progetto-firma-energetica/.

Monteveglio città in transitione Blog (e), Il Mercatino del riuso project available (in Italian) at: http://montevegliotransizione.wordpress.com/il-mercatino-del-riuso/.

Monteveglio città in transitione Blog (f), Pedibus project available (in Italian) at project http://montevegliotransizione.wordpress.com/scuola-sperimentiamo-il-piedibus/.

Monteveglio città in transitione Blog (g), Banca della Memoria project available (in Italian) at: http://montevegliotransizione.wordpress.com/la-banca-della-memoria/.

Monteveglio città in transitione Blog (h), Gruppo d'Acquisto Fotovoltaico e Solare Termico project available (in Italian) at: http://montevegliotransizione.wordpress.com/gruppo-energia/.

Transition Culture website, 2009. What it Looks like when a Local Authority REALLY Gets Transition... the Monteveglio Story.... Available at: https://www.transitionculture.org/2009/12/04/what-it-looks-like-when-a-local-authority-really-gets-transition-the-monteveglio-story/.

Transition Towns Network website: https://transitionnetwork.org/initiatives/map.

CHAPTER 24

Model for Integrated Urban Disaster Risk Management at the Local Level: Bottom-Up Initiatives of Academics

Marija Maruna, Ratka Čolić
University of Belgrade, Belgrade, Serbia

24.1 INTRODUCTION

In May 2014, catastrophic floods struck Serbia and the broader region. This disaster caused the most damage to Obrenovac, a town within the metropolitan area of Belgrade, the Serbian capital. The floodwaters completely inundated Obrenovac after embankments gave way along two of the three rivers surrounding the settlement (the Sava, Kolubara, and Tamnava). The already dramatic situation was aggravated by the fact that Serbia's largest coal-fired power plant is located in Obrenovac, on the very bank of the Sava; this power station is the largest single producer of electricity feeding into the Serbian national power grid. Although the power plant did not suffer significant damage, the urban area did. The floods killed 17 people and forced 25,000 local inhabitants to temporarily evacuate their homes. The damage ran into the billions of euros.

24.1.1 Background Characteristics of the Municipality of Obrenovac

Obrenovac is situated 29 km from Belgrade and is part of its administrative unity of the city. It has a moderate continental climate with an average annual temperature of 11°C and average precipitation of 640 L per square meter. Obrenovac began to develop as an electric power center for the former Yugoslavia in the second half of the 20th century, which caused a population influx and rapid urbanization. Today the town is the center of a mining, energy, and industrial zone of the utmost national importance, and the largest producer of electricity in southeastern Europe, with a population of over 70,000. However, the power plants burn 25 million tons each year and produce dozens of millions of tons of ash and thousands of tons of smoke particles. The resulting pollution has had a major adverse effect on the town's environment.

24.1.2 State of Resilience in the Municipality of Obrenovac

The May 2014 flooding in Serbia revealed the vulnerability of the urban disaster risk management (UDRM) system at all levels of governance. Although the floods were primarily the result of extreme precipitation, with 100-year recurrence for many river basins and even 1000-year recurrence for some, the damage was the result of interaction between natural and man-made factors. The floods revealed the inefficiency of the UDRM system, in particular in terms of basic infrastructure (incomplete flood defense systems along riverbanks, poor maintenance of regulated riverbeds and defensive embankments, lack of investment into anti-erosion works and afforestation). Institutions also failed; there was a lack of communication between authorities, early warning systems did not respond, spatial and urban planning was revealed as inadequate, illicit construction and landfills were tolerated close to urban watercourses, legislation was not mutually harmonized, and adequate instruments and human capacity were lacking to implement existing regulations (Čolić et al., 2015). Although major efforts have been made in recent decades to enhance UDRM in Serbia, together with the development of regional and national climate change protection policies, legislation remains fragmented and poorly aligned, whilst adequate instruments and human capacity to implement regulations are still lacking, in particular at the local level. It is now apparent that an integrated approach must be applied to solve this problem.

24.2 INITIATIVE BY THE INTEGRATED URBANISM MASTER'S PROGRAM TO BUILD UDRM CAPACITY AT THE LOCAL AUTHORITY

24.2.1 Role of the Integrated Urbanism Program in Building Professional Capacity

The Department of Urbanism (DU) at the Belgrade Faculty of Architecture has launched a new study program, entitled Integrated Urbanism, the first specialized master's degree course for urban planners in Serbia. The aim of this course is to enhance capacities of the planning profession in a postsocialist transition environment created by the shift to the market economy and democratic changes that took place in Serbia in 2000. The newly established socioeconomic framework, which led to a change in the concept of spatial intervention and required alterations to professional approaches of spatial and urban planning, revealed the inability of current practice to offer adequate answers (Vujošević, 2004; Lazarevic Bajec, 2009; Milovanović Rodić, 2015). The need arose to incorporate new knowledge and develop new practices that would be more efficient and effective at directing the spatial development of a society in transition.

Experiences of courses offered in the European context (Geppert and Cotella, 2010; Frank et al., 2014; Mironowicz, 2015) and recommendations for educating the planning profession (Kunzmann, 2004) were used in creating a 2-year master's program at the Faculty of Architecture, with additional critical adjustment to the peculiarities of the local environment (Nedović Budić and Cavrić, 2006; Lazarević Bajec, 2012). The DU felt that the development of new urban planning professionals constituted a change to the current planning paradigm and, consequently, required the inclusion of a variety of actors from practice that could readily accept this change in their day-to-day work.

The Department devoted particular attention to the role of academia and its responsibility for enhancing the profession, as the planning system is evidently facing a crisis. Academics have an obligation to act where proven routines are missing or unusable (Scholl, 2012). The social position of academia allows reassessment of modern theoretical concepts, value ideas, and principles, which consequently have an impact on changes to how the role and position of the urban planner are understood in the community (Milovanovic Rodić, 2015). In addition, in a transition society, characterized by poorly ordered institutions, vague procedures, underdeveloped instruments, and inadequate solutions on offer, academia constitutes a relatively neutral space in which to reexamine current practices and test new models as starting points for developing and enhancing the planning system.

The Department therefore devoted particular attention to developing ways to introduce new knowledge into urban planning in Serbia. A key principle here involved the establishment of closer cooperation between academia and practice in the education process to jointly solve complicated problems. Teaching had to be organized so as to ensure that the curriculum could gain legitimacy within the profession, as well as to build trust of the broader professional public in the importance of educating urban planners. One of the main objectives of this study program is networking with relevant institutions, organizations, and individuals in Serbia with the aim of exchanging experiences and knowledge creating a platform for collaborative learning (Maruna, 2015).

The production of students' final master works proved to be one of the more significant areas where these principles could be tested. Over the course of the entire semester, students, their mentors, and members of mentoring commissions had the opportunity to collaborate on final works, which are made up of two distinct units, a master's thesis and a master's design. These comfortable deadlines allow various stakeholders to participate creatively in the process at different stages. Students' assignments constitute a framework in which professional dialogue is established that focuses on the ability to apply topical issues from global planning discourse to solve local problems. To enhance the quality of discussion, professionals in various fields are involved: academics (as mentors and members of mentoring commissions), experts in the local context (those acquainted with specific problems that students' final works deal with), professionals (experts from the most-renowned professional institutions and organizations active in Serbia), and experts in the international context (as partnering academic institutions). Topics reflect the most recent European practice and are based on key documents (Maruna and Čolić, 2015).

The primary objective of the process of producing students' final master's theses and designs, set up as outlined herein, far exceeds its purpose as a tool for training young professionals. In essence, its intention is to enhance the capacity of the professional public as a whole through a carefully structured process of collaborative learning. Three main objectives achieved through this type of collaboration can be highlighted:

- Contemporary topics are introduced into the local professional discourse: Adoption of topical issues from the international context focuses discussion in local professional circles onto shared acquisition of new knowledge and establishment of a common professional terminology;
- Innovative methods of professional action are adopted: In parallel with the introduction of new topics, the profession takes up professional experiences developed elsewhere in the form of positive practices, as methodological guidelines of sorts that help create solutions;
- New instruments of professional action are introduced: These stem from the concept of urban governance and entail an integrated approach to planning that bases spatial development on actual resources and envisages the inclusion of numerous stakeholders.

This approach to learning and coming to understand new topics and professional approaches creates room for better comprehension of topical issues and problems that occur in day-to-day planning practice, and also allows joint professional action aimed at formulating appropriate responses and creating innovative solutions.

24.2.2 Overview of the Academic Initiative

"Urban disaster risk management at the local level" was the topic of the assignment given to students of the Integrated Urbanism course for their final master's works in academic year 2014–15. The suggestion to tackle urban disaster risk management, using the example of Obrenovac, was raised in the course of formal collaboration between the Faculty of Architecture and the Strengthening of Local Land Management in Serbia Project, implemented by Deutsche Gesellschaft für Internationale Zusammenarbeit between 2010 and 2015 (Müller et al., 2015). This international initiative assisted the establishment of official cooperation with the municipality of Obrenovac, whilst the master's degree course in Urban Management at the Technical University of Berlin was introduced as a partner program. The compatibility of these two study programs allowed parallel teaching over the course of the summer semester, with both sets of students sharing a single assignment using the example provided by the topical Serbian issue. The work took place in three stages: (1) parallel work by students of both programs in small groups designed to allow them to understand the topic, with communication via social networks; (2) joint workshop for students of both programs working in mixed groups in Belgrade, with direct communication with the local community of Obrenovac; and (3) joint development of recommendations, again facilitated by electronic means of communication. The final objective of the initiative was to establish recommendations for UDRM in the municipality of Obrenovac. Each stage consisted of multiple steps (Fig. 24.1).

The topics were based on *How to Make Cities More Resilient*, a report issued by the UN Office for Disaster Risk Reduction (UNISDR, 2012), and the subtopics raised in this paper: (1) Institutional and Administrative Framework, (2) Financing and Resources, (3) Multihazard Risk Assessment, (4) Infrastructure Protection, Upgrading, and Resilience, (5) Protect Vital Facilities: Education and Health, (6) Building Regulations and Land Use Planning, (7) Training, Education, and Public Awareness, (8) Environmental Protection and Strengthening of Ecosystems, (9) Effective Preparedness, Early Warning, and Response, and (10) Recovery and Rebuilding Communities. A group of mentors selected this document as it gives local authorities proven guidelines for developing resilient solutions based on enhancing local communities' abilities to resist, absorb, accommodate to, and recover from the effects of a natural hazard in a timely and efficient manner.

In the first stage of the effort, the students were divided into small groups and asked to develop individual aspects of the primary topic. The groups communicated via social networks [1]. This stage was designed to allow students to become acquainted with UDRM and the state of play in Obrenovac in this regard. Two workshops at the Faculty of Architecture in Belgrade allowed students to learn more about these issues: one included officials of the Obrenovac Municipality [2], and the other involved experts from reputable professional institutions, the line ministry (Ministry of Construction, Transportation, and Infrastructure), National Professional Licensing Body (Serbian Chamber of Engineers), National Planning Research Institute (Institute of Architecture and Urban and Spatial

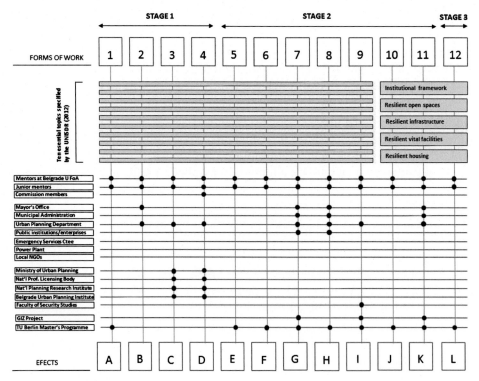

FIGURE 24.1 Flowchart of the bottom-up initiative. *Source: M. Maruna.*

Planning of Serbia), and the Belgrade Urban Planning Institute [3]. At the end of this first stage, the knowledge acquired was summarized and tested with the involvement of all stakeholders [4].

The second stage comprised an intensive 10-day workshop designed to develop basic UDRM recommendations for Obrenovac. Students from both Berlin and Belgrade attended the workshop, with the venue alternating between Belgrade and Obrenovac, depending on progress with construction of the shared body of knowledge. The task here was to adapt the recommendations of the UNISDR paper to the mode of governance in the municipality of Obrenovac. The first step involved an exchange of knowledge between students of the two faculties in Belgrade [5], which allowed the participants to identify problems and opportunities that made it possible to formulate questions to be posed to the local authority [6] (Fig. 24.2).

FIGURE 24.2 Exchange of knowledge between students of the two faculties in Belgrade — Step [5]. *Source: R. Čolić.*

FIGURE 24.3 Discussions with experts from Obrenovac – Step [7]. *Source: M. Maruna.*

The program cooperated closely with a number of relevant stakeholders in Obrenovac, such as the mayor's office, Urban Planning Department, municipal administration, public institutions/enterprises, Obrenovac Power Plant, Committee for Emergency Services, and local NGOs. This collaboration took place through visits and presentations [7], discussions, and interviews [8] (Fig. 24.3–24.5).

The next step was an SWOT analysis in which knowledge acquired in contacts with local stakeholders was summarized, with careful coordination with officers of the Obrenovac Urban Planning Department and risk assessment and crisis management experts from the Faculty of Security Studies [9]. In this step the participants developed a model for integrated urban disaster risk management (MIUDRM) for Obrenovac based on five key priority topics: Institutional framework, Resilient open spaces, Resilient infrastructure, Resilient vital facilities, and Resilient housing. Key topics were selected with reference to the sectoral division of governance at the local level. In this step, the students regrouped into new clusters to fully integrate the newly acquired knowledge and adjust it to the mode of operation of the Obrenovac local authority [10]. At the end of the workshop, the outputs were again cross-checked with the local authority at a presentation in Obrenovac [11] (Fig. 24.6 and 24.7).

In the third stage, the students continued working together in their individual topical groups, using electronic means of communication, to refine the tested concepts [12]. They additionally elaborated and transformed the MIUDRM for Obrenovac into a document that was finally publicly released for broader use (Maruna et al., 2015; Fokdal and Zehner, 2015).

FIGURE 24.4 Visit to Obrenovac institutions – Step [7]. *Source: M. Maruna.*

IV. CITIES DEALING WITH CLIMATE CHANGE: TRANSITION INITIATIVES

FIGURE 24.5   Interviews with experts from Obrenovac — Step [8].   *Source: R. Čolić.*

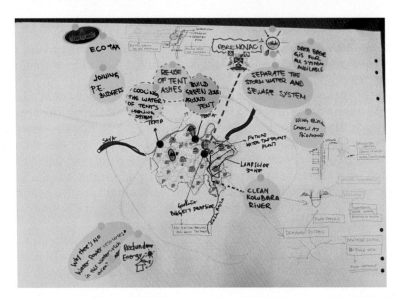

FIGURE 24.6   Development of the MIUDRM concept through priority topics — Step [10].   *Source: R. Čolić.*

FIGURE 24.7   Presentation of the MIUDRM concept in Obrenovac — Step [11].   *Source: M. Maruna.*

24.2.3 Achievements

Each stage of the collaborative learning process outlined herein resulted in a set of achievements that led, through the various steps involved, to the formulation of an MIUDRM adapted to the needs of the Obrenovac local authority. The entire process was an iterative effort consisting of three stages: introduction of a new topic into the professional context, testing of the topic with experts in a practical context, and adaptation of knowledge to the local context:

- Stage 1: Introduction of a new topic into the professional context — understanding the topic [A], becoming acquainted with the state of UDRM at the local level [B], discussing the topic with the broader professional community [C], and summarizing knowledge of the topic and establishing its significance for the local context [D];
- Stage 2: Testing the topic with experts in a practical context — establishing shared knowledge of students of both master's programs [E]; identifying problems, potentials, and key issues [F]; sharing knowledge with local authority stakeholders [G]; exchanging knowledge and working on the topic [H]; adapting knowledge [I]; developing the UDRM model [J]; and testing the UDRM model with local authority [K];
- Stage 3: Adapting knowledge to the local context — finalizing the proposed model [L].

The production of master's works as outlined here proved beneficial for all participants on a number of levels. First and foremost, the students were able to work on actual projects in cooperation with some of Serbia's most competent experts and so gain skills directly applicable to their future practice. Local authority professionals could collaborate with academics and colleagues from the country's most renowned institutions to develop innovative solutions for real problems they face in their day-to-day work. Finally, all of the stakeholders enhanced their professional capacities in mutual discussions focusing on the new knowledge and opportunities for its implementation in the Serbian context (Maruna et al., 2015). Academics were able to test contemporary theoretical approaches in a real context and so introduce innovations into their practice.

The final product of this effort was new knowledge, adopted from the contemporary international context and adapted to the needs and abilities of local practice. The initiative here came from academia, which used educational experiences and various learning and knowledge transfer methodologies to develop an innovative approach to solving a concrete problem in the selected local community. The resulting MIUDRM was directly adjusted to the governance structure of the municipality of Obrenovac and constitutes an innovative operating manual for the local authority.

24.3 CRITICAL ANALYSIS OF THE INITIATIVE

The development of the MIUDRM for the local authority was prompted by the problems and shortcomings of the current urban governance system, which came to the fore after the disastrous flooding in spring 2014. The MIUDRM was based on *How to Make Cities More Resilient*, a report released by the UN Office for Disaster Risk Reduction (UNISDR, 2012), which provided precise definitions of 10 key aspects that can serve as foundations for developing action plans. This ensured coverage of all aspects of the resilient cities concept.

The UNISDR guide is a document that details the latest approaches to climate change mitigation and adaptation adjusted in particular to the needs of local authorities. This allowed the study program to fully incorporate the new knowledge into UDRM practice at the level of the Obrenovac local authority and apply the most modern approaches to addressing climate change.

Collaboration in producing students' final master's works contributed to strengthening the network of stakeholders involved in developing the MIUDRM suitable for the Obrenovac local authority. This effort established a knowledge production network made up of multiple actors from different disciplines and sectors: students, mentors from both faculties, local policy makers, experts from local institutions and public companies, and representatives of the civil sector and the local community. On the one hand, mentors presented theoretical knowledge in the field of a new professional paradigm. On the other hand, this knowledge was accepted and tested through practical experience of renowned members of the consulting team, and in addition, experts from local administrations elaborated the knowledge and confirmed it in the local context (Maruna et al., 2015).

The initiative aimed at understanding the concept of UDRM and establishing guidelines for changing the professional approach to urban development from the perspective of resilience. The program created a conceptual model as the basis on which to reassess actions and introduce new solutions into urban governance in Obrenovac. The collaborative production of students' final works enabled the adjustment of key recommendations from the

UNISDR guide to the peculiarities of the local authority, existing sectoral organization, and traditional operating practices. In line with recommendations made in the UNISDR guide, urban planning is viewed as a constituent part of UDRM.

The issues faced by the existing urban governance system imposed on academia the leading role in transitioning toward a more collaborative planning process and governance. This example illustrates one of the actions academia has undertaken to accomplish this task. The disaster management planning initiative outlined in this chapter is just one example of the possible path of development of the entire urban governance system and its transition toward a new planning paradigm, but also provides insight into how a contemporary innovative solution can be used by local authorities in their efforts to address climate change.

References

Čolić, R., Lalović, K., Maruna, M., Milovanović-Rodić, D., 2015. Disaster Risk Management in Serbia and Flooding in Obrenovac Municipality. In: Fokdal, J., Zehner, C. (Eds.), Resilient Cities: Urban Disaster Risk Management in Serbia (Report on the Results of a Case Study Research Project (2015)). Berlin University of Technology, Berlin.

Fokdal, J., Zehner, C., 2015. Resilient Cities: Urban Disaster Risk Management in Serbia (Report on the Results of a Case Study Research Project (2015)). Berlin University of Technology, Berlin.

Frank, A., Mironowicz, I., Lourenco, J., Franchini, T., Ache, P., Finka, M., Scholl, B., Grams, A., 2014. Educating planners in Europe: a review of 21st century study programmes. Progress in Planning 91, 30–94. https://doi.org/10.1016/j.progress.2013.05.001.

Geppert, A., Cotella, G. (Eds.), 2010. Quality Issues in a Changing European Higher Education Area. AESOP Planning Education, No. 2. Leuven: AESOP.

Kunzmann, K.R., 2004. Unconditional surrender: the gradual demise of European diversity in planning. In: Keynote Address to the 18th Congress of the Association of European Schools of Planning (AESOP), Grenoble, France, 3 July.

Lazarevic Bajec, N., 2009. Rational or collaborative model of urban planning in Serbia: institutional limitation. Serbian Architectural Journal 1 (2), 81–106.

Lazarević Bajec, N., 2012. Design approach to education: learning across disciplines. STRAND: Role of Universities and Their Contribution to Sustainable Development. Available at: http://www.strand.rs/university/index.html.

Maruna, M., 2015. Proces izrade master rada kao poligon za dijalog i razvoj profesionalnih kapaciteta ['The master's thesis writing process as a forum for dialogue and professional capacity-building']. In: Maruna, M., Čolić, R. (Eds.), Inovativni metodološki pristup izradi master rada: doprinos edukaciji profila urbaniste [Innovative Methodological Approach to Master's Thesis Writing: A Contribution to Planner Education]. Arhitektonski fakultet., Beograd.

Maruna, M., Čolić, R., 2015. Inovativni metodološki pristup izradi master rada: doprinos edukaciji profila urbaniste [Innovative Methodological Approach to Master's Thesis Writing: A Contribution to Planner Education]. Arhitektonski fakultet., Beograd.

Maruna, M., Čolić, R., Fokdal, J., Zehner, C., Milovanović Rodić, D., Lalović, K., 2015. Collaborative and Practice Oriented Learning of Disaster Risk Management in Post Socialist Transition Countries. XVI N-AERUS Conference: Who Wins and Who Loses? Exploring and Learning from Transformations and Actors in the Cities of the South, Dortmund.

Milovanović Rodić, D., 2015. Edukacija za rehabilitaciju pozicije i uloge urbanista u procesima upravljanja razvojem grada ['Education to rehabilitate the position and role of urban planners in urban development processes]. In: Maruna, M., Čolić, R. (Eds.), Inovativni metodološki pristup izradi master rada: doprinos edukaciji profila urbaniste [Innovative Methodological Approach to Master's Thesis Writing: A Contribution to Planner Education]. Arhitektonski fakultet, Beograd.

Mironowicz, I. (Ed.), 2015. Excellence in Planning Education: Local, European and Global Perspective. AESOP Planning Education, No. 3.

Müller, H., Werman, B., Čolić, R., Fürst, A., Begović, B., Wirtz, J.C., Božić, B., Ferenčak, M., Zeković, S., 2015. Strengthening of local land management in Serbia, results of 6 years of German-Serbian cooperation, module 1: urban land management. AMBERO Consulting Representative Office in Belgrade, Deutsche Gesellschaft für Internationale Zusammenarbeit (GIZ) GmbH, GIZ Office in Belgrade, Belgrade.

Nedović Budić, Z., Cavrić, B., 2006. Waves of planning: a framework for studying the evolution of planning systems and empirical insights from Serbia and Montenegro. Planning Perspectives 21, 393–425. https://doi.org/10.1080/02665430600892146.

Scholl, B. (Ed.), 2012. HESP: Higher Education in Spatial Planning. Zürich: Vdf Hochschul-Verlag an der ETH Zürich.

UNISDR, 2012. How to Make Cities More Resilient: A Handbook for Local Government Leaders. UNISDR, Geneva. Available at: http://www.unisdr.org/we/inform/publications/26462.

Vujošević, M., 2004. The search for a new development planning/policy model: problems of expertise in the transition period. Spatium 10, 12–18. https://doi.org/10.2298/SPAT0410012V.

CHAPTER 25

Enhancing Community Resilience in Barcelona: Addressing Climate Change and Social Justice Through Spaces of Comanagement

Luca Sára Bródy[1], Lorenzo Chelleri[4], Francesc Baró[2], Isabel Ruiz-Mallen[3]

[1]Gran Sasso Science Institute (GSSI), L'Aquila, Italy; [2]Universitat Autònoma de Barcelona (UAB), Cerdanyola del Vallès, Spain; [3]Universitat Oberta de Catalunya (UOC), Barcelona, Spain; [4]Universitat Internacional de Catalunya (UIC), Barcelona, Spain

25.1 ADAPTING AND MITIGATING CLIMATE CHANGE FROM BELOW

Climate change has been usually recognized as a major issue to be addressed through top-down plans and strategies contributing to lower the exposures and vulnerability to different stresses, while planning through policies aiming step-by-step to decarbonize the economy and urban metabolism. In the light of these challenges, resilience entered the policy discourses supporting with its normative positivity any action or plan contributing to fight climate hazards (Rodin, 2014; TESS, 2016). A critical mass of multidisciplinary literature tackled the issue of how to operationalize resilience beyond its metaphorical meanings supporting adaptiveness (Pelling, 2011; Davidson et al., 2016), or tried to measure it for better assessing city resilience programs (Rose, 2004; Somers, 2009). Most of these concept applications, however, referred to disasters or climate risk reduction or mitigation (Serre et al., 2012). However, even if on the one hand the success of resilience recently has taken over the concept of sustainability (GSP, 2012), it has been receiving numerous criticisms from different scholars (Davoudi et al., 2012; MacKinnon and Derickson, 2013), mainly because of the trade-offs its operationalization imply (Chelleri et al., 2015; Meerow and Newell, 2016). Resilience of what, to what (Carpenter et al., 2001), where, and for whom (Vale, 2013; Meerow and Newell, 2016) are the recently posed questions when this label appears on city plans or projects. Indeed, from this critical perspective, the notion of community resilience (Berkes and Ross, 2012) is recently wishing to theoretically bridge both conceptual and practical misalignments between top-down and bottom-up initiatives. Even if community perceptions and contributions about risk and risk management are key elements within the specific literature on disasters (Cutter et al., 2008), unfortunately, in the urban agendas dealing with resilience in cities, communities usually play but a mere background role, in most of the cases, being only the audience and beneficiaries of the plans and institutional policies. As Mulligan et al. recently argued, by adding the word *community* to the already biased term *resilience*, the new concept of community resilience has been provided a technocratic discourse with strong public appeal and adopted by different nations in their national strategies for tackling climate change and hazards (Mulligan et al., 2016). The current gap and challenge, therefore, would be to recognize and account for community resilience taking into account values, beliefs, and the human agency nested with perceptions and the diversity of actions that people can express through grouping, associations, cooperatives, individuals, or enterprises acting within the city development and urban metabolism. There is a vast literature exploring urban informality, which in the last decades is no longer just a global south issue (Roy, 2005), which is challenging in global north cities the top-down management of spaces and uses (Certomà and Notteboom, 2017) because of the emergence of informal activities, functions, and self-managed services driven from groups of citizens (Cattaneo and Gavaldà, 2010; Camps-Calvet et al., 2016; Langemeyer et al., 2016). This chapter explores the emerging complexity of community-led initiatives in Barcelona, a European city well known for its ability for city branding after the 1992 Olympic Games (Balibrea, 2001; Degen and García, 2012). The potential tensions between community-led initiatives and city top-down plans (Eizaguirre et al., 2012) have been recently smoothed by a tentative framing of the comanagement of some public (vacant) spaces (Parés et al., 2014). This opened the door for social and governance innovations (González and Healey, 2005) through something that herewith we could define as community resilience, since it is leveraging a transition toward sustainability, enabling adaptive frameworks for action and city management.

25.1.1 Exploring Community-Led Initiatives in Barcelona: From Energy to Housing and Mobility

Barcelona embraces a rich variety of grassroots initiatives explicitly or implicitly related to climate change actions. These are mostly focused on the domains of energy, transport, and food, and going beyond the limits of the city, reaching even the national scale. The most known and successful experience is the case of *Som Energia* (we are energy), the first Spanish renewable energy consumers' cooperative, founded in 2010 in Girona (one of the four Catalonia provinces), and counting almost 30,000 members throughout the whole country. Som Energia offers its members the possibility of consuming electricity from energy sources that are 100% renewable, and at the same time developing its own renewable energy through different projects thanks to members' voluntary investments. The structure of this self-managed organization includes a technical and administration office employing 26 people, a governing council of 9 people, and more than 60 local groups composed of hundreds of volunteers. Participation and decision-making processes are organized through a general assembly that meets at least once a year, a summer school held every September, an online forum platform (called *Plataforma*) and periodic meetings of the local groups.

Also related to energy, but clearly particular to Barcelona, there is the comanaged Participatory Energy Plan (PEP) initiative of three city districts: Sant Martí de Provençals, La Verneda, and La Pau. Again in 2010, various community and public organizations (such as Eco-Union, Barcelona en Transició, IGOP Autonomous University of Barcelona, and Energy Agency of Barcelona) launched this initiative operating in the three mentioned working-class neighborhoods of Barcelona. PEP's main goals are (1) to promote the development of an urban and social environment embedding human values, (2) to move toward a healthier, participatory, and responsible neighborhood, (3) to disseminate a transversal and global knowledge on energy, (4) to creatively transform the neighborhood, and (5) to achieve inclusive and diverse citizen involvement. In order to fulfil these goals, the three PEP experiences promote and facilitate working groups that are: (1) OLEPEP (PEP's local energy observatory), (2) Pati La Pau (the "Peace Courtyard", hosting educational activities in the schools' playgrounds), (3) local economy (composed by the community network La Verneda Sant Martí), (4) community garden projects, and finally (5) PEP Dalera (activities related to mobility and urban cycling). PEP also organizes talks, conferences, and round tables to disseminate a holistic view on sustainable energy, as well as energy-related recreational/social activities, and community-empowerment activities. It is relevant to note that all the activities do not exclude a priori the involvement of institutions. PEP is a self-managed experience, looking forward to fostering a transition to a more just and sustainable society, so the linkages and networking with other neighborhood organizations and the City Council are also part of the plans.

From energy to housing, Barcelona has a dense network of experienced local professional cooperatives promoting the use of natural or recycled materials for housing. Among those, LaCol (www.lacol.coop) is probably the best known. LaCol ("col" meaning cabbage, recalling the fractal theory behind nature's organism and systems) is a cooperative of architects founded in 2009, working in the neighborhood of Sants, promoting low-energy constructions. Other initiatives regard mobility. Other relevant, self-managed community-led initiatives are related to mobility, including the Association for the Promotion of Public Transport (PTP Barcelona, www.transportpublic.org) and Biciclot (www.biciclot.coop) and Biciosxs (http://biciosxs.noubarris.org), which are communal bike garages located in different neighborhoods, offering free bike repair training and tools, and a barter of bike parts. The wave of sustainable mobility experiences in Barcelona is also linked to the huge investments, and strategy, of the municipality of setting up in the last decade a network of bike lanes and public bike parking (which, in 2017, counted 6,000 public bikes available for renting at 420 automatic parking sites, where 47,000 users can pick up or leave the bikes, or ride them using the 200 km of bike lanes built from 2007). However, the first experiences and associations promoting sustainable mobility are known as Trèvol, a historical cooperative founded in 1984, which focused on courier services. It was the first organization in Spain using bikes as courier transport, and nowadays also using electric vehicles for medium-distance services.

Other community-led initiatives deal also with food, through numerous consumer groups (or solidarity purchase groups). We counted at least 50 consumer groups in the city, usually composed of about 20–25 families at maximum. These self-organized groups generally buy seasonal food products from organic-certified local producers, contributing to the circular and 0 km economy.

25.1.2 Integrated Approaches from Cocreation to Squatting

Beyond groups tackling the diverse and specific issues of food, mobility, housing, or energy, there are several remarkable initiatives dealing with integrated and transversal approaches to climate change action through social transformation. An example is Smart Citizen Barcelona, a participatory educational platform cocreated by Barcelona citizens in 2012 through the digital fabrication center FabLab Barcelona (https://fablabbcn.org). Another is the center for art research and production, Hangar (https://hangar.org), aiming to produce, share, learn, and reflect on climate and pollution data in the city. Its final goal is the collective construction of the city for its own inhabitants. Interested citizens are invited to workshops and other educational activities on how to produce and use sensors and software for environmental monitoring. This is an international comanaged initiative that also exists in Amsterdam and Manchester. Digital technologies are also used in other bottom-up initiatives with environmental education and awareness-raising purposes at supra-local level, such as the pedagogical movement for renewing education, Rosa Sensat (http://www2.rosasensat.org/), through the online periodic publication Perspective Ambiental.

The other side of the coin is represented by several squatted initiatives, which are specific to the city of Barcelona and that also deserve attention because of their activities. The squatted farmhouse of Can Masdeu (www.canmasdeu.net) is a self-managed, community-based initiative that promotes and organizes social and educational activities on organic agriculture, renewable energy, bioconstruction, bike use/repair, and recycling, among others. These activities are open to the neighbors and other interested people, and also supported by a dense international network of universities and research centers thanks to its collaboration with the Autonomous University of Barcelona (which hosts one of the leading global academic groups from ecological economics, on degrowth). The Can Masdeu community garden project consists of approximately 50 small plots, and decision-making relies on a monthly assembly attended by gardeners living in the same farmhouse or in the neighborhood. Squatted urban gardens now expand beyond Can Masdeu to other areas and neighborhoods in Barcelona. There are at least 13 community-based squatted gardens in the city, many of them managed by informal institutions, in which management is driven by organic horticultural practices and often by biodynamic agriculture techniques. Decision-making on management usually relies on gardeners' assemblies. These self-managed community-led experiences of growing food following social, economic, and ecological sustainable practices in the city coexist with comanagement approaches promoted by the city council that seek to replicate them, as is shown in the next section.

25.2 FROM COMMUNITY-LED TO COMANAGEMENT: THE PLA BUITS EXPERIENCE

In light of the aforementiond active citizen groups and initiatives in Barcelona, the city council took the opportunity of the economic crisis after 2009 to frame an experimental program wishing to explore the pros and cons of comanagement, to be tested in some of the numerous vacant public property plots. The program, called Pla BUITS (Vacant Urban Spaces with Territorial and Social Involvement), offered temporary social and community use opportunities, inviting groups of public and private nonprofit organizations to propose activities fostering the public use value of those vacant plots. The program was at the same time aiming to avoid unwanted uses and social exclusion, offering an opportunity to invite different stakeholders in the regeneration and revitalization of marginalized spaces of the city center (Ajuntament Barcelona, 2016).

A first call for proposals was launched in October 2012, in order to award 12 vacant sites throughout 3 years (see Fig. 25.1) to the groups that scored the best in municipality evaluations based on proposals' self-sufficiency, feasibility, environmental sustainability, social impact, and the level of creativity and neighborhood involvement in the activities.

Almost 5 years after this experience, we can draw a preliminary assessment of the activities and outcomes of this social and governance innovation. During the 3 years (starting with the proposals in 2013 to be operative on the ground, and 2016) the main topics addressed by the initiatives' activities were related to gardening, sustainability, addressing vulnerable groups, and providing educational activities and workshops (see Table 25.1). In line with the extreme popularity gained in the last decades in Europe and North America, community gardening has been the most common activity across the projects. Indeed, gardens are relatively easy to set up, don't require large investments while offering a popular solution for vacant space management, and function along the way as a catalyst for activities that require more investment.

Four of the 12 projects address the issue of sustainability in the city. Two of them focus on easing traffic by either promoting biking or aiming to create a street free of car traffic. The other two projects are in search of sustainable alternatives in architecture and construction, also providing educational workshops.

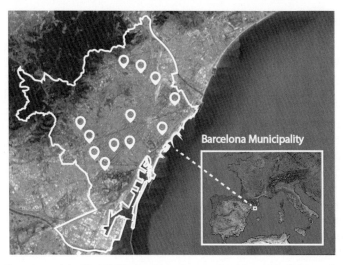

FIGURE 25.1 Map of the first 12 projects of the Pla BUITS program. *Source: Google map edited from authors.*

Vulnerable groups are addressed by various approaches, in four sites out of the 12. One project operates a social kitchen for people in need, the second provides job opportunities for the unemployed, the third offers recreational activities for people struggling with mental illness, and the last one aims to close up the generation gap by nurturing activities that connect the younger and older generations.

During the 3 years of development of the projects, two key points of reference have emerged in the collaboration between citizens and the municipality. One fundamental actor proved to be the initiator of the Pla BUITS projects in the Participation Unit of the Ecology, Urban Planning, and Mobility Department. One of the greatest strengths of the coordinator has been to sustain a close relationship with all the 12 projects, negotiating their needs and solving problems in this experimental phase. The other key figures were the district technicians, who partially managed the implementation of the projects in each district. The technicians throughout the 3 years learned how to negotiate and facilitate relationships between citizens and bureaucratic requirements. Based on interviews, this relationship was not easy in the beginning, but technicians turned out to be a very crucial mediating element in the process of developing the projects. Furthermore, in some cases, the organizing entity played an important role in acting as intermediary between the citizens and the municipality, creating an environment where initial prejudices and distrust toward administrative bodies could be released.

TABLE 25.1 Main Types of Activities Among the 12 Pla BUITS Projects

	Gardening	Sustainability	Vulnerable Groups	Education/Workshops
Bicipark Númancia		X	X	
Biobui(L) Txema		X		X
Can Roger	X		X	
Connect Hort	X			
El Portal de Sants				
Espai Gardenyes	X	X		X
Espai Germanetes	X	X		
Hort Aspanias	X			
L'illa dels 3 Horts	X		X	
La Ferroviária	X		X	
Porta'm an l'Hort	X			
Projecte Verd	X			

Administrative issues proved to be significant obstacles in carrying out rapid solutions to the problems of initiatives: base on the fieldwork, approval processes of proposed activities were slow, and participants criticized the complicatedness of paperwork that comes along with the projects, mostly for reasons of autonomy. However, Pla BUITS simplified enormously the licensing conditions of all the projects, creating an exceptional, multipurpose license that contained all the activities that were offered at the sites.

Finally, project leaders expressed issues with very limited budgets and economic resources. In the beginning of the first year and setting up of the projects, initiators had to finance the necessary changes that had to be made on the site that was available for the proposed activities. During the 3-year experimental period, the municipality adapted to the needs of project leaders and in the second round of setting up projects guaranteed the cleaning up and preparion of the sites for the winning proposals. However, funding and financial support from the municipality still remains unstructured.

Fortunately, at the end of the process, all the projects stated satisfaction and positive feedback with the first 3 years of operation, and mentioned that Pla BUITS generated certain advantages, compared with nonofficial appropriation of spaces. These include the possibility to involve more citizens (not just traditionally more radical ones, as squatters), and they receive better maintenance—if something goes wrong, they will not fix it, but the municipality provides help to fix the issue. And most importantly, Pla BUITS helps to create a dialogue between citizens and the municipality, generating an environment where they can discuss problems and decide about solutions together.

25.3 DISCUSSION: PROMISES FROM A NEW GOVERNANCE MODEL DEALING WITH COMANAGEMENT?

Navigating the initiatives and experiments dealing with cooperatives, associations, squatting, and finally comanagement of public vacant spaces, we can state that we have explored the potentialities of community-led initiative both for self-managing certain urban services and co-managing spaces.

In the Barcelona case the metaphors of smart, resilient, and transversal are used only partially from those initiatives, but maybe this is due to the fact that those metaphors, and slogans, are mainly used to label top-down policies and projects, aiming at fostering the city's marketing efforts. On the contrary, while not mentioning resilience, all the initiatives somehow (directly or indirectly) address societal transition toward sustainability, and therefore tackling climate change—related issues—shown through the case of energy, community gardens, housing and building materials, and mobility among the examples cited herein. Actually it is the Pla BUITS experience that builds and fosters the most the bridges between the grassroots potentialities and the institutional frameworks usually driving plans and actions, making relevant the grassroots actions without entering the city marketing game. The synergies and trade-offs among community-led initiatives and top-down practices have been actually only been partially explored through Pla BUITS, which is the inception for a future new model of governance, in which comanagement of city spaces and services could be a significant part of a more people-centred and sustainable planning and urban management policy. Indeed, as shown in the last section, not only the role of grassroots efforts but also, and mainly, the attitude and involvement of city institutions' technicians (beyond the politicians) are key in starting to bridge citizens' needs. This is but a long-term transition, in which the Barcelona Green Plan and the Climate Change Plan are related projects, all based on very top-down hard infrastructures, and huge investments depending on strategies.

To conclude, comanagement most of all has the potentialities to learn from the grassroots practices (Cattaneo and Gavaldà, 2010; Camps-Calvet et al., 2016; Langemeyer et al., 2016), in order to set synergies for transitioning toward a more decentralized and sustainable model of urban development, retaking the concept of community resilience, where values, beliefs, and human agency are crucial elements of change (Mulligan et al., 2016), and a necessary way to societal transition. What the Pla BUITS experience shows is that this transition requires flexibility both from city administration and the grassroots, and openness to overcome initial obstacles to be able to realize longer-term societal changes. These programs can only be successful if there is a strong commitment from all the actors involved to work for better policies and practices.

Acknowledgment

Authors are thankful to the Barcelona Municipality practitioners, providing support, documents and facilitating the linkages for interviewing stakeholders and responsible persons for developing the PLA BUITS section of the chapter. Also, I. Ruiz-Mallén gratefully acknowledges the financial support of the Spanish government's Research Agency through a 'Ramón y Cajal' research fellowship (RYC-2015-17676).

References

Ajuntament Barcelona., 2016. Pla BUITS. Available at: http://ajuntament.barcelona.cat/ecologiaurbana/en/bodies-involved/citizen-participation/buits-plan.

Balibrea, M.P., 2001. Urbanism, culture and the post-industrial city: challenging the 'Barcelona model'. Journal of Spanish Cultural Studies 2, 187–210.

Berkes, F., Ross, H., 2012. Community resilience: toward an integrated approach. Society & Natural Resources 26, 5–20. https://doi.org/10.1080/08941920.2012.736605.

Camps-Calvet, M., Langemeyer, J., Calvet-Mir, L., et al., 2016. Ecosystem services provided by urban gardens in Barcelona, Spain: insights for policy and planning. Environmental Science & Policy 62, 14–23. https://doi.org/10.1016/j.envsci.2016.01.007.

Carpenter, S., Walker, B., Anderies, J.M., et al., 2001. From metaphor to measurement: resilience of what to what? Ecosystems 4, 765–781.

Cattaneo, C., Gavaldà, M., 2010. The experience of rurban squats in Collserola, Barcelona: what kind of degrowth? Journal of Cleaner Production 18, 581–589. https://doi.org/10.1016/j.jclepro.2010.01.010.

Certomà, C., Notteboom, B., 2017. Informal planning in a transactive governmentality. Re-reading planning practices through Ghent's community gardens. Planning Theory 16 (1), 51–73.

Chelleri, L., Waters, J.J., Olazabal, M., et al., 2015. Resilience trade-offs: addressing multiple scales and temporal aspects of urban resilience. Environment and Urbanization 27, 181–198.

Cutter, S.L., Barnes, L., Berry, M., et al., 2008. A place-based model for understanding community resilience to natural disasters. Global Environmental Change 18, 598–606. https://doi.org/10.1016/j.gloenvcha.2008.07.013.

Davidson, J.L., Jacobson, C., Lyth, A., et al., 2016. Interrogating resilience: toward a typology to improve its operationalization. Ecology and Society 21 (2). https://doi.org/10.5751/ES-08450-210227.

Davoudi, S., Shaw, K., Haider, L.J., et al., 2012. Resilience: a bridging concept or a dead end? "Reframing" resilience: challenges for planning theory and practice interacting traps: resilience assessment of a pasture management system in northern Afghanistan urban resilience: what does it mean in planning practice? Resilience as a useful concept for climate change adaptation? The politics of resilience for planning: a cautionary note. Planning Theory & Practice 13, 299–333. https://doi.org/10.1080/14649357.2012.677124.

Degen, M., García, M., 2012. The transformation of the 'Barcelona model': an analysis of culture, urban regeneration and governance. International Journal of Urban and Regional Research 36, 1022–1038. https://doi.org/10.1111/j.1468-2427.2012.01152.x.

Eizaguirre, S., Pradel, M., Terrones, A., et al., 2012. Multilevel governance and social cohesion: bringing back conflict in citizenship practices. Urban Studies 49, 1999–2016.

González, S., Healey, P., 2005. A sociological institutionalist approach to the study of innovation in governance capacity. Urban Studies 42, 2055–2069.

GSP U, 2012. Resilient People, Resilient Planet: A Future Worth Choosing. United Nations, New York.

Langemeyer, J., Latkowska, M.J., Gómez-Baggethun, E., 2016. Ecosystem services from urban gardens. In: Bell, S., Fox-Kämper, R., Keshavarz, N., Benson, M., Caputo, S., Noori, S., Voigt, A. (Eds.), Urban Allotment Gardens in Europe, pp. 115–141.

MacKinnon, D., Derickson, K.D., 2013. From resilience to resourcefulness: a critique of resilience policy and activism. Progress in Human Geography 37, 253–270.

Meerow, S., Newell, J.P., 2016. Urban resilience for whom, what, when, where, and why? Urban Geography 1–21. https://doi.org/10.1080/02723638.2016.1206395.

Mulligan, M., Steele, W., Rickards, L., et al., 2016. Keywords in planning: what do we mean by 'community resilience'? International Planning Studies 1–14. https://doi.org/10.1080/13563475.2016.1155974.

Parés, M., Martínez, R., Blanco, I., 2014. Collaborative governance under Austerity in Barcelona: a comparison between evictions and empty urban space management. In: City Futures International Conference, Special Session: Collaborative Governance Under Austerity, Paris.

Pelling, M., 2011. Adaptation to Climate Change: From Resilience to Transformation. Routledge, London. ISBN:9780415477512.

Rodin, J., 2014. The Resilience Dividend: Being Strong in a World Where Things Go Wrong. Public Affairs.

Rose, A., 2004. Defining and measuring economic resilience to disasters. Disaster Prevention and Management: An International Journal 13, 307–314.

Roy, A., 2005. Urban informality: toward an epistemology of planning. Journal of the American Planning Association 71, 147–158.

Serre, D., Barroca, B., Laganier, R., 2012. In: Resilience and Urban Risk Management. CRC Press.

Somers, S., 2009. Measuring resilience potential: an adaptive strategy for organizational crisis planning. Journal of Contingencies and Crisis Management 17, 12–23.

TESS, 2016. Community Climate Action across Europe. Available at: http://www.tess-transition.eu/.

Vale, L.J., 2013. The politics of resilient cities: whose resilience and whose city? Building Research & Information 42, 191–201. https://doi.org/10.1080/09613218.2014.850602.

CHAPTER

26

Barriers to Societal Response and a Strategic Action Plan Toward Climate Change Adaptation and Urban Resilience in Turkey

Funda Atun
Politecnico di Milano, Milan, Italy

26.1 INTRODUCTION

Climate change (CC) is a complex phenomenon consisting of not only scientific but also political, economic, physical, and social dimensions. In the 21st century, although CC has been an accepted fact since the first conference held in Geneva in 1979 by the World Meteorological Organization, it still keeps its complexity and uncertainty as a global challenge (Zilman, 2009).

Considering uncertain political, organizational, and social aspects in cities highlights two unanswered questions: "How would the underlying vulnerability to climate change alter in the future in a complex system like Istanbul?" and "What would be the potential role of communities and bottom-up initiatives in CC adaptation strategies?"

Turkey accepted the Climate Change Strategy Document covering the years 2010–20. The action plan based on the report was completed in 2011. As regards to raising concerns in Turkey about CC, there are two main questions as to whether there are any trends in climate variability and extreme events in Turkey, and as a result, any (potential) damage and other (potential) impacts. According to the scientific results issued by the General Meteorology records in Turkey, temperatures are increasing everywhere in the country, especially the summer temperatures, and precipitation has increased in the northern parts. Because of such changes, sea level has risen and an increasing number of climate-related hazards have become prominent phenomena (Sen, 2013).

Especially in the implementation of climate change adaptation and mitigation policies, long-run strategies are highly crucial to achieve successful implementation. In normal conditions, city development plans and mitigation strategies are prepared in the present by forecasting future trends through looking at past trends. As the cities are highly complex, self-organizing systems and follow nonlinear patterns, the integrity of the plans and the implementation of the strategies defined in the plan cannot be guaranteed in the long run. In addition, to the scientific uncertainty of climate change, having an uncertain development pattern in cities, especially in a megacity like Istanbul, makes the situation more difficult to handle with the present resources.

Initially, I mentioned briefly the political dimension to show where Turkey stands in this picture. The next section includes some statistics regarding the physical dimension to better understand the relation between the increasing exposure and climate change. These include the increasing number of people, buildings, car ownership, the details about production of garbage, and consumption of electricity and water. Then we will discuss the societal dimensions of climate change, considering the results of a survey conducted in Istanbul in 2014 to better understand knowledge sharing between public and private actors, stakeholder involvement, and the reasons for the missing societal response in Turkey.

26.2 PHYSICAL DIMENSION: INCREASING EXPOSURE AND SYSTEMIC VULNERABILITY

Adger et al. (2005) states clearly that to understand the vulnerability of a system to climate change, there is a need to look at the exposure, physical setting, and sensitivity, as well as the system's ability to adapt to change. Vulnerability is a dynamic concept that can be formed over time and across the scales with a strong interaction and through policies and trends (Menoni et al., 2012). The changing trends of the economic policies in Turkey, especially in the 1950s and 1980s, have had distinct effects on Istanbul's economic, spatial, and social vulnerabilities against disasters, including earthquakes, floods, and climate change. Today the city system in Istanbul is highly vulnerable due to rapid population growth, rapid urbanization, low-quality housing supply, and traffic congestion (Atun and Menoni, 2014).

However, nowadays, the city has become a huge construction site to replace the vulnerable housing stock with newly built, modern high buildings and skyscrapers with the support of *Regeneration in the Disaster Risk Zones*, law no. 6306 (May 16, 2012) (Atun and Kundak, 2015). On the one hand, regeneration is a positive approach as the vulnerable houses are replaced with the disaster-resistant ones; on the other hand, the results has been increased exposure of buildings, infrastructure, and population.

With the increasing number of exposures, the consumption of electricity and garbage production increases as well. Statistical information about population, car ownership, production of garbage, and consumption of electricity and water are indicated in Table 26.1 for the year 2012, for both Turkey and Istanbul. The biggest concern is that both the consumption of electricity and garbage production are higher in the residential areas than industrial areas. Therefore, it is crucial for climate change mitigation that authorities in Istanbul make the public aware about the issues.

Indeed, in a city system, administrative and organizational decisions, such as legislation and development plans, have strong effects on its social, economic, and physical patterns. Therefore, as mentioned previously, due to the complexity of the phenomena, spatial aspects such as structuring elements, land uses, and infrastructures should be considered with economic and socioeconomic aspects, including both formal and informal sectors to achieve a sustainable and resilient city system.

26.3 SOCIETAL CONCERNS: STAKEHOLDER COORDINATION AND COLLABORATION

Climate change adaptation and mitigation cannot be dealt with by one group, or one authority, or only by individuals' efforts. Such a process needs societal actions with involvement of both profit and nonprofit stakeholders from all the related sectors, including governments and business on a global and national level, municipalities,

TABLE 26.1 Increasing Exposure With Numbers in Istanbul

		2012	Additional information
Population (inhabitants)	Turkey	75,627,384	In 2012, the population of Turkey had increased 1.2%, and Istanbul 2.7%, compared to the population in 2011.
	Istanbul	13,854,740	
Car ownership	Turkey	17,033,413	In Istanbul, car ownership is 18% of total car ownership in TR.
	Istanbul	3,065,465	
Production of Garbage (Ton)	Turkey	25,845	Istanbul produces ≈ 22% of the total garbage produced in Turkey.
	Istanbul	5,672	
Consumption of Electricity (MWh)	Turkey	194,923,349	In Istanbul, consumption of electricity in residential buildings (10,519,250) is more than the industrial consumption of electricity (9,560,861 MWh).
	Istanbul	33,084,558	
Consumption of water (1000 m^2/year)	Turkey	4,936,342	Consumption of water is highest in Turkey, with 18% percent of Turkey's total consumption, although there is no agricultural production in Istanbul.
	Istanbul	930,823	

Modified from: TUIK Selected indicators for Istanbul, 2013.

nongovernmental organizations (NGOs), and lay people in the local level. A successful outcome can be obtained as a result of a strong collaboration between these separate and/or somehow connected actors.

Herewith, to make myself clear, I would like to share the results of a survey conducted in 2014 as a part of the Know-4-DRR project in Istanbul to understand the level of collaboration across organizations, the level of information transfer, knowledge sharing between organizations about disaster risk reduction, and CC adaptation (for more information about the survey, see Norton et al., 2015). The survey focused on three prominent issues deserving to be mentioned here. These three issues concern "outreach of the information to public," "lack of coordination," and "implementation." The survey revealed several concerns of different organizations from public, private, and nongovernmental organizations on the topic.

Regarding "outreach of the information to public," respondents from public organizations mentioned their concerns that providing information to the public does not really help to increase public awareness (Norton et al., 2015). Before all else, having information available for public access does not necessarily mean that the public knows about it, and receives it in a sufficient and correct manner (Atun, 2014). That is the reason why there is no strong, notable bottom-up approach in Istanbul regarding CC adaptation.

Considering the general public, as Stoknes (2015) mentioned in his book *What We Think About When We Try Not to Think About Global Warming*, there are psychological barriers that people create when they have difficulties in understanding climate information; and consequently, they dismiss it by convincing themselves it is not important. Stoknes mentions the importance of showing the progress of actions through social media and storytelling to prevent people building barriers and increase their awareness about climate change.

The obscurity and inaccuracy of available information to the general public is mentioned also in the sixth final notice of the CC published in 2016 in Turkey by the Ministry of Environment and Urbanization. The report included the results of a survey that indicated the current awareness of the general public on the CC phenomenon. According to the results of the survey, 12.9% of the participants indicated that they do not have any idea about CC; 39.5% of the respondents stated that CC means "seasonal change"(Ministry of Environment and Urbanization, 2016).

Considering coordination among groups, several respondents stressed the lack of coordination between the various organizations (Norton et al., 2015). In Turkey, at the organizational level, since the occurrence of the 1999 Marmara Earthquake, there has been a positive tendency to focus on disaster risk reduction instead of just postdisaster response. However, the problem occurs in the involvement of several stakeholders that create connections between different groups. Trust and good coordination is strong between private and public organizations; however, the link between NGOs and other organizations needs to be strengthen.

In the survey, one respondent from a private organization pointed out his concern about the implementation of the regeneration projects to decrease the current disaster risk. According to the respondent, while the disaster risk reduction—related projects conducted in Istanbul, with the support of *Regeneration in the Disaster Risk Zones* law, aim to construct disaster-resilient modern living areas, they are not sufficient in considering the society-centric issues. The regeneration project areas were chosen from the socially depressed areas, as they are at the same time the most physically depressed residential areas (Norton et al., 2015). The newly built modern disaster-resilient housing stock in these areas is no longer affordable by the former inhabitants in the area. These results change not only the physical patterns, but also the social groups living in the area. It is true that the regeneration projects conducted in the high-level risk areas decrease structural vulnerability in the area, but at the same time, these projects may cause an increase of social vulnerability not related solely to the disaster as well (Norton et al., 2015).

In Istanbul within the general public, although there is little awareness of climate change, if any, there is such an attempt, as Stoknes (2015) mentions, where approximately 4 years ago, a small group[1] started a petition in change. org. That was supported by intellectuals, artists, academicians, and the general public, collecting 9,000 signatures, which is a spark that would be enough for a fire. They might need also to include the storytelling in their actions to transform their local actions into a societal response.

The most crucial factor for successful climate change adaptation and mitigation activities is to achieve an integrative strategic action plan considering various spatial and time scales through strong coordination and cooperation among diverse institutions, including governmental organizations, academic institutions, NGOs, the private sector, and the general public.

---

[1] http://www.iklimicin.org.

26.4 DISCUSSION AND CONCLUSION

Besides being complex, climate change is also a long-onset disaster where is difficult to be precise concerning future climate variability with today's knowledge. In the development history of cities, there are some milestones where the development rate increased tremendously in contrast to predictions. Therefore, we do not know whether there will be other milestones in the urban history, or, how the development rate will be in the future. This makes it difficult to define the economic cost of climate change. Therefore, it is difficult to do a cost-benefit analysis and to define and implement policies and strategies.

More than a century ago, Geddes (1915) mentioned the importance of an integrated view of life, and the fundamental unity and interdependence of culture and nature. He also mentioned that cultural transformation can be achieved only by locally adapted direct actions. Although there are actions in public and private organizations regarding CC adaptation, they are not successfully adapted and are insufficient for cultural transformation. The lack of collaboration between the various organizations is a well-known fact behind the unsuccessful attempts on CC adaptation and mitigation policies. Several reasons can be listed here. However, the question here is, although several organizations, both from public and private organizations, aim to achieve the same target, why does such a collaboration gap occur between organizations? Better collaboration depends mainly on the trust between organizations and that makes the trust-building activities more than necessary, especially between NGOs and public institutions, where a huge concern was observed during the survey in Istanbul in 2014 (Norton et al., 2015).

To conclude, Istanbul has a prominent role in the wider regional and national context for Turkey's economic and cultural development. Unfortunately, for a city like Istanbul, climate change is not an urgent need to be addressed on the first page of the political agenda. It will also take time to adapt the economic and social drivers to climate change, with the current level of awareness and perception of climate change, both in the business and social level. Although the strengths such as goodwill, awareness of some organizations, especially in the governmental level, and some intentions exist, both in the business and nongovernmental organizations, the weakness is the missing collaboration between these organizations and several initiatives. Therefore, these intentions have not been sufficient for a city of around 15 million inhabitants and more than 1 million migrants. The current disaster reduction strategies and their implementation are strategically significant for achieving the sustainable development in Istanbul, however, the continuity of this trend in the future cannot be guaranteed due to potentially urgent matters that could emerge at any time. The migrants, the unstable situation and tensions in the Middle East and being subject to several other natural disasters are the current threats before a successful implementation of climate change adaptation and mitigation strategies in Turkey.

References

Adger, W.N., Arnell, N.W., Tompkins, E.L., 2005. Successful adaptation to climate change across scales. Global Environmental Change 15, 77–86. https://doi.org/10.1016/j.gloenvcha.2004.12.005.

Atun, F., 2014. Improving societal resilience to disasters. A case study of London's transportation system. In: Springer Briefs in Applied Sciences and Technology. Springer. ISBN:978-3-319-04653-2 (book). Available at: http://www.springer.com/gp/book/9783319046532.

Atun, F., Kundak, S., 2015. Before and after HFA. Retrospective view of progress in disaster risk reduction system in Turkey. In: Input Paper Prepared for the Global Assessment Report on Disaster Risk Reduction 2015. UNISDR, Geneva, Switzerland. Available at: http://www.preventionweb.net/english/hyogo/gar/2015/en/home/documents.html.

Atun, F., Menoni, S., 2014. Vulnerability to earthquake in Istanbul: an application of the ENSURE methodology. ITU Journal of the Faculty of Architecture, Special Issue on "Cities at Risk". ISSN: 1302-8324 11 (1), 99–116. Available at: http://www.az.itu.edu.tr/azvol11no1web/09-AtunMenoni-1101.pdf.

Geddes, P., 1915. Cities in Evolution. Williams and Norgate, London.

Menoni, S., Molinari, D., Parker, D., Ballio, F., Tapsell, S., 2012. Assessing multifaceted vulnerability and resilience in order to design risk-mitigation strategies. Natural Hazards 64 (2057). https://doi.org/10.1007/s11069-012-0134-4.

Ministry of Environment, Urbanization, 2016. Turkiye Cevre ve Sehircilik Bakanligi 2016. Turkiye Iklim Degisikligi 6. Bildirimi. Available at: https://www.csb.gov.tr/db/destek/editordosya/Turkiye_Iklim_Degisikligi_Altinci_Ulusal_Bildirimi.pdf.

Norton, J., Atun, F., Dandoulaki, M., 2015. Exploring issues limiting the use of knowledge in Disaster Risk Reduction. Tema. Journal of Land Use, Mobility and Environment 8 (Special Issue ECCA 2015), 135–154. https://doi.org/10.6092/1970-9870/3032.

Sen, Ö.L., 2012/13. Mercator-IPC Fellow, a Holistic View of Climate Change and its Impacts in Turkey. Available at: http://ipc.sabanciuniv.edu/publication/a-holistic-of-climate-change-and-its-effects-in-turkey/?lang=en.

Stoknes, E., 2015. What We Think About When We Try Not To Think About Global Warming. In: Toward a New Psychology of Climate Action. Chelsea Green Publishing.

Zilman, 2009. A history of climate activities. World Meteorological Organization, Bulletin 58 (3). Available at: https://public.wmo.int/en/bulletin/history-climate-activities.

Website

Iklim icin harekete gec: http://www.iklimicin.org.

Further Reading

TUIK, 2013. Selected Indicators for Istanbul. (TUIK seçilmis göstergelerle İstanbul 2013). Available at: http://www.tuik.gov.tr/ilGostergeleri/iller/ISTANBUL.pdf.

CHAPTER 27

Victims or Survivors: Resilience From the Slum Dwellers' Perspective

Deepika Andavarapu[1], David J. Edelman[1], Nagendra Monangi[2]

[1]University of Cincinnati, Cincinnati, OH, United States; [2]Cincinnati Children's Hospital, Cincinnati, OH, United States

27.1 PEDDA JALARIPETA (LARGE FISHING VILLAGE)

Of 1.7 million residents in the city of Visakhapatnam, India, 770,971 (44.1%) live in slums (Government of India, 2011). Land is a valuable and scarce resource in Visakhapatnam with the sea to the west and hills to the east (Fig. 27.1). Low-skill work opportunities in the city such as in the navy port and steel industry attract rural migrants to Visakhapatnam, further exacerbating the land scarcity and thereby increasing slums.

The Pedda Jalaripeta (PJ) slum is the largest (6000 residents) and oldest slum in the city (GVMC, 2009). However, three of the top tourist destinations in Visakhapatnam are in proximity to the PJ slum, and luxurious five-star hotels and high-rise buildings surround it. Pedda Jalaripeta, means "large fishing village," and could be characterized as a low-income fishing community where 75% of the adults still depend on fishing as their livelihood (Immanuel and Rao, 2012; Central Marine Fisheries Research Institute, 2005). More than 90% of the residents earn less than US$1.25 a day, which is below the international poverty line (GVMC, 2009). Nearly 88.5% of Pedda Jalaripeta residents are Hindus and belong to the Jalari caste (GVMC, 2009). Despite their limited resources, the residents have survived various disasters, such as a fire in 1983, and eviction/gentrification threats since early 2000. The homogeneity of the community and its historical existence as a fishing village for decades result in high reserves of bonding and bridging capital in the community.

27.1.1 Pedda Jalaripeta Social, Cultural, and Economic Identity

An important feature of Pedda Jalaripeta's social, cultural, and economic identity is its beach; the section of Visakhapatnam beach along PJ slum mainly functions as a fishing beach, which is an ecological and economic resource and also acts as the social gathering space for men. This is where men store their boats, prepare to go fishing, and repair their nets, but also where they rest and go to consume (large volumes of) alcohol (Fig. 27.2).

With regard to its social organization, it is also important to highlight the value of the Grama-Sabha for the slum. Grama-Sabhas are ancient forms of village assemblies in India. These organizations strengthen the communities' bonding and bridging capital through dispute resolution and managing religious events (Aldrich, 2011). Pedda Jalaripeta's Grama-Sabha has existed since before the community was notified as a slum in 1969, and continues to be active to date. For example, it organizes 10–12 religious festivals such as Polaramma-Jatara, Nookalamma-Jatara, Sri Rama Navami (each of these festivals is aimed at appeasing a Goddess (Polaramma or Nookalamma) or God (Ram)) year round. As is discussed following, the Grama-Sabha has recently taken on an important role in the redevelopment of the slum following a major fire.

Ceremonies and construction activities impose economic burdens, but residents are typically willing to bear them if they help them build their internal and external social networks. The resulting bonding capital is critical for Pedda Jalaripeta, since during the offshore fishing ventures, fishermen rely on friends and family in the nearby boats for first-aid and emergency evacuation. These networks built on the ground save lives on the sea (Focus group, March 30, 2014). One of our respondents commented:

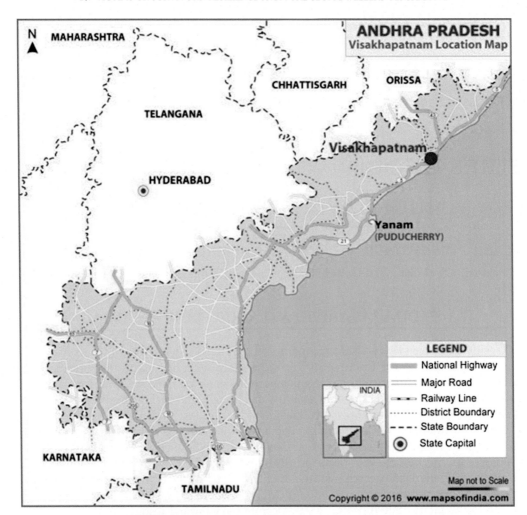

FIGURE 27.1 Map of Visakhapatnam. *Mapsofindia.com.*

FIGURE 27.2 Fishermen mending the nets. *Authors.*

FIGURE 27.3 Ramalayam temple. *Authors.*

> If we are not united, our lives will be very difficult. If we are united here, then when we go in to the sea we help each other out and save each other. So that is the reason for our unity. If someone falls in to the sea, the other person who is fishing far away needs to feel the need to go and save this person. Unity is our strength; otherwise we cannot survive. That's why we build temples and conduct festivals. Our safety on the sea is dependent on our unity on the ground. ***Informal Interview with Retired Senior Fisherman on March 22, 2015***

In 2006, Grama-Sabha organized a donation campaign for building the Ramalayam temple (Fig. 27.3). The slum residents donated 3 days of fishing catch, equivalent to 1,500,000[1] Indian rupees. The residents continue to donate 1% of their income voluntarily for the temple's upkeep and maintenance. Following is an excerpt from the Grama-Sabha president about this effort:

> Author: do you think everyone donates (1% of their income)?
> Grama-Sabha President: In June/July and January/December we (the temple) get donations around 32,000 or 35,000 rupees. The other months we get anywhere from 15,000 to 20,000 rupees. Those 4 months are season for getting fish so our donations in temple are higher in those 4 months. So, yes I believe that the fishermen donate 1% of their income for the temple upkeep

Over the past 40 years, despite multiple challenges, the PJ slum has retained its social, cultural, and economic identity as a fishing village. It successfully avoided undesirable alternative states, including the loss of local control, loss of fishing, gentrification, and eviction. This chapter details how the community has dealt with multiple challenges they faced over the years. Combating disasters and retaining the community character is relatively difficult for poor urban neighborhoods. Many such neighborhoods in New Orleans, Louisiana, failed to revive after Hurricane Katrina. Only communities with linking networks to external agencies survived Katrina (Elliot et al., 2010). This chapter, therefore, evaluates the role of social networks in the ongoing resilience of the PJ community. Social networks can be a powerful defense and survival mechanism in a poor urban neighborhood with meager tangible physical or financial resources.

27.2 DATA SOURCES AND COLLECTION

Between September 2013 and March 2014, preliminary data about the PJ slum were collected through a variety of data bases including GVMC household surveys (GVMC, 2009); the Department of Fisheries socioeconomic survey on the fishermen in the PJ slum (Immanuel and Rao, 2012); Action Aid (an international nongovernmental

---

[1] $24,105 at a conversion rate of $1 = Rs. 62.22.

organization [NGO] active in the community) project reports outlining their programs and successes in the PJ slum (Phillipose, 2013); and other scholarly articles about the city of Visakhapatnam and the PJ slum. In February 2014, University of Cincinnati's Institutional Review Board reviewed and approved the PJ slum research protocol. The fieldwork in the PJ slum was conducted from March through July 2014. During that time, 54 interviews (44 with the slum dwellers, 10 with government employees) and three focus group discussions were held. Local leader perspectives were canvassed through multiple key informant interviews. Additionally, ethnographic observations were recorded on a daily basis in a journal. During this time, an average 15 h per week were spent on interviews and participant observation.

27.3 SURVIVING DISASTERS THROUGH COLLABORATION: FIRE AND GENTRIFICATION/EVICTION

As mentioned earlier, Pedda Jalaripeta retains its social, cultural, and economic identity as a low-income fishing village where 70% of the adults still depend on fishing as their livelihood (Immanuel and Rao, 2012). The PJ slum has successfully avoided alternative undesirable states, including the loss of local control, loss of fishing, gentrification, and eviction. In this section, two disasters that the PJ slum faced in the past 30 years will be discussed.

27.3.1 Fire

On a cold December night in 1983, an accidental fire destroyed 600 of the 800 huts in the PJ slum. The accident was a fast-moving variable, but the community was able to recover from the loss primarily due to its linking capital (a direct link to the political leadership). This vertical linkage to the government provided the capital to recover from the fire and transform the slum into a new community. Thirty years after the redevelopment, the community continues to be desirable and loved by a younger generation, which is willing to invest time, energy, and resources to preserve and protect the community's identity as a fishing village.

The Greater Visakhapatnam Municipal Corporation (GVMC) redeveloped PJ in 1984 under the ongoing Slum Improvement Program, funded by UK's Department for International Development (Abelson, 1996). Slum redevelopments are often controversial since they can be poorly built (Perlman, 1976; Patel, 2013), or in case they regenerate, they run the risk of being gentrified by middle- or high-income residents (Peattie, 1982; Patel, 2013).

The PJ is a slum redevelopment success story since it continues to provide quality affordable housing 3 decades after its inception (GVMC, 2009). A local fisherman, who was the elected representative (MLA) of the PJ community, was the reason for PJ's successful redevelopment. According to interviews with slum residents and government officials, the MLA used his political clout and appointed the slum's Grama-Sabha as the neighborhood committee in charge of monitoring the redevelopment and allocating the lots.

27.3.2 Gentrification/Eviction

After the fire in 1983, Pedda Jalaripeta faced a second disaster. In early 2000, the slum witnessed a multitude of tourist development projects in and around their community, which not only threatened to gentrify it, but also challenged PJ community's livelihood (Equations, 2008; Phillipose, 2013). If the seashore was redeveloped as a tourist destination, the PJ residents could potentially lose their livelihood (fishing). The resulting gentrification could also evict nearly 65% of the PJ community residents who built their houses on encroached government land (and therefore possess no legal right to the land). In order to avoid the gentrification/eviction, the community accessed capital in the form of knowledge and training from the NGO Action Aid to file legal petitions against the private developers and government agencies.

Gentrification is not typically considered a threat especially when compared to potentially deadly consequences of a tsunami or a fire. However, the PJ residents considered the tourism developments and gentrification as a key threat to the village, primarily since these projects threatened the community's source of livelihood (fishing) and had the potential to evict a large majority of the community (Equations, 2008; Phillipose, 2013). Previous research has shown that tourism-oriented development along the fishing beaches has a detrimental effect on fishermen (Benansio et al., 2016; Derman and Ferguson, 1995). Fishermen lose access to fishing grounds, and their fishing gear is often destroyed due to tourist activities such as diving, snorkeling, boat riding, and others. Employment in the tourism sector is often not an option for these fishermen because of their low level of education (Derman and Ferguson, 1995).

The residents, especially the younger fishermen, thwarted the gentrification threats through media outreach and legal filings when necessary. Action Aid, an international aid organization, provided these fishermen with the knowledge and resources in their battle against unpredictable threats including eviction or gentrification. This example elicited an important process of building community strength and resilience, and it shows a different way of "bouncing forward" not by drawing on the old, established structures of leadership as in the case of the fire and development after that, but instead by drawing on the skills, knowledge, and motivation of the younger generation.

27.4 COLLABORATING, EMPOWERING, AND EDUCATING

The resilience metaphor in the context of a slum can be questionable, since slums are typically seen as vulnerable, and slum residents are considered victims. The Pedda Jalaripeta case, however, illustrates that slums can also be resilient, and slum dwellers can fight and survive not only natural disasters but also man-made disasters. PJ survived two disasters over the past 30 years and continues to fight tourist developments that threaten its quality of life. The fact that the community persistently holds onto its values is due to its desirability for a new generation. PJ provides several spatial (affordable housing, public-private space, and access to the beach for fishing) and social advantages (bonding and bridging capital through cultural and religious activities and vertical links to external agencies) to its residents. Overall, PJ's resilience reflects its livability and vibrancy. Thirty years after redevelopment, the community continues to be desirable and loved by a younger generation willing to invest its time, energy, and resources.

Several other scholars have observed that bonding and bridging capital, while critical to coping, do not sufficiently help residents to transform and successfully adapt to disasters in the long run (Putnam, 2001; Islam and Walkerden, 2014; Aldrich, 2011; Elliott et al., 2010). Similarly, the PJ community's transformative capacity enticed mobilizing linking capital with politicians and external agencies. Specifically, linking networks, first with the MLA during the 1980's fire, and then with Action Aid starting in the 1990s. The link between slum residents and the government improved PJ's physical infrastructure. That physical transformation, in turn, provided the space and scope for human enrichment, preserved the residents' access to the fishing beach, and retained the social capital within the community. Action Aid provided knowledge and resources to educate and empower the youth in the community to fight for their human, cultural, and community rights (Phillipose, 2013; Equations, 2008). These rights, along with the physical upgrading, help shield the community against unexpected future threats including gentrification, displacement, and demolition.

Through this case study, it has been demonstrated that resilience on the ground is a long and laborious process. However, the PJ slum residents collaborated with the government and NGOs to improve the physical and social infrastructure of their slum. The residents invested time, money, and labor to rebuild their community from the ground up. The PJ case is an exception in that the community was able to access linkage capital at two different times in its history. Accessing linkage capital is not possible for all urban slums, especially when slums are seen as a hindrance to urban development (Bhan, 2009). The urban poor's existing hardships are further exacerbated when governments and municipalities reject the subalterns' claims on urban space (Datta, 2016). Informal, small-scale businesses, including traditional fishermen, are being fought against and displaced, which leads to the marginalization of the informal economy and contributes to its restructuring. Today, aesthetics and beautification have become key criteria of urban governance, and marginalizing and physically displacing the traditional elements of the urban economy are seen as important to modernizing cities.

References

Abelson, P., 1996. Evaluation of slum improvements: case study in Visakhapatnam, India. Cities 13 (2), 97–108.
Aldrich, D.P., 2011. The externalities of strong social capital: post-tsunami recovery in Southeast India. Journal of Civil Society 7 (1), 81–99.
Benansio, J.S., Wolff, M., Breckwoldt, A., Jiddawi, N., 2016. Have the fishing communities of Zanzibar Island benefited from increasing tourism development? Journal of Development and Agricultural Economics 8 (5), 95–107.
Bhan, G., 2009. This is no longer the city I once knew. Evictions, the urban poor and the right to the city in millennial Delhi. Environment and Urbanization 21 (1), 127–142.
Central Marine Fisheries Research Institute, 2005. Available at: http://www.cmfri.org.in/.
Datta, A., 2016. The Illegal City: Space, Law and Gender in a Delhi Squatter Settlement. Routledge.

Derman, B., Ferguson, A., 1995. Human rights, environment, and development: the dispossession of fishing communities on Lake Malawi. Human Ecology 23 (2), 125–142.

Elliott, J.R., Haney, T.J., Sams Abiodun, P., 2010. Limits to social capital: comparing network assistance in two New Orleans neighborhoods devastated by Hurricane Katrina. The Sociological Quarterly 51 (4), 624–648.

Equations (Equitable Tourism Options), 2008. Right to Information and Tourism. Available at: http://www.equitabletourism.org/stage/files/fileDocuments311_uid10.pdf.

Government of India (2011), August 2, 2013. Indian Census Data on Slums 2011. Available at: http://www.census2011.co.in/census/city/402-visakhapatnam.html.

GVMC, 2009. Survey of Households in Greater Visakhapatnam Municipal Corporation. In: Accessed through Andhra Pradesh Mission for Elimination of Poverty in Municipal Areas (MEPMA). Available at: http://www.apmepma.Gov.In/HHPS/Hh_Rep_Slum.Php?Ulbid=109.

Immanuel, S., Rao, G.S., 2012. Social status of hook and line fishermen in Visakhapatnam. Fishery Technology 49 (2), 204–209.

Islam, R., Walkerden, G., 2014. How bonding and bridging networks contribute to disaster resilience and recovery on the Bangladeshi coast. International Journal of Disaster Risk Reduction 10, 281–291.

Patel, S., 2013. Upgrade, rehouse or resettle? An assessment of the Indian government's Basic Services for the Urban Poor (BSUP) programme. Environment and Urbanization 25 (1), 177–188.

Perlman, J., 1976. The Myth of Marginality. University of California Press.

Peattie, L.R., 1982. Some second thoughts on sites-and-services. Habitat International 6 (1–2), 131–139.

Putnam, R.D., 2001. Bowling Alone: The Collapse and Revival of American Community. Simon and Schuster.

Phillipose, B.P., 2013. 'Sea our life, coast our right'. Learnings from Visakhapatnam's Fisher community. In: Critical Stories of Change. Action Aid. Available at: http://www.actionaid.org/sites/files/actionaid/critical_stories_of_change_i.pdf.

CHAPTER 28

Bottom-Up Initiatives for Climate Change Mitigation: Transition Town in Newcastle

Giuseppe Forino, Jason von Meding, Graham Brewer
University of Newcastle, Callaghan, NSW, Australia

28.1 INTRODUCTION

Newcastle hosts Transition Newcastle (TN), a long-term established community group that promotes initiatives of transition toward a low-carbon society, by targeting sustainability and environmental issues, including climate change. TN is one of the Transition Town groups that is part of a larger global network called the Transition Network. This chapter aims at briefly presenting TN as a grassroots initiative active in Newcastle and its relation with climate change mitigation. Data were retrieved from TN's website[1] and from two interviews, conducted in July 2016 with two TN members, Albert and Mark.[2] These data allowed us to explore the connection between TN's initiatives and climate change mitigation. Conclusions will reflect on challenges and opportunities for TN in the climate change discourse in Australia and Newcastle.

28.2 TRANSITION AND TRANSITION TOWN

A radical and systemic transformation is required by reducing the consumption of fossil fuels and the production of greenhouse gases emissions by shifting to renewable energy and more sustainable and efficient resource management (Seyfang and Haxeltine, 2012 and references therein). Transition is one of the concepts to describe such transformation (Haxeltine and Seyfang, 2009). Particularly, "transition towards sustainability" is a broad concept including those grassroots initiatives which address several environmental challenges, including climate change, to move towards a low-carbon society (Seyfang and Haxeltine, 2012). Transition involves economy-wide sectors such as energy efficiency, sustainability, technology, transport, manufacturing, agriculture, and tourism (Chappin and Ligtvoet, 2014; De Pryck et al., 2014). Transition describes changes in everyday life practices of environmental management and of political-economic approaches by individuals and communities (Seyfang and Smith, 2007; Seyfang and Longhurst, 2013).

Within such a scenario, the Transition Network represents one of the most long-term, established, and widely recognized grassroots, bottom-up initiatives at the global level. The Transition Network was born in England in 2006 and is currently in more than 40 countries with more than 1100 local transition initiatives. The Transition Network aims at creating resilient local communities by dealing with simultaneous challenges related to the shrinkage of fossil fuels, climate change, and economic crisis. The Transition Network exemplifies the potential of social movements to create alternatives to mainstream fossil fuel–oriented, neoliberal economies (Seyfang and Haxeltine, 2012; Feola and Him, 2016). A fundamental characteristic of the Transition Network that distinguishes

---

[1] http://transitionnewcastle.org.au/.
[2] Names have been changed in accordance with anonymity requirements by the research ethics committee of the University of Newcastle (Australia).

it from other social movements is the attempt of performing societal change "here and now," for example by challenging the status quo and promoting new practices, institutions, and forms of socioeconomic organization, and alternative systems of food, transport, and energy provision. These forms of transitions represent innovative experiences by networked individuals, communities, and organizations generating bottom-up solutions for sustainable development. These initiatives are distinguished from mainstream green business as they operate with a bottom-up perspective and are contextualized in civil society arenas, experimenting often radical social and technological innovations in order to reflect alternative worldviews and systems of values (Seyfang and Haxeltine, 2012).

Nevertheless, these innovations present some challenges. Grassroots innovations do not always operate as smoothly as idealized or as inclusive and supportive communities of practice. Furthermore, like many organizations relying on volunteers, such initiatives often struggle with ensuring and sustaining participation over time and ability of promotion within the community (Feola and Him, 2016). Meanwhile, the scarcity or lack of secure financial resources can inhibit their effectiveness. Also, these grassroots innovations do not always mirror the diversity (e.g., ethnic or religion) of local communities, consequently struggling to establish strong links with the larger community (Feola and Nunes, 2014).

28.3 TRANSITION NEWCASTLE

According to the TN website, TN is part of the global Transition Network and aims to create ways of living that (1) promote environmental and social sustainability, (2) build strong, connected communities, (3) value relationship and quality of life over wealth and material goods, (4) and conserve the finite global resources. Albert says that "TN started…in 2007 and was formed from a slight different basis of a Climate Action Newcastle (founded in 2005) group, because it was involved in the climate emergency and peak oil." According to Albert, "climate change was the initial focus of TN, but such focus is not very high profile in the TN now. It is not something we personally promote but it was in the early days and is internationally in the movement." TN aims at strengthening community education and actions toward a more sustainable Newcastle, particularly by advocating around an anticonsumerist message, and at building social capital at the level of household and street-scale network. Therefore, according again to Albert, TN promotes "different ways of being. We don't tackle the global economic system but the idea is that we would." As Mark advocates, in fact, "there are movements which are rebels and push the boundaries, others who talk about legislation and policies…. But… change can be part of the everyday life, and I think TN is doing this…I think we are an easy, safe way for people to start exploring these issues. I believe that the idea behind TN is that you are creating the alternative you want." Therefore, TN attempts raising awareness among the people in their everyday life about reducing their emissions and thinking about the future of the world. "The action involved people changing individual household, energy usage, how we became more conscious about our ecological footprint, how should we do as community" (Albert). Toward this goal, TN is currently promoted some initiatives, briefly presented next.

28.3.1 Newcastle Upcycle

Newcastle Upcycle focuses on changing how society considers waste. It aims at reimagining a waste product as a starting point for something new or useful (a process called "upcycling"), rather than a product that needs to be disposed of and that adds to the waste stream. Therefore, TN is working to create a thriving upcycling community in Newcastle through workshops gathering a vibrant and creative community with the goal of giving new life to products. Within Newcastle Upcycle, participants gain and share skills and confidence in creating new things from old, reinvigorating, reinventing, and adapting objects.[3]

28.3.2 Transition Street

Transition Street aims at allowing the involved streets in Newcastle to explore different forms of energy and water usage, food, waste, and transport over a period of 6 months. A practical workbook (Transition Newcastle, 2015) provides the basis for discussion within the streets. The workbook has chapters on energy, water, food, transport,

---

[3] See the TN website.

and waste/consumption and aims at reflecting on how the current society considers resources and the environment, and on how creative responses can be developed in ways to help the streets become more sustainable. Each month is dedicated to a specific theme, by encouraging to think about what communities can do together to reduce their environmental footprint. Through the Transition Street program, participants learn how they can save money while improving their sustainable footprint.[4]

28.3.3 Nourishing Newcastle Urban Tucker Stall

A further initiative is the Nourishing Newcastle Urban Tucker Stall (NNUTS) at the Newcastle Farmers Market (at the Broadmeadow showground) on the 1st and 3rd Sunday of each month. Such an initiative aims to support local urban growers who provide a range of homegrown products.[5]

28.4 TRANSITION AND CLIMATE CHANGE MITIGATION: CHALLENGES AND OPPORTUNITIES FOR TRANSITION NEWCASTLE

These briefly presented initiatives are examples of bottom-up practices aiming at the reduction of consumption and waste production. As we have seen, in the initial phases of TN, climate change was part of the discussion. However, after the initial years, it was realized that climate change was "a symptom of environmental unsustainability issues" (Mark); therefore, climate change and related mitigation just became one part of the broader issue of sustainability. Promoting the reduction of production and consumption is ultimately related to the reduction of greenhouse gas emissions. In this way, rather than directly targeting climate change and mitigation, all these initiatives frame them within a picture of environmental unlimited growth and the reduction of reliance on fossil fuels. For example, the workbook by Transition Newcastle (2015) recognizes that climate change is likely to have effects on water resources and that it is likely seeing more droughts and floods due to the increasing climate change. Improving the sustainability of individual and community water footprints is an imperative for ensuring access to water for everyone into the future.

Nevertheless, as recent researches on the Transition Network have found (Feola and Nunes, 2014; Nicolosi and Feola, 2016), challenges exist for TN in terms of being effective within the wide mitigation action. As previous chapters on Australia in this book have already demonstrated, the extent at which local communities and related initiatives can influence the demand of fossil fuels by public policies and in a fossil fuel–oriented government is still limited. At the federal scale, for example, the promotion of, and the large-scale public investment in, renewable energy advocated by different community groups cognate to the TN such as the Climate Action[6] is constantly under political attack in 2016, the Australian government attempted to dismantle the Australian Renewable Energy Agency (ARENA), and removed AUD1.3 billion from its budget (Hopkin, 2016). Additionally, both Albert and Mark recognize that finding time and energy for TN's initiatives is hard, as TN is totally run by volunteers, mostly working full time. For example, talking about the Transition Streets, Albert notes that "at the beginning a lot of people was excited, but then most people…fell away because people is so busy day to day with their families, their jobs." Therefore, it is hard to sustain these activities, and generally "unless there is someone driving, it's very hard sustaining a community no profit or a volunteer organization" (Albert).

These also represent some of the challenges for the creation of partnerships with other organizations, which are still limited in Newcastle. Albert argues that "There is very little interaction between community group…The big stakeholders (*such as corporations*) have a lot of money so they give money to part of the community who relies on them in the short term." Therefore, partnerships and interactions are mainly "between government, big corporations and exporters, as they collude to each other." In terms of community groups, TN's agenda is different from others' agendas, however, TN has some overlaps with local groups advocating environmental and social justice goals. For example, TN provides some support to Climate Change Action Newcastle or Lock the Gate, grassroots groups that advocate climate change issues or support farmers against the expansion of coal mining in agricultural areas.

---

[4] See the TN website.

[5] See the TN website.

[6] http://www.climateaction.org.au/index.php/our-projects/87-smart-energy-expo.

TN also tried to work with the Newcastle City Council and to raise its voice in Council meetings. For example, Albert told that the Council had put a lot of money into the consultation phase of the Community Strategic Plan (The City of Newcastle, 2015), but "at the end they (*the Council*) are doing what they usually do. (*They*) do the consultation but the community has not enough power." TN tried to claim its voice into the consultation. As Albert argues, "We (*TN*) had numbers of meetings with the Council at different times. At one stage, the Mayor had a consultancy committee to have environmental voice, but it did not do any difference." However, Mark claims that, when he told the Council that TN was keen to work together, the reaction by the Council surprisingly was, "Look, you do your things, we do ours. (*The reaction*)…probably may be related to that specific person who gave me the answer, however, that person had a quite high role for which a different response should have been provided."

In conclusion, innovative transition initiatives such as those promoted by TN can represent important input for the formulation and implementation of mitigation strategies and public policies toward environmental innovations (De Pryck et al., 2014). TN, for example, could be a key contribution toward a post-carbon society in the Hunter Valley and Newcastle (Evans, 2008), able to move from the reliance on fossil fuels and to reinvent new everyday practices and economies that are environmentally and socially sustainable. In this way, the idea of transition such as the one developed by the TN can provide an understanding of how niche activities promoting low-carbon experiments can eventually challenge the existing regime of production and consumption, offering interesting opportunities for climate change mitigation (Saikku et al., 2016). However, transition needs more space within urban policies and planning in Newcastle to be considered an effective metaphor. In fact, TN faces a lot of issues related to its volunteer basis, to the different agendas by other organizations, and basically to the isolation among grassroot groups in Newcastle, which does not allow creating effective and collaborative partnerships. Particularly, creating institutional spaces for these initiatives worldwide (Feola and Nunes, 2014) is necessary to increase opportunities for influencing multi-level policies and actions (Feola and Nunes, 2014) in order to influence the policies and actions by different levels of governments.

Acknowledgments

Giuseppe Forino thanks Dr. Geoff Evans for a preliminary discussion about the contents of this chapter.

References

Chappin, E.J.L., Ligtvoet, A., 2014. Transition and transformation: a bibliometric analysis of two scientific networks researching socio-technical change. Renewable and Sustainable Energy Reviews 30, 715–723.

De Pryck, K., Forino, G., Da Cunha, C., Remvikos, Y., Gemenne, F., 2014. D7.1: Analytical Framework of the Decision-making Process on Adaptation. RAMSES EU Project. Available at: http://www.ramses-cities.eu/fileadmin/uploads/Deliverables_Uploaded/RAMSES_D7.1.pdf.

Evans, G., 2008. Transformation from "carbon valley" to a "post-carbon society" in a climate change hot spot: the coalfields of the Hunter Valley, new South Wales, Australia. Ecology and Society 13 (1), 39.

Feola, G., Him, M.R., 2016. The diffusion of the transition network in four European countries. Environment and Planning A 48 (11), 2112–2115. https://doi.org/10.1177/0308518X16630989.

Feola, G., Nunes, J.R., 2014. Success and failure of grassroots innovations for addressing climate change: the case of the transition movement. Global Environmental Change 24, 232–250.

Haxeltine, A., Seyfang, G., 2009. Transitions for the People: Theory and Practice of 'Transition' and 'Resilience' in the UK's Transition Movement. In: Tyndall Centre for Climate Change Research, Working Paper, vol. 134. Available at: http://library.uniteddiversity.coop/Transition_Relocalisation_Resilience/Transition_Network/Transitions%20for%20the%20People.pdf.

Hopkin, M., 2016. Australian Renewable Energy Agency Saved but with Reduced Funding – Experts' React. Available at: https://theconversation.com/australian-renewable-energy-agency-saved-but-with-reduced-funding-experts-react-65334.

Nicolosi, E., Feola, G., 2016. Transition in place: dynamics, possibilities, and constraints. Geoforum 76, 153–163.

Saikku, L., Tainio, P., Hildén, M., Antikainen, R., Leskinen, P., Koskela, S., 2016. Diffusion of solar electricity in the network of private actors as a strategic experiment to mitigate climate change. Journal of Cleaner Production 142, 2730–2740. https://doi.org/10.1016/j.jclepro.2016.11.00.

Seyfang, G., Smith, A., 2007. Grassroots innovations for sustainable development: towards a new research and policy agenda. Environmental Politics 16 (4), 584–603.

Seyfang, G., Haxeltine, A., 2012. Growing grassroots innovations: exploring the role of community-based initiatives in governing sustainable energy transitions. Environment and Planning C: Government and Policy 30 (3), 381–400.

Seyfang, G., Longhurst, N., 2013. Desperately seeking niches: grassroots innovations and niche development in the community currency field. Global Environmental Change 23 (5), 881–891.

The City of Newcastle, 2015. Newcastle 2030. Our vision for a smart, liveable and sustainable city. Newcastle Community Strategic Plan. Available at: http://www.newcastle.nsw.gov.au/Newcastle/media/Documents/Strategies%20Plans%20and%20Policies/Plans/Newcastle-2030-V4.pdf.

Transition Newcastle, 2015. Transition Streets. How to Change the World, One Street at a Time. A Sustainability Program for Local Neighbourhoods in Australia. Available at: http://transitionnewcastle.org.au/wp-content/uploads/2016/09/TS-Newcastle-Region-3rd-Edit-April-2015.pdf.

SECTION V

CROSS-CUTTING ISSUES: HINTS FOR INTEGRATED PERSPECTIVES

CHAPTER

29

Integrated Knowledge in Climate Change Adaptation and Risk Mitigation to Support Planning for Reconstruction

Scira Menoni
Politecnico di Milano, Milano, Italy

29.1 CITIES AS COMPLEX ENTITIES

Cities are complex three-dimensional spaces in which social, political and economic organizations interact in different ways and at multiple levels with buildings, infrastructures, production and service facilities, open areas. These interactions reflect the cultural features and the degree of technological development of cities and their inhabitants. **Menoni and Atun (2017)**

Cities differ from one another both qualitatively and quantitatively. Quantitatively as to the number of inhabitants, dimension, number of businesses; qualitatively as to their specialization domain, political and governance arrangements, diversity and richness of economic assets that make some cities more central than others, more vital than others. Through the specialization of assets and functions, cities aim to attract more job opportunities, wealth and gain relevance internationally. Some specializations are permanent, some others, increasingly in the last decades, are temporary, linked, for example, to a specific event a city is going to host for a given period of time, such as becoming the capital of culture in Europe, running an international exhibition, sport games, the Olympic Games. Cities that acquire a high level of specialization orient their assets and services around a key function, such as a port, the financial district, universities, or fairs. The development around a core function or an area makes each city differently vulnerable both physically and systemically to the disruption provoked by an "external" stress such as an earthquake or a flood, either permanently or temporarily.

Contemporary cities need to be understood as nodes acting at multiple scales in space and time. Local, regional, national, and global scales are both interconnected and in tension with each other. The clear-cut dichotomy between central and peripheral has lost its explanatory capacity of the complex interrelationships that relate cities through complementary functions and services, and with their "surroundings." Contemporary cities may be central in a global perspective, or within a nation or a region, depending on the type of functions and services they offer. Cities that are at the margins of global networks can be still central to a region. Mobility has changed forever the spatial dimension of cities, selecting those that are easier to reach with respect to those that are peripheral to high-speed means of transportation. The fact that distances count differently in a globalized world means that regions and cities are closer or farther depending on their position and centrality within transport routes and systems. This has implications also for the way in which aid can be provided to areas stricken by a disaster and on how fast recovery can be achieved.

More than 54% of the world population lives today in cities, 75% in Europe, reversing any prior historic trend, implying differential development dynamics and requiring multilayered governance, resulting from a deeply transformed geography. The tremendous changes implied by the recent urbanization dynamics are able to explain alone the increasing trend of damage witnessed in the last decades as a consequence of natural disasters. Following the 2012 IPCC Report, Pielke (2014) has shown that increased damage and losses provoked by natural hazards are largely explained by increased exposure in hazardous areas and increased vulnerability of exposed assets and cannot be rigorously attributed to climate change. This statement is reinforced by Neumayer (2011) using data from Munich-Re.

Cities as we experience them today are the result of both planned and unplanned actions carried out at different spatial scales by different actors, from individuals to economic, political stakeholders and interest groups. Still, the role of planned intervention holds a very relevant role in providing more livable cities with a balanced mix of functions in different areas, access to services and amenities, provision of green spaces. Cities' complexity requires highly skilled managers to keep cities functioning in ordinary times and able to prevent and respond to multiple stresses including natural disasters and man-made incidents. The role of land use and urban planning in disaster risk prevention has been increasingly stated in internationally relevant documents and agreements, last but not least by the Sendai Framework. Yet, significant obstacles still hamper the full integration of risk prevention into ordinary town and city planning.

29.2 DEALING WITH HAZARDS AND RISK IN CONTEMPORARY CITIES: A MULTIDISCIPLINARY CHALLENGE

Given the complexities that characterize contemporary cities on the one hand and the potential risks that may threaten them on the other, it should be quite evident that a multidisciplinary team with necessary and diverse expertise will be in charge of planning and managing them. In most cases significant fragmentation of competences among different offices dealing at the municipal level with a variety of aspects including transportation, lifeline management, safety issues, climate change adaptation, ordinary planning, ordinary administrative procedures to concede building permits and development rights, is still the rule. Such fragmentation occurs at many levels, from administrative to political, and stems from lack of knowledge and information sharing among relevant experts and organizations. In a study on how knowledge on climate change and adaptation is managed at the local level in Sweden, Glaas et al. (2010) found that often even intersectoral committees created to bridge between different agencies and offices are not well integrated in the ordinary organizational structure of municipalities. Their mandate is not always clear and confined to officers with a technical background, failing to incorporate all potentially interested stakeholders.

Another important divide exists between planners and disaster specialists. Wamsler et al. (2013) noticed that "whilst it is generally recognized that the role of spatial planning for adaptation [to risks and climate change] should be strengthened, the practice is still not well developed." An important reason for this situation relates perhaps to the disciplinary background of most planners. Those educated in policy science focus generally on the social, political, and cultural aspects of risk perception and mitigation, while those educated in architecture and design focus mainly on the features of buildings and the urban fabric. Both generally lack the technical expertise to fully understand the type of stress natural hazards may exert on infrastructures and cities and also the very fast dynamics that may irreversibly change an apparently stable landscape in a very short time.

A second reason relates to how scientific information and knowledge on hazards and risks is provided to planners. Starting from hazard maps and indicators, planners would need different representations depending on the scale at which planning decisions are taken (regional or local) and on the type of locational and development or redevelopment choices to be made (see Margottini and Menoni, 2018). Furthermore, the focus on individual hazards and risks that is usually taken by experts is problematic, especially in areas that are prone to multiple hazards and risks. The coupling of natural and na-tech (incidents triggered by natural hazards) is an eventuality that has to be taken into consideration particularly in large and highly industrialized urban agglomerations. Modeling multihazard and multirisk conditions is very difficult for scientists and engineers and is still a matter of research and of few pioneering experiences. After an earthquake, for example, all attention goes to "seismic" proofing buildings and the urban fabric, disregarding the fact that the affected city may be exposed also to other hazards, such as floods or landslides. However, having to face both the consequences of climate change and of natural hazards requires a multihazard and multirisk perspective. Such perspective would help planners and city managers to consider the potential for cascading effects due to hazards that may trigger one another and to balance prevention investments considering the relative severity of all existing present and future threats.

The fragmentation among scientific communities has also led to a certain separation between science and information produced to support climate change mitigation and adaptation and knowledge to prevent and overcome damage due to natural hazards. In this regard, planners would certainly greatly benefit from a stronger cooperation between the two scientific communities: first, because some "traditional" hazards may be exacerbated and/or modified by climate change; second, because measures for making communities more resilient to threats due to climate change as a hazard and hazards that may or not be influenced by climate change are often much more interconnected than often acknowledged; and third, because often the same officers are responsible for implementing both policies of disaster risk reduction and climate change adaptation.

29.3 RESILIENCE: WILL THIS CONCEPT PROVIDE THE NECESSARY LINK BETWEEN DISASTER SCIENTISTS, CLIMATE CHANGE EXPERTS, AND PLANNERS?

Resilience is a concept that has encountered a certain fortune in recent years among climate change and disaster researchers as well as among planners. Whilst the exact meaning of the term is neither obvious nor identical for different practitioners and across scientific communities, it still holds the potential for overcoming some knowledge barriers and preparing the ground for a much more convergent effort toward making communities, cities, and the built environment able to better cope with climate change, natural and man-made risks.

What "resilient cities" means in practice is not yet clear, nor what is the circular loop between resilient and resistant cities. It may be suggested that "resilience" refers mainly to the capacity to bounce back and recover, referring to the postevent phase, while "resistant" refers mainly to the measures that have been taken before the event to strengthen physical assets and to prepare communities' and organizations' responses. However, as stated by Campanella and Godshalk (2012), "ideal resilient cities are sustainable networks of strong and flexible physical systems, natural environments, human communities, and economic enterprises," suggesting that in order to be resilient a city needs to have addressed correctly at some point the issue of becoming also more resistant. According to Geis (1996), "the concept of a disaster resistant community is just the first step in developing sustainable communities." There are several definitions and even calls for more resilient cities, but very little profound understanding about what this actually means and how in practice the goal of being more resilient can be attained. The "Resilient cities campaign" launched by the International Strategy for Disaster Risk Reduction (ISDR) in May 2010, certainly had the merit to put the issue on the table, however, it has given too little guidance to public administrators and decision makers regarding what to do concretely for their own city.

Disasters and large accidents have been traditionally considered as exceptional and unexpected events in cities, even in cases where hazards were well known and fragilities anticipated long before the occurrence of extreme events. Cities had to recover from the shock and reconstruct; in some instances the issue of creating or rebuilding urban fabric more resistant to future similar events was considered and attended by decision makers, designers, architects, and engineers. Structural measures to protect cities from floods, fires, and even earthquakes developed along with disaster experience and provided a rather sound set of practices that are still conceptually valid today. Most measures are meant to reduce the frequency and/or severity of the expected hazard; yet some others focus on building regulations and on the layout of roads, addressing therefore vulnerable exposed systems. Apparently cities proved to be rather resilient in their history, meaning by resilience the ability "to survive a traumatic blow to physical infrastructure, economy, or social fabric" (Campanella and Godshalk, 2012). An important question to be asked, though, is whether or not modern cities are comparable to ancient ones, whether or not the resilience manifested in the past relies on the same set of capacities that are necessary today to recover from an extreme event.

The climate change community often considers resilience as a synonym, as complementary or as a component of adaptation capacity, whilst the disaster community has been increasingly using it to depict the capacity to recover, to overcome the trauma provoked by a catastrophe and to rebuild in a better way (transforming losses into opportunity of improving the preevent vulnerable conditions). Despite the differences, the two communities seem to converge on considering resilience as a key component making communities better able to manage uncertainties and to prepare for change. The latter may be consequent to temperature rise, changes in relevant meteorological conditions, or to the fact that the environment and the urban fabric, to which one was used to, has been lost due to the level of destruction provoked by a "natural" extreme. In both regards and when applied to cities, the term has the important connotation of learning from and after events, embedding such knowledge in recovery and reconstruction, blending different dimensions of recovery together including symbolic, emotional, and societal beyond the material effort of rebuilding what has been destroyed.

In more recent years, research efforts and publications show greater attention being devoted to the recovery and reconstruction phases (Olshansky, 2017). Partially this is because it is increasingly recognized that disasters cannot be avoided, as the history of the last 50 years has clearly shown, just when science and technologies have reached unprecedented results in understanding and coping with extreme phenomena. According to Guénard and Simay (2011), there has been a shift from research focused mainly on risk prevention approaches toward studies acknowledging the need to prepare to cope with catastrophes, that are to a certain extent inevitable. Without fully embracing the rather radical view of the latter authors, it can be still held that scientists are more aware nowadays of the impossibility to fully anticipate or control the consequences of very large disasters. From a rather aseptic consideration of risks in terms of probabilistic numbers regarding events that can be fully envisaged, a more realistic recognition of the complexity of both natural phenomena and exposed vulnerable human systems has arisen. As a result, being able to recover and reconstruct in a resilient way has been brought back on the agenda of decision makers, and there is renewed interest in models to help decision makers understand how postevent actions may change the condition of risk and vulnerability in the future, in case another extreme event occurs (Miles and Chang, 2006).

Some authors (Olshansky, 2005; Quarantelli, 1999; Comerio, 1998) also suggest that there is difference between recovery after a disaster and a catastrophe: the latter entails a wider and more spread level of destruction. Longer time and larger effort to bring back the situation to normal is required after a catastrophe, also because a partial or even total loss of references and sense of community identity that was embedded in places occurs due to massive destruction. Instead, disasters would imply a less-disruptive impact, which requires more affordable timeline and resources to recover. Having said that, it is clear that the capacity to recover is influenced not only by the extent of the devastation but also (mainly) by the available resources, both financial and human. Also, the window of opportunity in terms of improvement and changes may stay open for a longer time after a catastrophe, as the total reconstruction that is required leaves more options for alternatives. As the authors warn, however, this is true only to a limited extent, as the larger the level of destruction the faster needs to be at least the response to satisfy the most basic needs of sheltering and return to services and work. As Chang and Olshansky (2009) put it, "the central issue in post-disaster recovery is the tension between speed and deliberation: between rebuilding as quickly as possible and considering how to improve on what existed before."

29.4 KNOWLEDGE SUPPORTING BETTER AND MORE ADAPTIVE RECOVERY AND RECONSTRUCTION

If learning is an essential component of a resilient response, one needs to be clear about the type of knowledge, data, and information that is most useful to support postdisaster recovery. First some discussion should be spent on the latter term as it stands with respect to reconstruction. Different interpretations coexist currently. In some instances (Arendt and Alesh, 2015), recovery is addressing the entire process of rebuilding, healing the community, relaunching economic activities and competitiveness whilst reconstruction is considered limited to the physical rehabilitation and reconstruction of the destroyed assets. According to other interpretations (De Ville de Goyet, 2008), recovery and reconstruction are two consequent phases following the impact and the emergency response. The former term depicts the efforts aimed at restoring minimal services and functions in place, while the latter refers to a more solid, stable, and durable restoration of activities, places, and both private and public facilities. The former is shorter in time, encompassing the first 2 to 3 years, whereas the latter has a longer time horizon that, for very severe events, may extend over more than 10 years. The interpretation by phases is due to the pioneering work of Haas et al. (1977) who warned that phases do not follow one another automatically but require instead efficient administration, provision of resources, both human and material, to be massively conveyed to the affected regions. More recently, researchers and practitioners have highlighted the importance of predisaster preparedness not only, as usually considered, to manage the emergency but also, and most importantly, to provide a set of instruments and provisions to support a more effective, efficient, and faster recovery. Smith (2012) stated that whatever is standard and known to be similar in most disasters' aftermath should be prepared in advance, such as ordinances and fast tracks for financial instruments to address the pressure coming from the affected communities and still compliant with regulations and norms. Schwab et al. (1998) goes even further suggesting that having a plan "in case of," providing alternative pathways of development in safer zones for an affected city, is a condition that can not only speed but also support a much more resilient reconstruction. In the absence of the latter, the strife between the need to follow administrative procedures and the wish to provide a better planned future for the disaster zone is not going to be solved, failing to comply with the imperative call for "a vision" that Kartee Shah (2001) opposed to "administration" for the reconstruction of Gujarat devastated by the 2001 earthquake, in a memorable letter to the chief minister of the state at the time (Fig. 29.1).

A plan blending a creative vision for the future but grounded on reliable knowledge and information cannot be improvised in the tense time of the crisis. Not only must ordinances and kits of tools and provisions be preorganized

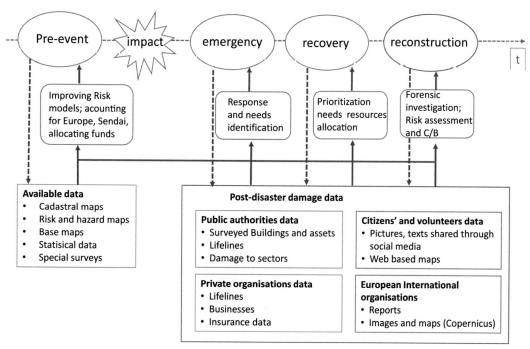

FIGURE 29.1 Pre- and postdisaster data useful for recovery planning. *Author's elaboration. From Mejri O., S. Menoni, K. Matias, N. Aminoltaheri (2017). Crisis information to support spatial planning in post disaster recovery.* International Journal of Disaster Risk Reduction, 22, pp. 46–61.

but also datasets and information systems that use such a vision must be developed. In very broad terms, two types of data and information are necessary: (1) On the one hand, ordinary data such as statistics, thematic maps, reports of various kinds that are produced at predefined time intervals and that support usually a variety of planning and city management purposes. Current advances in information technology and geographic information systems have remarkably augmented the capacity to produce such a knowledge base with satisfactory quality and updates whenever necessary at much faster rates and lower production costs than ever before. (2) On the other hand, as the event may have significantly disrupted the built environment and may have provoked a chain of second and higher-order damage (Rose, 2004), it is fundamental to first collect and second analyze damage and losses data in order to understand how the environment has changed and what are the needs to be prioritized.

In recent years there has been a growing interest in postdisaster data, to overcome the present deficient situation in which only anecdotal evidence is provided for the damage to sectors and systems, and where limited data is collected in a fragmented way mainly for compensation purposes, by a variety of disconnected stakeholders (Molinari et al., 2014). A better structured and coordinated procedure for postdisaster data collection and analysis has been promoted as a pillar of the identification of the most urgent and relevant needs (PDNA, see GFDRR, 2013), and to permit a critical analysis of the causes of the damage. The latter has been labeled "forensic disaster investigation" (Oliver-Smith et al., 2016), aimed at learning from the event and improving the situation with respect to the past, by reducing preexisting levels of vulnerability, exposure, and risks. Interestingly enough a forensic investigation method has also been developed by an insurance company, reinforcing the idea that analysis of damage and loss data can teach significant lessons on how exposed assets become more or less prone to damage and on the overall consistency and reliability of currently used risk models (Zurich Insurance Company Ltd., 2015). Our own research has shown that whatever information system is developed to better organize and subsequently query damage and loss data needs to be tailored to the different purposes that such data could positively serve and requires a much more sophisticated understanding of what damage means for different sectors (ranging from lifelines to agriculture, to commercial and industrial activities, and to cultural heritage).

Yet the two broad groups of data, that of "before" and that of "after," the data that referred to the "normal" and that referred to the "disaster" conditions, must be used together in order to support the vision for the recovery, for deciding what can and has to be maintained as was before and what requires substantial transformation and improvement. This in light of the social and economic conditions of the area that set the arena of the possible and the attainable, even given that external help and financial support will be provided.

29.5 AN EXAMPLE OF THE POSSIBLE APPLICATION OF A MULTIRISK, ATTENTIVE-TO-CLIMATE CHANGE RECOVERY AFTER A DEVASTATING EARTHQUAKE: THE CASE OF THE 2016 CENTRAL ITALY EVENT

The seismic swarm that has and is still affecting Central Italy since August 24, 2016, provoked very severe damage to the built environment, to lifelines, and to cultural heritage in an area that is for a large part mountainous, characterized by a marginal economy as compared to metropolitan areas and coastal regions. The scene where reconstruction will have to be designed displays all the relevant challenges that have been briefly addressed in the previous sections. By being for its most part a mountain area of the Apennines, this area is exposed to a variety of risks, including floods, avalanches, and landslides; the rather unfortunate coincidence of two moderate shakes and of a very severe snowstorm that paralyzed the power sector for more than a week (and even longer in the most remote places) in January 2017 has to be evidence of the need to take a multirisk approach in the reconstruction effort. Some of the municipalities within the epicenter area have been affected also by floods in the last 4 to 5 years, as a consequence of more intense precipitation that has immediate response effects in small catchments. Incorrect land use choices of the past in locating industrial and service facilities and larger lifelines in areas exposed to high levels of multihazards could be revised also in consideration of the potential exacerbation of some hazards (including hydrogeological and also forest fires and drought) as a consequence of climate change. Climate proofing new constructions and to-be-rebuilt infrastructures has to be seen as a key investment while tackling the relative vulnerabilities both physical and systemic to a variety of hazards besides the seismic one.

The two sets of data mentioned previously are both key to developing a strategy for recovery. On the one hand there is the need to prioritize and plan recovery considering the social, economic, and community perspective ahead of those areas that for most part are considered "inner" as defined by an important strategic report that was prepared in 2014 by UVAL (Public Investment Evaluation Unit) and made the object of a specific national policy. According to the report, inner areas are characterized as: "(a) being at some significant distance from the main essential service centres (education, health and mobility); (b) they are rich in important environmental (water resources, agricultural systems, forests, natural and human landscapes) and cultural resources (archaeological assets, historic settlements, abbeys, small museums, skills centres); (c) extremely diversified, as the result of the dynamics of varied and differentiated natural systems, and specific and centuries' old anthropization processes."

On the other hand, data and information regarding damage and losses constitute an essential baseline for learning about the damage causes and identifying needs and requirements for achieving a safer environment, in the meantime livable and at reasonable economic costs, considering the real demographic and economic patterns traceable in many municipalities before the occurrence of the disaster.

In this regard, two noticeable data management efforts can provide significant support with respect to both data groups. On the one hand, the Italian Statistical Office has produced invaluable reports and data extraction for all the municipalities included in the declaration of State of Emergency. At first a full report was issued after the August 24, 2016, first episode for the 17 affected municipalities; later, after the severe shakes of October 25 and 30, 2016, the data provision was extended to all of the newly involved municipalities, the number of which has dramatically grown to 140. The extracted datasets consider demographic indicators, permitting, for example, to establish a ranking regarding the aging condition of the different areas, economic parameters related to the number of activities, employees, type of production, agricultural production, income, etc., and also relevant indicators characterizing the type and quality of existing residential buildings. Initial analysis of such data permits, for example, to distinguish within the affected area a more vital, productive zone that stretches in the north from Perugia in the Umbria Region to the coast of the Marche Region and inner, mountain municipalities. The first is characterized by an order of magnitude larger number of economic activities, and in the second area the latter drops down to less than 10 and even five businesses (see Fig. 29.2).

A second set of data that would deserve further analysis and disaggregation is contained in the report sent by Italy to ask and obtain monies from the Solidarity Fund that the European Union is providing according to the New Union Community Mechanism approved in December 2013. It is interesting to analyze, for example, the share of damage to different sectors represented in Figs. 29.3 and 29.4.[1]

It is rather evident that the largest amount of damage has been suffered by residential buildings, which is easy to explain given the larger exposure and vulnerability of the latter to earthquakes. The figures point also at the relatively high share of damage to lifelines, in particular to the transportation system severely affected by the

---

[1] FSUE: Report sent by the Italian Government to DG-REGIONAL and URBAN POLICY to apply for the Solidarity Fund on the 16th May 2017.

29.5 AN EXAMPLE OF THE POSSIBLE APPLICATION OF A MULTIRISK

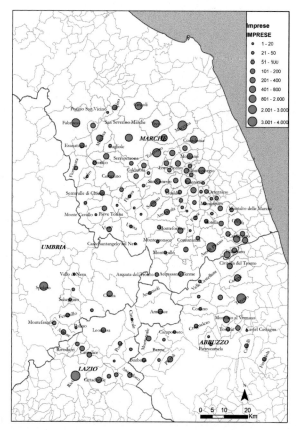

FIGURE 29.2 Number of businesses in the more central areas affected by the 2016 Central Italy earthquake. *Author's elaboration on data from Italian Statistical Office.*

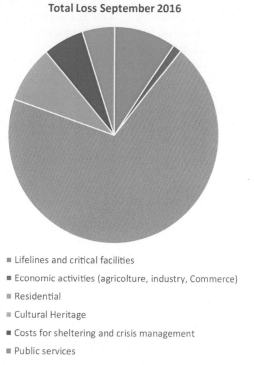

FIGURE 29.3 Share of damage to sectors after the August 24, 2016, earthquake. *Author's elaboration on data extracted from Fondo Solidarietà Europeo (European Solidarity Fund) (FSUE) Report.*

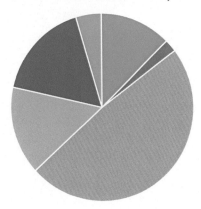

FIGURE 29.4 Share of damage to sectors after the August 24, 2016, earthquake. *Author's elaboration on data extracted from FSUE Report.*

many landslides triggered by the seismic shakes, as well as to cultural heritage sites, not only individual monuments but also entire historic towns such as Castelluccio di Norcia or Amatrice, which were completely devastated.

While the aggregated information provided in the report is certainly not sufficient to carry out a forensic investigation of damage, it is useful to identify the priorities in the different areas, considering also the differential development and degree of economic activities and characteristics of the present population of the latter. The challenge is certainly to use in a clever way data already produced and provided in tailored forms; yet other sets of data and information will have to be integrated. The latter include, for example, the flood hazard maps compliant with the Flood Directive, the avalanches and landslides zonation, as well as the results of the ongoing microzonation studies that will provide the expected level of amplification at a rather detailed scale (10 × 10 m) on which rebuilding decisions and resistance levels to be guaranteed in the future will be set.

Blending all this information and data is necessary to provide the knowledge base regarding the physical, social, and economic aspects of the affected territories in light of what available regionalized scenarios and projections of climate change show for Italy. Recovery strategies will need to envisage both technical and economic objectives as well as constraints and opportunities of the quantitatively small community that has been affected, bearing in mind the most significant risk of those areas—abandonment and loss of cultural and social values.

References

Arendt, L., Alesh, D., 2015. Long-Term Community Recovery from Natural Disasters. Taylor & Francis Group, LLC.
Campanella, T., Godshalk, D., 2012. Urban resilience. In: Randall, C., Weber, R. (Eds.), The Oxford Handbook of Urban Planning. Oxford University Press.
Chang, S., Olshansky, R., 2009. Chapter 2. Planning for disaster recovery: emerging research needs and challenges. In: Blanco, H., Alberti, M. (Eds.), Progress in Planning, vol. 72, pp. 195–250.
Comerio, M., 1998. Disaster Hits Home. New Policy for Urban Housing Recovery. University of California Press.
De Ville de Goyet, C., 2008. Recovery and reconstruction in the aftermath of natural disasters. In: Aimn, S., Goldstein, M. (Eds.), Data Against Natural Disasters : Establishing Effective Systems for Relief, Recovery and Reconstruction. The World Bank, Washington D.C.
GFDRR, 2013. Post-disaster Needs Assessment, Volume A, Guidelines. https://www.gfdrr.org/sites/gfdrr/files/PDNA-Volume-A.pdf.
Geis, D., July 10, 1996. Creating Sustainable and Disaster Resistant Communities. Working Paper. The Aspen Global Change Institute, Aspen, Colorado.
Glaas, E., Jonsson, A., Hjerpe, M., Andersson-Skold, Y., 2010. Managing climate change vulnerabilities: formal institutions and knowledge use as determinants of adaptive capacity at the local level in Sweden. Local Environment: The International Journal of Justice and Sustainability 15 (6), 525–539.
Guénard, F., Simay, P., Mai 23, 2011. Du risque à la catastrophe. À propos d'un nouveau paradigme. La Vie des idées. http://www.laviedesidees.fr/Du-risque-a-lacatastrophe.

Haas, J., Kates, R., Bowden, M., 1977. Reconstruction Following Disasters. Cambridge University Press.

Margottini, C., Menoni, S., 2018. Hazard assessment. In: Bobrowsky, P. (Ed.), Encyclopedia of Engineering Geology. Springer forthcoming.

Mejri, O., Menoni, S., Matias, K., Aminoltaheri, N., 2017. Crisis information to support spatial planning in post disaster recovery. International Journal of Disaster Risk Reduction 22, 46–61.

Menoni, S., Atun, F., 2017. Cities and DRR. In: Educen Culture & Urban Disaster. A Handbook. The Netherlands, ISBN:978-94-6343-600-7.

Molinari, D., Menoni, S., Aronica, G., Ballio, F., Berni, N., Pandolfo, C., Stelluti, M., Minucci, G., 2014. Ex post damage assessment: an Italian experience. Natural Hazards and Earth System Sciences 14, 901–916.

Miles, S., Chang, S., 2006. Modeling community recovery from earthquakes. Earthquake Spectra 22 (3), 439–458.

Neumayer, E., 2011. Normalizing economic loss from natural disasters: a global analysis. Global Environemntal Change 21 (1), 13–24.

Oliver-Smith, A., Alcántara-Ayala, I., Burton, I., Lavell, A., 2016. Forensic Investigations of Disasters (FORIN): A Conceptual Framework and Guide to Research, (IRDR FORIN Publication No. 2). Integrated Research on Disaster Risk, Beijing, 56 pp.

Olshansky, R., 2005. How do Communities Recover from Disaster? A Review of Current Knowledge and an Agenda for Future Research. 46th Annual Conference of the Association of Collegiate Schools of Planning Kansas City. October 27.

Olshansky, R. (Ed.), 2017. Urban Planning After Disasters: Critical Concepts in Built Environment. Routledge, New York.

Pielke, R., 2014. The Rightful Place of Science: Disasters and Climate Change. Consortium for Science, Policy & Outcomes, Tempe, AZ.

Quarantelli, E.L., 1999. The Disaster Recovery Process: What We Know and Do Not Know from Research. Preliminary Paper 286. Disaster Research Center, University of Delaware, Newark, Delaware.

Rose, A., 2004. Economic Principles, Issues, and Research Priorities of Natural Hazard Loss Estimation. In: Okuyama, Y., Chang, S. (Eds.), Modeling of Spatial Economic Impacts of Natural Hazards. Heidelberg, Springer.

Schwab, J., Topping, K.C., Eadie, C.C., Deyle, R.E., Smith, R.A., 1998. Planning for Post-Disaster Recovery and Reconstruction. Fema Report no. 483/484.

Shah, K., December 2001. Earthquake rehabilitation: a 24 strategy for shelter and settlements development. Letter to the Chief Minister Government of Gujarat, Sachivalaya, Gandhinagar, March 1st 2001. Urbanistica (Italian Review of Urban Planning, Also in English) 117.

Smith, G., 2012. Building a theory of recovery: institutional dimensions. International Journal of Mass Emergencies and Disasters 30 (2), 147–170.

Wamsler, C., Brink, E., Rivera, C., 2013. Planning for climate change in urban areas: from theory to practice. Journal of Cleaner Production 50, 68–81.

Zurich Insurance Company Ltd., 2015. The Perc Manual. Learning from Disasters to Build Resilience: A Simple Guide to Conducting a Post Event Review.

Further Reading

Kammerbauer, M., 2013. Planning Urban Disaster Recovery. Spatial, Institutional, and Social Aspects of Urban Disaster Recovery in the U.S.A. – New Orleans after Hurricane Katrina. Verlag und Datenbank für Geisteswissenschaften, Weimar.

Wamsler, C., 2006. Mainstreaming risk reduction in urban planning and housing: a challenge for international aid organisations. Disasters 30 (2), 151–177.

CHAPTER 30

Boundaries, Overlaps and Conflicts Between Disaster Risk Reduction and Adaptation to Climate Change. Are There Prospects of Integration?

Kalliopi Sapountzaki

Harokopio University of Athens, Athens, Greece

30.1 CLIMATE CHANGE AND DISASTER RISK: ARE THEY RELATED?

How much and what type of disaster risks are climate-related? Is this relationship more evident in certain regions of the world? Do climate-related disaster risks show an upward trend? Is climate change (CC) accountable for increases and scaling up of climate-related disaster risks? Are the causal factors/origins of climate-related disaster risk, i.e., the hazard, exposure, and vulnerability factors affected by CC? How are these factors expected to change in the future? In the first section of this chapter about the prospects of integration of disaster risk reduction (DRR) and CC adaptation (CCA) the author attempts to address above these questions by looking:

1. Backward, to address existing trends in weather-related extreme natural processes and respective disasters in the decades since 1970 with a focus on the recent 20-year period 1995–2015, and
2. Forward, into the global and regional development projections and CC scenarios regarding climate variability and extremes to examine whether there is a possibility for suggesting future trends in CC hazards, exposure, and vulnerability.

At first it is necessary to recall the two different approaches to CC: the scientific and the institutional. The scientific originates predominantly from the Intergovernmental Panel on CC (IPCC), which in the respective IPCC Glossary (2013) specifies that:

> Climate change refers to a change in the state of the climate that can be identified (e.g., by using statistical tests) by changes in the mean and/or the variability of its properties, and that persists for an extended period, typically decades or longer. Climate change may be due to natural internal processes or external forcings such as modulations of the solar cycles, volcanic eruptions and persistent anthropogenic changes in the composition of the atmosphere or in land use.

On the other hand, the institutional approach to CC is represented basically by the definition of UNFCCC the United Nations Framework Convention on Climate Change (UNFCCC), which is the main international treaty on CC. The latter identifies CC with the part of it that is attributable to human activity. In particular, Article 1 of UNFCCC (1992) reads: "CC means a change of climate which is attributed, directly or indirectly, to human activity that alters the composition of the global atmosphere in addition to natural climate variability observed over comparable time periods."

Identification of climate-related hazards is a basic precondition for addressing trends in climate-related disaster events in the past. CRED and Munich Re (2009) classify hydrological, meteorological, and climatological extreme events as weather-related hazards. The first category, i.e., the hydrological, includes floods and mass movements (wet) that may cause as secondary hazards rockfalls, landslides, avalanches, and subsidence. Meteorological hazards refer to storms (tropical storms, extratropical cyclones, and local/convective storms), while the third category covers the cases of extreme temperatures (e.g., heat waves), drought, and wildfires.

Already since 1950 changes have been observed with respect to extreme climatic and meteorological events. Extreme events are one component of climate variability. Extreme is defined as an event that is characterized by a climatic or meteorological parameter value that is below or over a threshold, i.e., a value that surpasses the lower or upper limit of a series of observed values of the parameter. Extreme meteorological and climatic events are an issue preoccupying researchers of both fields of CC adaptation and disaster management. Both fields examine the relationship of these phenomena with disasters and disaster risk. The field of adaptation to CC focuses basically on hydrometeorological and oceanographic phenomena and consequent disasters like floods, heat waves, and droughts.

According to the report "The Human Cost of Weather-Related Disasters 1995–2015" (CRED and UNISDR 2015) there has been a significant increasing trend from decade to decade in the number of weather-related disasters worldwide. The average number of weather-related disasters per year in the decade 2005–14 (335) is 14% higher than the respective number in the period 1995–2004 and more than double of that in the decade 1985–94. This upward trend indicates an increase either of the frequency of occurrence of extreme hazardous events or the exposure/vulnerability factor. However, the fact that the number of geophysical disasters remained almost steady in the 30-year period indicates that increase of the hazard factor has been determinant for the upward trend of weather-related disasters.

Worth referring to are the trends of specific types of weather-related hazards and disasters, those that predominate in each one of the three categories, i.e., (1) floods, (2) storms, and (3) drought, extreme temperatures, and wildfires. In the period 1995–2015, floods indicate the following trends:

- There are likely more land regions where the number of heavy precipitation events has increased than where it has decreased. The frequency and intensity of heavy precipitation events has likely increased in North America and Europe (IPCC, 2014).
- The number of floods per year globally shows a 35% increase in the decade 2005–14 compared to the decade 1995–2004. However, it should be taken into account that floods are strongly influenced by a variety of human activities impacting catchments and making the attribution of detected changes to climate change difficult (IPCC, 2014).
- While floods strike in Asia and Africa more than in other continents, they pose an increasing danger elsewhere (in South and North America and Europe, Fig. 30.1) (CRED and UNISDR 2015).
- The features and types of disastrous floods have changed with flash floods, acute riverine, and coastal flooding becoming increasingly frequent (CRED and UNISDR 2015). Urbanization has significantly increased flood runoffs.

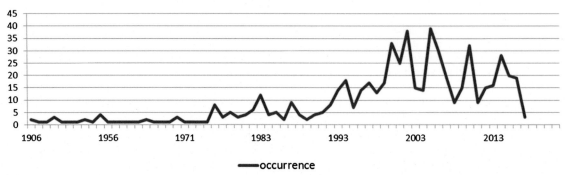

FIGURE 30.1 Total number of flood disaster events, Europe 1906 and 2016. *Elaboration by the author on the basis of data from http://emdat.be/emdat_db/*

With respect to storms (including hurricanes, cyclones, and storm surges) as hazard, there is low confidence that long-term changes in tropical cyclone activity are robust. However, it is certain that intense tropical cyclone activity has increased in the North Atlantic since 1970 (IPCC, 2014). On the other hand, it is likely that experienced storm surges as a result of extreme sea levels have increased since 1970, being mainly the result of mean sea level rise. Overall, storms as disaster events show the following trends (CRED and UNISDR 2015):

- Storm disasters presented the second highest frequency of occurrence in the period 1995 to 2015. Asia (in particular south and southeastern regions) and small island regions/states are highly exposed and the most affected by storms.
- The progress made in weather forecasting in recent years (leading to highly reliable predictions of extreme precipitation and storms 48 h before their outbreak) has made it possible to reduce exposure and vulnerability by means of appropriate warning and preparedness measures.

IPCC (2014) suggests that "it is likely that the frequency of heat waves has increased in large parts of Europe, Asia and Australia" since the mid-20th century. Besides, the report remarks with certainty that "human activity has contributed to the observed global scale changes in the frequency and intensity of daily temperature extremes." Heat waves as disasters present the following features (CRED and UNISDR 2015):

- In the period 1995—2015 extreme temperature events were the third type of disasters in terms of death toll (27% of all deaths attributed to weather-related disasters), with the overwhelming majority being the result of heat waves.
- Under the reservation that deaths from extreme heat are underreported in some countries, it seems that 92% of deaths from heat waves occurred in high-income countries, with Europe representing the majority (90%).
- Europe warms up quicker than the global average temperature (EEA, 2012).

According to IPCC (2014) "there is low confidence in observed global-scale trends in droughts, due to lack of direct observations, dependencies of inferred trends on the choice of the definition for drought, and due to geographical inconsistencies in drought trends." There is also low confidence in the attribution of changes in drought over global land areas since the mid-20th century, due to the same observational uncertainties and difficulties in distinguishing decadal scale variability in drought from long-term trends. Drought affects Africa more than any other continent.

The prrevious statements indicate uncertainty as an essential problem in assessing and attributing extreme event trends of the past to causal factors, and much more so in anticipating future causal relationships IPCC (Mastrandrea et al., 2010). The European Environment Agency defines uncertainty as the degree to which a parameter value (e.g., the future characteristics of the climatic system) is unknown (EEA, 2008). This may be due to lack of information or the controversy about what is known and what is not known. Uncertainty may originate from measurement errors, scientific inadequacies/difficulties, and/or unpredictability of human action. Almost certain is considered an anticipation/scenario characterized by a probability of occurrence reaching 99%—100%, most probable is a probability of 90%—100%, while unlikely is a probability of 0%—33%. In general, reliability and validity of a finding is judged on the basis of its scientific documentation and the degree of consensus over it.

While the anticipated hazards/impacts of CC present spatial imbalances and sectoral diversity, some of them are considered cross-sectoral and highly probable for broad regions. More specifically, the following CC hazards and disaster risks present high probability of occurrence in the decades to follow (IPCC, 2014): risks to health and impacts on livelihoods originating from extreme temperature and heat waves; sea level rise and coastal floods; urban floods and long periods of urban exposure to extreme heat waves; systemic risks of operational interruption of lifelines and critical public services due to extreme weather phenomena; lack of water security and loss of agricultural income; and risks of loss of ecosystems and biodiversity. Water, ecosystems, food, human health, and coastal regions will be increasingly affected by global warming.

Projections for increasing climate-related disaster risks in the next decades are founded not only on the expected increase of the intensity, duration, and frequency of extreme meteorological processes due to CC but also on exposure changes. Upward trends in exposure are not only due to urbanization but also due to the spatial expansion of climate-related hazards. An indicative example of a rapidly increasing climate-related disaster risk owing to mutual reinforcement of hazard and exposure is heat waves in the cities (through the heat island effect). Consequently, climate-related disaster risk increases at the global level due both to intensification and always higher frequency of CC hazards and to exposure expansion and increase of the absolute value of the exposed gross domestic product per capita. According to UNISDR (2011) however, whereas exposure increases vulnerability lowers in the OECD countries.

Reduction of the manmade component of CC is possible by means of appropriate CC mitigation policies reducing its range and rate. These will subsequently lower the possibilities of manifestation of extreme meteorological phenomena. However, reversal of current trends is achievable in the long term only and it is involved as well with spatial imbalances. Additionally, the respective efforts for consensus at the international level still face major governance difficulties. Therefore, there is a need for a combined approach aspiring to both mitigation and CCA, where part of the latter is mitigation and management of climate-related disaster risks. It is obvious then that CCA and DRR overlap in their (theoretical) scope and objectives; the pending query, however, is whether the respective policy sectors support each other in operational terms through synergies, coordination, and/or integration into development and other programs. The following section concentrates on the analysis of the terminology, approaches, and tools of each sector to address convergences and inconsistencies facilitating or complicating integration.

30.2 DRR AND ADAPTATION TO CC: CONVERGENCES AND MISMATCHES IN TERMINOLOGY, PROCESSES, AND TOOLS

DRR is widely considered as a systematic approach to identifying, assessing, and primarily reducing the risks of disaster. While in the past (before the 1970s) the former paradigm of disaster management was considered and practiced as a separate policy sector focusing basically on hazard mitigation and emergency management, since the 1970s the new DRR paradigm has obtained a much broader scope. More specifically, it has been acknowledged that there is a potential for DRR measures and practices in just about every sector of development (and humanitarian work). This is because since then disasters have been considered as socioeconomic and political in origin. Mercer (2010) emphasizes the multidisciplinarity as a result of the "all hazards" component of DRR and the linkages of disaster risk not only with hazard(s) but also with the wider exposed context (social, political, economic, and environmental) that formulates vulnerability and coping capacities. However, DRR as a policy practice is distinctive for and anchored to each, single locality/region bedeviled by a specific hazard or compound of hazards. DRR has been developed basically as a series of practical applications differing from one hazard to another and one community/region to another. The cross-cutting issue of the various DRR local policies is vulnerability reduction. As Mercer (2010) suggests, DRR has a long history *of being successfully implemented at the local level* by means of locally appropriate tools *to reduce community vulnerability to environmental hazards*. It has been acknowledged, though not very successfully in practice, that a major path to territorial vulnerability reduction is mitigation spatial planning, which Balamir et al. (2012) call "planning to avoid," with land use mitigation planning as the spearhead of this policy domain.

The widely acknowledged relationship between DRR and development policies is reflected in the definition of DRR shared by UNISDR and UNDP (2004):

> The conceptual framework of elements considered with the possibilities to minimize vulnerabilities and disaster risks throughout a society, to avoid (prevention) or to limit (mitigation and preparedness) the adverse impacts of hazards, within the broad context of sustainable development.

This definition indicates (1) the emphasis of DRR on vulnerability reduction regardless of the hazard confronted; (2) the introduction of the challenge of embeddedness of DRR measures into every level of development policy making (from transnational to local); and (3) the acknowledgement that DRR is an essential and fundamental aspect of sustainable development. The attention given by the definition to vulnerability is evidence of the interest for the root causes of disaster risk and the conviction that these are to be found in development processes.

In the context of DRR, vulnerability accounts for the characteristics and circumstances of a community, system, or asset that make it susceptible to the damaging effects of a hazard (UNISDR, 2009). This definition identifies vulnerability as a characteristic of the element independent of its exposure. However, the word is often used more broadly to include the element's exposure. While it is a single property, vulnerability is multifaceted, i.e., a "whole" with a number of dimensions or facets (physical, economic, social, systemic, cultural, organizational, institutional, territorial, ecological). Each facet is intrinsically related to every other facet although the nature of these relations varies, i.e., some are closer or stronger than others. These relations are played out in time and space (ENSURE, Del. 2.1, 2011) Vulnerability as a precursor of disaster risk may be measured by susceptibility to loss and the capacity to recover (Cutter, 2006). The close linkages between the social and economic vulnerability facet are demonstrated by the common use of the term *socioeconomic* vulnerability. Socioeconomic vulnerability reflects the processes that "deprive people of the means of coping without incurring damaging losses that leave them physically weak, economically impoverished, socially dependent and psychologically harmed" (Bankoff, 2001, p. 25). Vulnerability is capable of being transferred or "externalized" (one agent may off-load vulnerability to another) and of being transformed (i.e., change in composition). The processes that lead to vulnerability may operate at different scales so that we may recognize vulnerability at the individual, community, region, and state levels.

DRR is a cyclical procedure of distinct, yet overlapping, consecutive stages. Sapountzaki et al. (2011) suggest that while these stages, i.e., Response-Recovery Prevention-Preparedness (the RRPP chain), should function as a continuum, fragmentation and introversion of the respective policies and agencies, i.e., civil protection, development/spatial planning, and sectoral planning (e.g., forest planning in the case of forest fire risk), result in breaks in the chain. Indeed DRR alone faces internal problems of coordination between the involved actors and minimal attention to prevention and the respective role of development and spatial planning. It suffers besides from lack of concrete responsibilities and synergies between the involved actors and sometimes from duplicated measures and funding. According to INCA, EC project (2011) there is a widespread demand in Europe to bridge spatial, functional, and operational gaps and to correct divergence in approach, competence, and perspective between civil protection, spatial planning, and other administrations in charge of risk prevention by a collaborative process with concrete results to make measures and actions of risk mitigation efficient, effective, strategically aligned, and sustainable. According to the same project, "we are in need of more efficient governance and flexibility in risk prevention and response actions."

DRR is accomplished through hazard modification, vulnerability modification, and/or resilience enhancement. This is translated into management or risk governance choices on a policy package to include the most appropriate from a wide spectrum of structural and nonstructural measures. These range from environmental engineering works and emergency warning systems, to preparedness programs that change human behavior in crisis situations, to appropriate spatial and development plans navigating development away from hazard zones. Except for (hazard resisting) building construction codes issued at national or transnational levels (e.g., EU level), most of the other measures are planned and undertaken at the regional and local levels.

CCA as defined by IPCC (2007) is "an adjustment in natural or human systems in response to actual or expected climate stimuli or their effects, which moderates harm or exploit benefit opportunities." CCA is the process of adjusting to new conditions, stresses, and natural hazards resulting from CC with the aim to minimize losses and maximize benefits. Adaptation is both a postimpact and preimpact response; in the second case it comes about in anticipation of expected impacts. CCA can be either a spontaneous, autonomous process/action undertaken by individual agents (private or social) according to their perceptions and capacity (the so-called adaptive capacity or resilience) or a planned and collectively agreed initiative/intervention. There is a widespread agreement that CCA just as DRR should be integrated into development and spatial planning. CCA actions may be part of national or regional adaptation strategies or step-by-step processes at the community or personal level.

A number of European countries have already carried out CC impact assessment studies and the respective national adaptation strategies. The character and form of each strategy and the measures included vary according to the scientific assumptions considered, the particularities of CC impacts in each country, as well as the specific social, economic, and political context. In general terms, the profile of national strategies is affected by the answers to the following questions (Swart et al., 2009):

1. What are the adopted scenarios for future CC and the prevailing development model?
2. What are considered as key-vulnerability sectors and attract attention?
3. Does CC represent primarily a risk or important emerging opportunities?
4. What is CCA perceived as, a local, national, or international issue?
5. Which is the prevailing paradigm in the public dialogue for adaptation? Or there is no public dialogue on CCA?

With regard to the first question, some scenarios are based on future climate projections where the latter are based in turn on assessment of the trajectory of future greenhouse gas emissions. Some others are connected with policy objectives, such as the objective *of the EU for limiting the global temperature increase to 2°C* (compared to the preindustrial period). A third category of scenarios consists of two arms: the climatic and the economic-demographic-technological. In such cases the adaptation strategies are based not only on future climate forecasts but also on projections regarding national and regional economic development and demographic changes. These advanced scenarios acknowledge that adaptation costs and capabilities depend on development standards and spatial allocation of population and activities.

The second question is indicative of the importance of vulnerability both as a concept to understand CC impacts and as a tool to address key groups, sectors, and systems in need of adaptation and to assess adaptive capacity and options accordingly. According to IPCC (2007), *"Vulnerability to climate change is the degree to which geophysical, biological and socio-economic systems are susceptible to, and unable to cope with, adverse impacts of climate change"* (see also Füssel and Klein, 2006). In the report of the OECD (2006) "Adaptation to CC: Key Terms," it is acknowledged—on the basis of the Third Assessment Report by IPCC, 2001—that *"Vulnerability is a function, of the character, magnitude and rate of climate variation to which a system is exposed, its sensitivity and its adaptive capacity.* The term *vulnerability* may therefore refer to the vulnerable system itself, e.g., coastal cities; the impact to this system, e.g., flooding of coastal cities; or the mechanism causing these impacts. Key vulnerabilities are associated with many climate-sensitive systems: food supply, infrastructure, health, water resources, coastal systems, ecosystems, global biogeochemical cycles, etc.

Even within the community of CC there are two distinct and different vulnerability conceptions. The first identifies vulnerability in connection to the final, residual CC impacts after adaptation, i.e., those that have not been tackled by means of adaptation measures. The second uses as reference point the beginning of the adaptation process and considers vulnerability as a general feature/property of the societies originating from a diversity of social and economic processes. In the final stage approach, adaptive capacity (or resilience) is crucial for the resulting vulnerability level, while in the initial stage approach, vulnerability is the basic factor determining adaptive capacity organization for optimum adaptation. To use an example, according to the final stage approach, only part of the aged population in Europe remain vulnerable to heat waves after adaptation measures (private or public) at least in relation to the resources at their disposal. On the other hand, according to the initial stage approach, all aged people are vulnerable to heat waves due to their advanced age and therefore predisposition to disease (Villagrán de León, 2006).

The third question above indicates that CCA has a wide scope that exceeds the part of DRR referring to CC hazards; it covers as well issues of exploitation of opportunities emerging from CC (e.g., reduction of energy demand for heating in Northern European countries). Hence, the two sectors overlap and both of them are wider than their common part (Fig. 30.2).

The fourth question above indicates clearly that CCA is a multiscalar issue. However, as Birkmann and Teichman (2010) point out, most of the adaptation strategies designed up to now refer to entire countries/regions, and in any case there is no such thing as a hierarchy of adaptation plans. The causes of the frequent lack of such plans at the local level lie first and foremost in the lack of local, downscaled data of CC effects or localized forecasts of CC extreme events (Birkmann and Teichamn, 2010), as well as in low public perceptions about the necessity of local-scale climate policies. This constitutes a major mismatch between DRR and CCA; while the first is practiced mostly at the regional and local levels, the second is performed basically at the national and transnational levels. On top there exists an institutional mismatch; while CC issues and responsibilities are assumed by environment ministries and

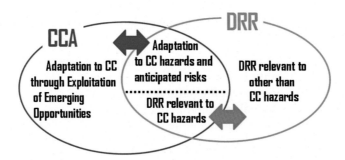

FIGURE 30.2 Scope of DRR and CCA and overlaps. *Author's elaboration.*

meteorological services, DRR lies within the responsibilities of the ministries of defense or interior (for civil protection), sectoral ministries (depending on the agent at risk), and development or spatial planning ministries (only if the latter are institutionally involved in DRR).

Regarding temporal scales, CCA strategies require principally long-term perspectives "far beyond any political election period" (Birkmann and Teichman, 2010). In fact, short-term-oriented adaptation measures that contradict long-term CC mitigation and adaptation objectives are usually referred to as "maladaptations." On the other hand, DRR has a dual temporal profile. Theoretically, as a sustainability objective, it should be based on a long-term prevention perspective considering both the response and recovery stages after disasters as windows of opportunity for comprehensive vulnerability reduction and resilience enhancement. In practice, however, most of DRR operates as a short-term preparation or postdisaster response to restore the predisaster conditions. This is due to neglect of safety issues in routine development planning, the short-term character of funding schemes oriented to specific disaster events, and the pressing needs and backward-looking forms of demand from disaster victims.

The fifth question above reveals the critical governance issues involved in a public dialogue on CCA: the issue of knowledge and uncertainty about CC and adaptation results, the issue of CC information dissemination and communication to the experts and nonexperts, as well as issues of public perception. An obvious problem rests with the difficulties in communication between the climate and risk scientists and practitioners (particularly those responsible in public administration) due to differences in the respective terminologies and concepts. An additional cause of uncertainty already mentioned is the lack of knowledge, data, and information on local effects of CC, trends in extreme weather events, social and economic census data, etc. (Birkmann and Teichman, 2010). Finally, uncertainty is increased further by the lack of norms and clear consensual visions to determine ultimate objectives of adaptation, monitoring indicators, and allocation of adaptation costs. Under conditions of uncertainty, perceptions of the lay public and responsible administrations on the necessity to undertake CC adaptation programs are low (particularly at the local level), compared with those on the necessity for DRR actions.

Theoretically, the major convergence of DRR and CCA policy sectors is their common concern for vulnerability reduction (Schipper, 2009), despite the fact that the notion of vulnerability has different connotations in the respective scientific fields. In particular, vulnerability in the context of DRR is considered against all versions of hazards whereas in the context of CCA it is considered against climatic variability and extreme events; vulnerability in the first case refers to exposed elements (population, assets, economic activities, infrastructure, territories) whereas in the second it refers to biophysical-social systems; vulnerability in the first case is a function of susceptibility to loss and coping capacity, whereas in the second it is climate variation, sensitivity, and adaptive capacity. However, the two fields share a vulnerability component of great significance for adaptations to reduce susceptibility to loss and sensitivity or upgrade adaptive capacity and resilience—the nonclimatic factors accountable for vulnerability, i.e., socioeconomic, territorial, and institutional. This is the reason why development-spatial planning is the ideal policy platform for integration of the two fields. Table 30.1 summarizes the basic features of DRR and CCA functioning as catalysts or obstacles to synergies between and integration of the two fields.

30.3 THE NEED FOR AND PROSPECTS OF INTEGRATION OF DRR AND ADAPTATION TO CC

The need for integration of DRR and adaptation to CC is an imperative just as is the need to mainstream climate adaptation into sectoral policies and the need to coordinate civil protection, spatial planning, and sectoral policies (pertaining to specific types of risk management) to rehabilitate the continuum of the disaster risk management cycle. The reasons are always comparable:

- To avoid situations where the respective processes of hazard information collection, vulnerability assessment, and planning for resilience and adaptation run in parallel without any linkages, feedback, and interactions;
- To avoid situations of fragmented and duplicated funding to achieve one and the same objective;
- To shift the emphasis of measures and funding from emergency short-term responses to disasters and short-term maladaptation to climatic stresses toward long-term disaster risk prevention and long-lasting adaptations to changing climatic stresses;

TABLE 30.1 Scope of DRR and CCA Policy Sectors and Catalysts/Obstacles to Integration

	Components/Features	DRR	CCA	Catalysts/Obstacles to Integration
Terminology (Key terms)	Hazard	"...a dangerous phenomenon, substance, activity or condition associated with disaster risk" (UNISDR, 2009). DRR is concerned with all hazard cases	It is concerned with CC hazards only	• Easy communication between DRR and CCA experts and practitioners • Easy communication of known hazards to the public • From the hazards point of view, DRR is a wider field including CCA
	Exposure	The elements (and their value) present in hazard zones	The nature and degree to which a system is exposed to climatic variations/extremes	• Communication difficulties due to different conception of "exposed agent" and "stressor"
	Vulnerability	"...susceptibility to loss in case of exposure to hazards...It is a function of susceptibility and coping capacity"	"...the degree to which geophysical, biological and socio-economic systems are susceptible to, and unable to cope with, adverse impacts of CC...A function of the climatic hazard, sensitivity and adaptive capacity"	• Difficulties in communication due to different conception of the vulnerable agent and vulnerability determinant parameters
	Resilience	"The ability of a system, or society ...to resist, absorb, accommodate to and recover from the effects of a hazard ...., including through the preservation of its basic structures & functions."	"...the intrinsic capability of a system to adapt to the stresses and impacts of CC (existing or anticipated)"	• Confusion with the term *adaptive capacity*, because they are very close to each other. Their basic difference rests with the stressor which, in the case of CC, is not always a hazard
	Adaptive Capacity	The term *coping capacity* is used instead	"The ...capability of a system to adapt to (to alter to better suit) climatic stimuli or their effects/impacts" (IPCC, 2001)	• Confusion between the terms *adaptive* and *coping capacity*
	Adaptation	Not used	The process of adjusting to a changing climate and its impacts (CC hazards)	• A term familiar only to CC scientists and practitioners
Knowledge, Information, Uncertainty		• Clear norms, measures, and indicators leading to DRR and community protection up to acceptable levels of risk on the basis of probabilistic and deterministic scientific approaches	• Knowledge gaps and lack of information on local effects of CC, trends of extreme weather events, etc. • High uncertainty due to knowledge gaps and the lack of norms and clear consensual visions on objectives of adaptation	• Low public perceptions do not favor CCA policies at the regional and local level • Risk-based approaches to CCA may motivate CCA (OECD, 2015a) • Approaches to DRR and CCA based on spatial planning is a path to coordination of the three policy fields
Scales of reference and performance	Spatial	*Prevention*: primarily at the regional and local scale (except hazard-resistant national-level building standards); *Emergency plans*: Structured all along the administrative hierarchies	Ideally, all scales from the international to the local. In practice, mostly the national and transnational scale	• Inconsistency between DRR and CCA spatial scales • Embeddedness of DRR and CCA in the development and spatial planning system may secure CCA and DRR implementation at all scales

30.3 THE NEED FOR AND PROSPECTS OF INTEGRATION OF DRR AND ADAPTATION TO CC

	Aspect	DRR	CCA	Integration / Spatial Planning
	Temporal	Ideally, a perpetual cyclical process; in practice mostly short-term responses restoring predisaster conditions	Ideally, a long-term recurrent process involved with multiple planning horizons (the near future, i.e., 2020s, and the long term, i.e., 2050s); in practice often short-term reactive maladaptations	• Inconsistency between the theoretically optimal long-term processes and actual short-term practices • Spatial planning is a platform for proactive and preventive practices and synchronization of DRR and CCA
Procedural elements	Scope and objectives	*Scope*: All hazards *Objectives*: To reduce vulnerability to all hazards and multihazards and boost resilience	*Scope*: CC hazards and impacts *Objectives*: To reduce vulnerability and boost adaptive capacity	• Common objectives favor unification of the respective governance processes in what might be called "Risk + CC Impact Governance"
	Operational features of the process	A standalone perpetual cyclical policy of consecutive, overlapping stages facing problems of articulation with policies addressing normal period needs.	• A cross-sectoral policy involving multiple planning horizons with different degrees of uncertainty • The ideal process is to mainstream adaptation into almost every case of policy making	• Emergency planning is a field at the crossroads of policies for enhancement of both coping capacity/resilience (DRR) and climatic adaptive capacity (CCA) • Spatial planning (SP) is a platform for multisector and territorial vulnerability reduction
	Internal coordination	Problems of coordination with SP despite its critical role in vulnerability reduction	Mainstreaming adaptation into every case of policy making may cause problems of coherence of climate-related actions	• DRR is a more coherent policy suffering, however, from a prevention deficit • CCA has started as an all-sector policy lacking internal coherence
	Normative framework	More or less concrete, commonly known, and socially acceptable norms	Uncertain, ambiguous, and difficult-to-communicate normative frameworks	Need for risk-based communication to the public to establish convincing norms for both DRR and CCA. The UNISDR "Resilient City" campaign is an example
	Linkages with development and SP	Development and SP is an ideal field for long-term preventive DRR through vulnerability reduction	Development and SP is a basic means for mitigating the nonclimatic factors of vulnerability to CC	There is a need for a major turn in SP to combine equally and effectively objectives of "Planning to Achieve" and "Planning to Avoid"
Responsible institutions and Issues of governance		• Agencies falling within the competence of ministries of interior, defense, sectoral and spatial planning	• Environment ministries, meteorological services • Every policy sector in case of successful mainstreaming of CCA (OECD, 2015b)	Solutions to integration differ and depend on the preexisting institutional landscapes. The spatial planning and emergency/preparedness structures are the spearheads for integration.

Author's elaboration.

- To consider the cumulative impact of multihazard sources (climatic and nonclimatic) and multirisk situations and formulate vulnerability reduction and resilience/adaptive capacity enhancement policies accordingly;
- To avoid situations of waste of administrative resources owing to unreasonably fragmented areas of competence;
- To avoid lack of accountability owing to duplicate or overlapping administrative competences;
- To avoid misalignments between measures taken to serve DRR on the one hand and CCA on the other.

Integration of DRR and CCA is mutually beneficial, because:

- The scope of DRR is expanding as the basic policy to address and adapt to new and intensified climatic stresses; also because social and political acceptance of DRR costs improve and DRR gains currency as a primary objective within the scope of development and spatial planning.
- The urgency of CCA is boosted further (as a path to climate risk mitigation) and relevant public perceptions (about the need for adaptation) heighten, thus leading to proactive initiatives at the private, social, and institutional level; also because CCA obtains a solid normative framework and a familiar operational procedure (i.e., modified versions of DRR) toward objectives.
- Performance of "generic vulnerability" analysis to serve simultaneously CCA and DRR aims allows both policy fields to enjoy economies of scale.

The remarks in the last column of Table 16.1 referring to catalysts and obstacles to integration of DRR and CCA lead the author to propose specific paths to integration: integration on a theoretical/terminology basis, procedural basis, institutional basis, the basis of public information to change predominant public perceptions on CC risks and adaptation, and on the basis of governance toward common objectives (vulnerability reduction and strengthening adaptive capacity).

Integration on a theoretical basis requires resolution of inconsistencies in the two terminologies and introduction of a common extended one to embrace all hazards and risks (climatic and nonclimatic and combinations) and include all versions of coping and adaptive capacity and resilience with parallel clarification of differences in concept and use. From this perspective, critical queries focus around (1) the definition and bounding of vulnerable and exposed entities (e.g., "are these complex systems or simply elements of a distinct character, economic, social, technical…?") and (2) the issue of generic vulnerability, i.e., the part of vulnerability that is detached from the hazard and that is common for all hazard/stressor cases (climatic and nonclimatic). A step forward in theoretical integration is represented by mixed approaches combining methodological elements from both the CC and the DRR scientific fields; an indicative example is risk-based approaches to adaptation planning (OECD, 2015a). The respective framework proposes four steps for managing climatic risks: (1) identifying the risks, (2) characterizing the risks, (3) choosing and exploring adaptation policies to address the risks, and (4) responding to evolving risks through an iterative process of feedback and learning. What is especially interesting is first that the framework acknowledges the multirisk landscapes created by the CC process and the evolving, uncertain character of climate hazards and risks, and second, that the framework is based on previous ones built by OECD and taking advantage of methods of risk governance, resilience systems analysis, and approaches to adaptation to single climatic risks (like water risk).

Integration on a procedural basis is about addressing operational stages and the respective facilitating structures and policies that are common in DRR and CCA. Preparedness and emergency planning agencies accommodate reinforcement of both coping capacity for DRR and adaptive for CCA. On the other hand, development and spatial planning agencies and processes might accommodate long-term vulnerability reduction for the sake of both CCA and DRR should professionals and the concerned community adopt it as a high-priority objective. According to the Dutch National Adaptation Strategy 2007 entitled "Make Space for Climate," climate proofing involves the spatial development of the entire country. The Strategy relates primarily to spatial measures although raising awareness and identifying knowledge gaps are also concerns. It is not by chance that the ministry for spatial planning is also the ministry for climate as its title is Ministry for Climate and Spatial Planning.

Regarding integration on an institutional basis, ministries and public agencies with parallel competences for the environment, climate, and spatial planning is a means to realize coordination between CCA and DRR and of both with spatial planning.

Integration on the basis of governance of consensual community objectives is an approach that is favorable at the regional and local levels. In 2006–07 Mercer (2010) conducted research in three rural indigenous communities in Papua New Guinea (PNG) (Singas, Kumalu, and Baliau in the provinces Morobe and Madang) to show that (1) CC is only one factor amongst many contributing to community vulnerability, reduction of which is the basic issue at stake, and (2) CCA should be embedded within DRR. PNG is a multihazard territory prone to a wide range of environmental hazards (volcanic eruptions, landslides, earthquakes, floods, cyclones, tsunamis, wildfires, droughts, and frost) and suffers as well from technological, biological, health risks, and crises related to civil unrest and conflict. Mercer (2010) remarks that PNG is considered highly vulnerable to CC due to rural population's mainly subsistence livelihoods, and rapid environmental degradation due to extensive deforestation. Regarding CC and DRR policies, Mercer (2010) observes:

> In PNG developing climate change and DRR policies has been very much top-down, frequently initiated with limited community involvement. This is an all too familiar story occurring worldwide as evidenced by recent reports analyzing progress towards the Hyogo Framework for Action 2005–2015, both at the global and local levels...

This research resulted in a process framework indicating how indigenous and scientific knowledge might be integrated for DRR. The framework was subsequently revised to incorporate CC concerns (Mercer et al., 2010), and it was developed as a community tool to reduce vulnerability to hazards. As Mercer (2010) put it, "the terminology of CCA and DRR meant nothing to PNG communities." The participatory approach adopted was the "guided discovery" process where "community members were facilitated to identify, facts, problems and solutions for themselves with minimal interference from facilitators… The process enabled the communities to situate the hazards within their wider development context." According to Mercer, CCA is not a stand-alone issue especially at the local level suffering from actual impacts of CC; it is a part of a DRR approach where all aspects of vulnerability contributing to community risk are of concern, and DRR is a part of a wider sustainable development policy. Should Mercer's remarks be correct, we should acknowledge that by inventing and pushing forward CCA as a new stand-alone top-down policy, we have created an unnecessary problem we are now struggling to resolve—integration of DRR and CCA.

References

Balamir, M., Colucci, A., Sapountzaki, K., 2012. Aspects of Learning in Planning to Avoid. AESOP 26th Annual Congress, "Planning to Achieve/Planning to Avoid". 11–15 July 2012, Ankara, Turkey.

Bankoff, G., 2001. Rendering the world unsafe: "vulnerability" as Western discourse. Disasters 25 (1), 19–35. https://doi.org/10.1111/1467-7717.00159.

Birkmann, J., Teichman, K., 2010. Integrating disaster risk reduction and climate change adaptation: key challenges-scales, knowledge and norms. Sustainability Science 5, 171–184. https://doi.org/10.1007/s11625-010-0108-y.

CRED-Centre for Research on the Epidemiology of Disasters and Munich RE-Munich Reinsurance Company, Disaster Category Classification and peril Terminology for Operational Purposes, 2009, Working Paper, Available at: https://elsevierbookweb2.proofcentral.com/index.html?token=f088044504153e12b0f823c76633e1dc&typen=AU.

Cutter, S., 2006. Hazards, Vulnerability and Environmental Justice. Earthscan, Sterling, VA.

EC project ENSURE, 2011. Enhancing Resilience of Communities and Territories Facing Natural and Na-tech Hazards. Del. 2–1 Relations between different types of social and economic vulnerability, (Parker, D.J., Tapsell, S., Handmer, J., Kidron, G., Omer, I., Benenson, I., Bakman, Y., Zilberman, T., Costa, L., Kropp, J., Molinari, D., Bonadonna, C., Gregg, C., Menoni, S.). EC Contract No. 212046. Available at: http://cordis.europa.eu/publication/rcn/14275_en.html.

EC project INCA, 2011. Linking Civil Protection and Planning by Agreement on Objectives. Civil Protection Financial Instrument of the European Communities. Available at: http://www.project-inca.eu/.

EEA – European Environment Agency, 2012. Climate Change, Impacts and Vulnerability in Europe 2012 –An Indicator-based Report. EEA Report, No. 12/2012. Available at: https://www.eea.europa.eu/publications/environmental-indicator-report-2012.

EEA – European Environment Agency, 2008. Impacts of Europe's Changing Climate – 2008 Indicator-based Assessment. Joint EEA –JRC – WHO Report. EEA Report, 4/2008.

Füssel, H.M., Klein, R.J.T., 2006. Climate change vulnerability assessments: an evolution of conceptual thinking. Climate Change 75 (3), 301–329. https://doi.org/10.1007/s10584-006-0329-3.

IPCC, 2001. Climate Change 2007: 3rd Assessment Report.

IPCC, 2007. Climate Change 2007: 4th Assessment Report.

IPCC, 2014. Climate change 2014: synthesis report. In: Pachauri, R.K., Meyer, L.A. (Eds.), Fifth Assessment Report of the Intergovernmental Panel on Climate Change. Contribution of Working Groups I, II and III. IPCC, Geneva, Switzerland.

IPCC, 2013. Annex III: Glossary [Planton, S. (Ed.)]. In: Stocker, T.F., Qin, D., Plattner, G.-K., Tignor, M., Allen, S.K., Boschung, J., Nauels, A., Xia, Y., Bex, V., Midgley, P.M. (Eds.), Climate Change 2013: The Physical Science Basis. Contribution of Working Group I to the Fifth Assessment Report of the Intergovernmental Panel on Climate Change. Cambridge University Press, Cambridge, United Kingdom and New York, NY, USA. Available at: http://www.ipcc.ch/pdf/assessment-report/ar5/wg1/WG1AR5_AnnexIII_FINAL.pdf.

IPCC (Mastrandrea et al.), Guidance Note for Lead Authors of the IPCC Fifth Assessment Report on Consistent Treatment of Uncertainties, 2010, Available at: http://www.ipcc.ch/pdf/supporting-material/uncertainty-guidance-note.pdf.

Mercer, J., 2010. Disaster risk reduction or climate change Adaptation? Are we reinventing the wheel? Journal of International Development 22, 247–264. https://doi.org/10.1002/jid.1677.

Mercer, J., Kelman, I., Taranis, L., Suchet-Pearson, S., 2010. Framework for integrating indigenous and scientific knowledge for disaster risk reduction. Disasters 34 (1), 214–239. https://doi.org/10.1111/j.1467-7717.2009.01126.x.

OECD/IEA, 2006. Adaptation to Climate Change: Key Terms. Available at: http://www.oecd.org/env/cc/36736773.pdf.

OECD, 2015a. Framing risk-based approaches to adaptation planning. In: OECD, Climate Change Risks and Adaptation: Linking Policy and Economics. OECD Publishing, Paris. https://doi.org/10.1787/9789264234611-en.

OECD, 2015b. Tools to mainstream adaptation into decision-making processes. In: OECD, Climate Change Risks and Adaptation: Linking Policy and Economics. OECD Publishing, Paris. https://doi.org/10.1787/9789264234611-en.

Sapountzaki, K., Wanczura, S., Casertano, G., Greiving, S., Xanthopoulos, G., Ferrera, F., 2011. Disconnected policies and actors and the missing role of spatial planning throughout the risk management cycle. Natural Hazards 59 (3), 1445−1474. https://doi.org/10.1007/s11069-011-9843-3.

Schipper, L., 2009. Meeting at the crossroads?: Exploring the linkages between climate change adaptation and disaster risk reduction. Climate and Development. 1 (1), 16−30. https://doi.org/10.3763/cdev.2009.0004.

Swart, R., Biesbroek, R., Binnerup, S., Carter, T.R., Cowan, C., Henrichs, T., Loquen, S., Mela, H., Morecroft, M., Reese, M., Rey, D., 2009. Europe Adapts to Climate Change: Comparing National Adaptation Strategies. Report No 1, PEER (Partnership for European Environmental Research), Helsinki. Available at: http://www.peer.eu/.

UNFCCC-United Nations Framework Convention on Climate Change, 1992. UN. Available at: https://unfccc.int/files/essential_background/background_publications_htmlpdf/application/pdf/conveng.pdf.

UNISDR-United Nations Office for Disaster Risk Reduction, 2009. Terminology. UN. Available at: http://www.unisdr.org/files/7817_UNISDRTerminologyEnglish.pdf.

UNISDR−International Strategy for Disaster Risk Reduction, 2011. Global Assessment Report on Disaster Risk Reduction (GAR2011)−revealing Risk, Redefining Development. UN. Available at: http://www.preventionweb.net/english/hyogo/gar/2011/en/home/download.html.

Villagrán de León, J.C., 2006. Vulnerability − a Conceptual and Methodological Review.

Further Reading

IFRC-International Federation of Red Cross and Red Crescent Societies/the ProVention Consortium, 2007. Tools for Mainstreaming Disaster Risk Reduction: Guidance Notes for Development Organizations [Ch. Benson and J. Twigg]. ProVention Consortium, Switzerland. Available at: http://www.proventionconsortium.net/themes/default/pdfs/tools_for_mainstreaming_DRR.pdf.

Schipper, L., Liu, W., Krawanchid, D., Chanthy, S., 2010. Review of Climate Change Adaptation Methods and Tools. MRC Technical Paper No. 34, Mekong River Commission, Vientiane. Available at: http://www.mrcmekong.org/assets/Publications/technical/Tech-No34-Review-of-climate-change.pdf.

CHAPTER

31

The Contribution of the Economic Thinking to Innovate Disaster Risk Reduction Policies and Action

Giulia Pesaro

Poltecnico di Milano, Milan, Italy

31.1 URBAN SYSTEMS AND DISASTERS FROM AN ECONOMIC PERSPECTIVE

An urban area can be seen as a system of components, an aggregate formed by combining several elements, subjects, objects, activities, and use functions for the built and open environments. Following an economic approach, such elements represent the system of resources an urban area itself may rely on to function and continuously evolve or, in other words, the territorial capital on which the lives and activities of both inhabitants and economic subjects may count on to continue to develop and, optimistically, produce new innovative activities and components. In a disaster risk reduction (DRR) perspective, such elements and capital resources are the values exposed to disasters, and their characteristics and quality degrees represent the vulnerability of the territorial capital when facing disasters.

The shapes, characteristics, local specificities, and availability of the whole of the urban resources constituting a certain territorial capital are the result of a complex system of actions/reactions/nonactions, which are, in turn, the results of combinations and interactions of elements and their dynamics. Combinations and interactions are stratified over time and integrated in territorial "objects," activities, and resources such as to produce a specific and, to some extent, unique territorial capital, different from many others. This happens particularly where the natural ecosystems, natural and cultural landscapes, the cultural heritage, and the community model and quality are involved.

A specific territorial capital is, therefore, also the result of a variety of decision-making processes, developed over time by a huge variety of different subjects/stakeholders, with defined systems of interests, needs, and goals. The capital of resources that characterizes an urban area may assume different roles according to the variety of different activities and expected productions that are the effects of such decisions. This means that, starting from a similar territorial capital, the characteristics and the image of the area may deeply differ, producing different exposure and vulnerability profiles. The same elements themselves, in different systems, may assume different specific potentials and values and are more or less valuable in relation to their availability, renewability, sustainability, and, of course, direct economic and financial values. Moreover, they not only sustain the usual, every day, functioning of the related area but also have potential functions and roles for future developments, often not completely exploited or visible just looking at the state-of-the-art settings.

As a consequence, if a disaster occurs, the view offered by the economic thinking is that of a reduction in the territorial capital as the result of damages and losses suffered, which means a reduction of available values integrated in the damaged or lost resources—a reduction that can be so high as to produce the definitive death of a territorial area because of the lack of major values, components, and means needed for the maintenance of the specific production and consumption models as before the catastrophe (Pesaro, 2007). Here some additional assumptions are needed in order to better understand the implications of such an approach. Here it is important to briefly mention that, in an economic reasoning perspective, damage and loss have different meanings (Van Der Veen et al., 2003). The meanings are mainly related to the capability to measure or assess, by way of money as the unit of measure, the reduction of part or the entirety of the components—and the related values—constituting the capital of the territorial resources and/or of their functionality if a disaster occurs. It is possible to assess damage in direct value terms using money as the unit of measure, even if the assessment process is often complex and not evident. Losses, on the contrary, are very difficult to assess and quite often never using a clear value measure unit, as they refer to components that cannot be replaced or have intrinsic, frequently very high, values that can hardly be quantified, like the natural environment, the cultural heritage, the community elements, the social relationships. Losses, therefore, may refer to resources that cannot be substituted by others and once lost will not be available anymore (at least in the short and medium term).

The greater is the vulnerability, or fragility, of the territorial components exposed to risks and their values and the lower is their substitutability, renewability, or recoverability, the greater will be the total value of the expected damages and losses. The application of the economic approaches and assessment tools can therefore offer an effective support to explain to what extent and how a disaster will affect the capital of local resources. This is in terms of direct and indirect values of the whole amount of resources involved in the disaster, looking to the capability of the system to continue producing over time goods and services, values, and revenues for final consumers and production means and to reproduce (if possible) the lost resources. This means that damages and losses not only have to be accounted for directly but also in terms of indirect and systemic impacts according to the cause-and-effect chain that characterizes the damage and loss dynamics over time and involves huge territorial systems, even very far away from each other but linked by economic or geopolitical relationships. For instance, the tsunami that hit Japan in 2011 produced important effects and damages to US companies because of the commercial links between the two countries (Nanto et al., 2011).

If both potential damages and, mainly, losses are very high, looking at the whole of the direct, indirect, and systemic impacts, the best possible resilience strategy is risk mitigation and reduction. The ex ante action to reduce or at least mitigate damage and losses are the best possible solutions not to lose not only the investments needed to replace territorial capital but mainly the resources that cannot be quantified and are not replaceable, as well as the systemic impacts, whose dimensions and effects are rather difficult to identify and intervene on. This means that, from an economic perspective, a smart territory is a territory able to invest in preparedness, damage mitigation, and risk reduction as the best possible strategies to use the available financial means (considering both public and private territorial subjects). Moreover, this means it is necessary to work on different elements and dynamics using, in a strategic way, a complex and richer toolbox, drawing attention to civil protection and emergency management as well as on ex ante mitigation measures and, when possible, such as for floods, on risk sources.

Following this reasoning, DRR can be considered the premise to look at urban environments and their components from a different perspective, which is that of the "system of values at stake" or exposed to uncertainty and systemic dynamics. These are, of course, not only related to risks themselves but also to other territorial events and developments, which, individually, have to be adequately understood in order to push the system toward a "stronger equilibrium."

If, for instance, we consider the urban environments, it is increasingly important to be able to recognize the whole of the urban components and dynamics that, at different levels and intensity, shape the capital of urban values and positively or negatively influence their performances—particularly concerning the functions and functioning of the built environments and urban services when exposed to disasters. As an example, in the Mediterranean region, many important urban areas are prone at the same time to several different natural hazards and other territorial risks, whose disruptive potential might be amplified by three main elements. The first one is the average age of the built heritage and the related core characteristics in terms of architectural styles, building materials, and state of conservation, on which basis restoration and regeneration investments should be done in order to make a city or urban area stronger in coping with disasters. A second element is the density of the built environments, which creates the need for huge amounts of economic and financial means in order to intervene so not to produce inequalities or just move exposition and vulnerabilities from an area to another. Finally, the presence of multicultural communities and a variety of production and consumption models can produce an increase in vulnerability and a greater difficulty in action. This is due to the different perceptions people have toward preparedness and emergency itself or the capability to follow rules, guidelines, procedures, and other instructions concerning the use and maintenance of the buildings and the urban infrastructures when an event occurs. These elements make it even clearer how much the social components, together with the economic ones, may become key elements to make an urban community more or less exposed and, mainly, vulnerable.

31.2 COPING WITH DISASTER RISK: HOW ECONOMIC UNDERSTANDING MIGHT ENHANCE THE DRR ACTION

Following an economic perspective, the damage produced to a territory by risks like major natural hazards means a loss of development resources for a territorial system as a whole, in terms of damage if the value of the loss can be measured in money and if the damaged resources can be substituted by others. This means drawing attention to civil protection and damage management in the short period and, at the same time, carefully stating action goals and tools to enhance prevention and ex ante risk and damage mitigation in the long run. This is in order to minimize the loss of total territorial resources in case a major natural event occurs.

Over time, it has become clear that the intervention of the state or, more generally, of the public bodies in DRR and resilience matters is central to obtain both an effective and efficient emergency and recovery program and an action able to reduce risk and damage in a prevention perspective. In urban areas, rich of exposed and often vulnerable elements but also where the economic and financial resources have to be shared among a variety of different uses, the risk issues can find obstacles. Action toward the enhancement of prevention and preparedness might therefore be slowed down because of the uncertainty that characterizes the risks and, therefore, because of the low perception and willingness to pay for producing safety when the benefits of such investments are not visible because the risks are not visible. On the other hand, safety and resilience seem to be elements able to increasingly condition decision-making with reference to the choice for where to establish new residential investments and the headquarters and production units of firms and multinational companies. A smart urban area should therefore better consider the production of safety and resilience as a resource itself for enhancing the attractiveness of strong and productive activities and people.

Interventions for the enhancement of the DRR action system and for the production of safety and resilience depend of course on the infrastructures and bodies in charge for the governance and management of risks, especially when referring to natural hazards—a governance system making decisions, selecting actions, and finding investments and financial means for prevention interventions, organization, and management of the consequences of natural disasters (civil protection), reestablishment, and mitigation of the impacts of damage on the territory. Moreover, these actions have to consider the whole of the exposed and damaged components of a community at the same time (nonrivalry), without excluding anyone, no matter what gender or social conditions (nonexcludability). If the economic approach is applied, the conditions are recognizable at which it is possible to assume that the production of safety from risks, with particular regard to natural hazards, can be legitimately regarded as a public good. This is a particularly important assumption, as it implies that the public bodies have to lead the action for producing and enhancing safety and resilience, for instance, facing the impacts of climate change dynamics such as floods and heat waves in urban areas. Actually, another implication of defining community safety and resilience as public goods is that the market mechanism itself is not able to autonomously produce such services and resources in the private markets and therefore the direct intervention of the state or public bodies, at the national as well as regional or international levels, is indispensable. Public strategies, action plans, and regulations and an increasing amount of tools and mitigation measures are therefore tools needed to push the urban systems acting in order to find resources and financial support to intervene for climate change adaptation and mitigation, also increasing the involvement of the individual private stakeholders. This is a system that, over the years and based on experience and knowledge building, has increased in both number and variety of stakeholders and in resources and capabilities available to support decision-making processes becoming more and more important and complex and involving many urban actors. This system, in synthesis, has become increasingly effective and efficient—more able to obtain the expected results and more able to maximize the productivity of the resources used to produce safety and resilience.

Among others, one of the main economic-based tools to support decision-making for the selection of risk and damage mitigation measures is the cost-benefit analysis, a concept that underlines the idea that direct, often public, investments (costs) can be undertaken to obtain safety as mitigation of damage and risk (benefits). In this way, investments to produce a public good and savings in terms of damage and losses if an event occurs become savings in terms of territorial capital and of the related production of value flows over time—particularly referring to nonrenewable and nonreplaceable resources and to environmental, social, and cultural values exposed and vulnerable to risks (Pesaro et al., 2016).

In light of the increasing importance and recognition of the economic dimensions of DRR, since the early 1990s there has been increasing attention to economic-based assessment methods and solutions. These mainly refer to damage accountability, from the one side, and, more recently, to the contribution of the economic thinking and toolboxes in enhancing action and funding for risk and damage mitigation facing, in particular, natural disasters. Thus, there is an increase in the contribution of economists in multidisciplinary working groups, integrating other study fields with their disciplinary approaches.

One result has been a renewed interest in and an increasing number of applications of cost-benefit analysis (CBA) applied to natural hazards as it makes it clearer and easier to understand the relationship between investments in mitigation measures and the related effects. Moreover, looking at urban and territorial systems, CBA helps in better understanding exposition and vulnerability in a more direct and clear economic way. Exposition is the portion and typology of the components of the territorial capital (elements and resources) potentially involved if an event occurs. Vulnerability refers to both the quality and characteristics of the exposed elements, which might increase damage and losses due to an event, and the nature and importance of the chain of interconnected impacts at the system level—direct, indirect, and systemic impacts, which might amplify the losses of the territorial area involved and increase the difficulties in recovery and rebuilding.

One main element has to be emphasized when considering the operational implementation of CBA. Even if this assessment tool is becoming popular, the methods, elements, and indicators needed to adequately identify the whole of the costs and benefits to be considered are still under development. Looking at damage data and mitigation measures provides the ability to produce effective images of both values prone to risks and investments needed for mitigation measures. Moreover, in light of the great variety of local conditions influencing the impacts of events, even just floods, it allows a better understing on territorial subjects and objects with reference to exposure and vulnerability. A second aspect refers to the identification of the elements needed to develop an effective cost-benefit analysis itself. This is from a methodological and operational perspective, where effective means not only able to sustain real decision-making toward the enhancement of regional resilience but also based on the whole of the needed data required to design an effective and complete image of the values at stake and the chain of related and indirect impacts (see, among others, the results of the European Commission—funded project IDEA, as reported in Pesaro et al., 2016).

31.3 CBA AS AN ECONOMIC-BASED TOOL TO ENHANCE DECISION-MAKING PROCESSES IN DRR ACTION AND TERRITORIAL RESILIENCE

The use of CBA analysis in order to enhance the implementation of a disaster mitigation measure demonstrates the concept of intervention profitability, which refers to the capability of investments in mitigation measures to obtain the expected outcomes in terms of risk prevention and/or damage mitigation. It is a method to compare the whole of the direct costs associated with each typology of mitigation measure and the whole of the results of it, measured as the total value of the avoided damage and losses plus the benefits coming from the increase in safety and territorial quality. Such a method can therefore sustain the decision makers by making available knowledge about the positive results attainable by way of the expenditure referring to a certain mitigation measure to be implemented. In this conceptual framework, CBA is seen as an ex ante decision-making tool, which calls for the capability to develop the assessment process in an ex ante perspective. It is important to emphasize this point as the prevention concept is involved.

The assessment methods and tools to quantify some of the damage categories are central elements when referring to the economic approach to DRR and resilience matters. The debate developed during the last 2 decades agrees that damages suffered by populations and built environments have been much more clearly assessed than in the past for almost all events. Less evidence is available, in contrast, with respect to damage to other damage categories like the economic sectors and the whole of the local level territorial resources, such as the cultural heritage, the natural environment and the community's identity, and social models (some of these elements are discussed in Mechler, 2016). This is mainly due to the incidence of indirect and systemic damage (Cochrane, 2004a,b), which is often underestimated (Shreve and Kelman, 2014) but also on the losses suffered by the territorial systems looking at intangibles, public and common resources, and cultural and historical heritages (Pesaro, 2005).

In economic sectors, for instance, indirect and systemic damage may be huge and the time needed for restoration or reconstruction and may become crucial factors for the capability of a whole territorial system to start again with its "everyday life." These factors should be better recognized and deepened, the costs being related to the cascade of impacts coming from business interruptions (direct or because of the interruption of lifelines and other territorial infrastructures and services; Rose and Huyck, 2016), rebuilding and reconstruction investments and time needed, substitution of machineries and production materials, and injuries to workers. This may result in a deep loss of competitiveness and reduction of market shares—so high, sometimes, to cause economic subjects to stop their activities. It is therefore easy to understand how indirect and systemic losses might weight in terms of future development and resilience of the territories hit by natural disasters.

Looking at the whole of the resources a territorial area can rely on to sustain its production and consumption models and qualitative and quantitative growth, the economic perspective suggests to account for not only direct quantitative damage, using money as its measure unit, but also indirect and systemic damage and losses even if not easily quantifiable and measurable—for instance, when the lost resources are public, not renewable, or have a very long regeneration time, such as, as already mentioned, for cultural heritage, natural environment, cultural landscapes, community identity, and social models. Monetization models have been developed in the field of environmental economy, first, and in risk and damage matters afterward, but still, monetary evaluation remains very difficult (see, among others, Meyers et al., 2013). This can reduce the feasibility of CBA in DRR, as it is not possible to take into account the whole of the values prone to risk from a direct quantitative monetary perspective, to be compared with the direct and clear costs of the mitigation measures. When the reference area is rich in cultural heritage, for instance, or when the indirect costs cannot be adequately detected, CBA reliability itself weakens as a decision-making tool for selecting the most effective mitigation measure among the possible ones. The loss of nonrenewable values or potentially very high indirect losses might actually suggest implementing expensive solutions even if not directly comparable to the dimensions assessed for the avoided damage. The evidence itself of the existence of such "definitive" potential loss of resources might make the expenditure for the reduction of risk and damage even more desirable and crucial, enhancing the desirability and acceptability of expenditure (investments) in prevention actions and territorial and community resilience projects and solutions. As a matter of fact, from an economic perspective, the costs payed to reduce losses as much as possible mean resilience, because it means maximizing the protection of the whole of territorial resources, engines, and basis for future development or precious and irreplaceable heritage from the past. Moreover, avoided damage also means lower emergency costs and time to recover from the disaster impacts, values very difficult to forecast in advance, also due to the variability of events themselves. Finally, costs in risk and damage mitigation measures may mean benefits coming from the enhancement of territorial and community resilience itself, seen as a strength of a certain area and making it more attractive for people and economic subjects.

On the other hand, even if the incidence of the nonmonetizable values influences the potentials of CBA, it remains an important tool to rely on. In a decision-making framework, the main question is how to choose among different possible action/intervention alternatives using an economy-based toolbox, whose strength also lies in the use of quantitative measure units, able to offer clear-to-read results to support and address a selection process that, many times, takes place at the policy level. Moreover, it enables the possibility to look at the results envisaged for different mitigation measures not only for what concerns the technical/technological and operational performance, linked to the physical capability to face hazards and the related events, but also in terms of investments' effectiveness.

This takes us also to another crucial element, that is the suffering of finance that characterizes many territorial areas and countries, mainly affecting the public subjects but having effects also on private spending options. Such suffering creates the conditions for the development of tools to support decision-making processes among investments, comparing different mitigation measures models, different expected effectiveness results due to local specificities, different impacts of the undertaken measures, and possible negative externalities (see, for instance, mitigation measures producing important interferences with natural environment and cultural landscape).

The results of a CBA might be effective also in light of a common problem decision makers often have to face, that is, the perception of direct and quantifiable costs related to mitigation measures implementation. Quantified costs, even when related to damage prevention, are much clearer and easier to perceive than the possibility of a reduction of potential damages and losses in an uncertain and not visible or known future. If uncertainty about an event is great, there is also a lower perception of the value of the avoided future damages. A dynamic which may produce heavy negative impacts because uncertainty is greater when referring to highly devastating events (high return time and severity of natural disasters). The more today's mitigation costs are clear but not directly comparable to future damage decrease, the more difficult it is to obtain consensus on expenditures (both public and private) in time of peace. This is a core matter, as the predisposition to investments in mitigation measures is a milestone in enhancing the implementation of DRR policies and in the involvement of territorial subjects. In places where floods are more likely to happen and could be considered in terms of "ordinary" territorial events, the territorial area involved develops mitigation measures as a "normal" way to react, and the perception about the advantages of prevention becomes higher. If looking at the "extreme" case of Venice, it is clear how preparedness is central to maintain the city. Flood events are perceived as normal and take a specific name, the so-called *acqua alta*. These are expected floods and forecasting services are normally provided to the urban community and the tourists. Looking at the behaviors of public authorities, residents, shopkeepers, and businesses (especially those linked to tourism), expenditures in defense measures and preparedness are normal and the benefits of interventions and investments in safety are clear.

Some of these measures have even become local trademarks, like the bulkheads and shutters for the front doors, moving up from the ground floor to close the shop and house entrances, and the gangways to walk over the water in open spaces. Venice's prevention measures also include other less-visible actions. In restaurants, there are places to put tables and chairs in case the water comes inside, electrical devices and plugs are positioned above the floor level, walls are protected with water-resistant coverings, etc. Advice is also normally given to tourists when there is *alta aqua*, like to take cash with them as the flooding might close down cash machines and shops' credit card devices.

31.4 CONCLUDING REMARKS

The economic approach, particularly if integrated in CBA, is a way to read an urban area and its strength and weaknesses. Of course the innovation is in the perspective, as a value and a vulnerability assessment are attached to each territorial component. Clearly, deep knowledge of the urban area and the related developments and dynamics are crucial to obtain a realistic knowledge framework for DRR and urban resilience enhancement.

This is even more important when referring to risks, like floods, where the mitigation action can be applied to the risk source itself. This allows the system to act in both risk and damage mitigation and reduction. The action on risk mitigation appears more effective because the investments work on the reduction of the flood risk itself and of the exposition, no matter how vulnerable the exposed subjects and objects are. Working on the damage reduction side means, in contrast, the capability to recognize and intervene on exposure and vulnerability of a great variety of individual situations, and identify the potential loss of territorial resources/values that could be not replaceable with others or be easily restored/rebuilt. Finally, it is important to consider the economic subjects' viewpoint if a disaster occurs. The events might be more and more disruptive and the damage might be so high, compared to the territorial and social specificities and working framework, that there might no longer be conditions to continue producing goods and services to sustain the life of the concerned area. Mountain and internal areas are, for instance, very prone to such additional risks (Botzen et al., 2017). Mitigation measures therefore must be identified and mitigation costs also have to be evaluated according to such potential dynamics.

Once the importance of making the ratio between mitigation effectiveness and implementation costs is more visible and understandable, the next step is related to how to concretely recognize the "list" of costs and benefits suitable for describing the risk of a certain territorial area and how to evaluate and monetize them. This is because there is great variation in local conditions influencing the impacts of events, even just floods, on territorial subjects and objects, with reference to exposure and vulnerability. Moreover, the accounting of potential damage has to be related to a realistic dimension and severity of the potential risk affecting the area, which means related to the probability of an event with different possible destructive consequences. The experts in the economic appraisal should therefore work closer with risk and damage scientists, engineers, architects, and other experts able to recognize the exposed elements and to identify and understand their vulnerabilities according to the severity of events, thus producing an image of the components of the potential damage to be finally evaluated in CBA.

The evidence of the economic and territorial effectiveness of the measures to mitigate risk and damage is even more important as the growth, in number and values, of territorial objects (buildings, infrastructures, and other physical elements), subjects and activities exposed to risks is increasing. Why is it therefore so difficult to share consensus on mitigation measures (apart from the difficulties in modeling the potential damage to be compared to)? The dialogue with territorial active subjects and stakeholders reveals to be very important, as their viewpoints and demands are central components in the CBA design. Costs and benefits to be compared are, as a matter of fact, connected to the related system of stakeholders, their activities, resources, and expectations in terms of development and quality of life and work. Linked to this, another question arises: who is paying for what and for whose advantage? This particularly refers to public/private expenditures and public/private impacts of such expenditures, which, in the presence of suffering finances, should consider emerging trade-offs between mitigation and compensation. Over time, apart from the evidence that the "pure" compensation for the entire suffered damage is no longer sustainable for public systems alone, the importance of mitigation and prevention measures is increasing. This is also because of the value of human lives and the existence of nonrenewable and highly valuable territorial resources and heritage, which are increasingly considered and perceived as fundamental resources and values. Mitigation measures become therefore also means of territorial resilience enhancement.

In some cases, like flood risk, it is possible to intervene to mitigate the risk itself or the exposure or its vulnerability. Exposure assembles the exposed elements (territorial subjects, objects, and activities) to the hazard in terms of quantity and values in danger compared to the event characteristics. Vulnerability addresses the factors, characteristics, and other conditions that can make the exposed elements more fragile and prone to potential damage. The effectiveness of mitigation measures, might they be structural/physical or nonstructural, is therefore assessed in terms of capability to reduce the (economic) impact of events working on the reduction/mitigation of risk, exposure, and vulnerability.

One of the limits of the economic tools based on damage assessment, like CBA, is that they often are still too complex and demand scientific knowledge to be implemented. It is therefore difficult to imagine an extensive use of such tools as "normal" assessment tools to support decision-making in public administrations and the state. This is particularly evident when CBA or even just the damage assessment (avoided damage in CBA) is meant to be used in ex ante situations, where real damage data may not be available and where, therefore, other data sources should be made available. Such datasets, often developed for other territorial assessment purposes, might not be directly or easily usable in ex ante simulation for expected damage modeling. The predictive capabilities of CBA in assessing mitigation measures without the evidence from previous events and when trying to shift attention from the micro- to the mesoscale could therefore be weakened. Still, the increase in the availability of case studies and operational methodologies suggests that relevant improvement could be done in order to better learn how to asses and interpret such evaluation methods, what kinds of professional profiles are needed from an interdisciplinary perspective, and, finally, how to make them become more diffused.

From the benefit-side perspective, when referring to the accounting for damage as the avoided costs, problems have been observed in real data collection activities (as an example, see Pesaro et al., 2016), particularly looking at economic subjects. For households, available data and data collection methods and tools seem able to catch the high majority of the damages really suffered. For the economic subjects, in contrast, improvements are still needed. This is mainly related to the high variability and differences of these kinds of subjects and of their activities and built environments, which, at least at present, make it more difficult to identify indicators, collect and organize the needed data, and produce the measures. There are also problems in providing base mapping and reliable images of land uses needed to develop territorial assessments at the over-local level, especially when trying to obtain models of ex ante damage assessment for a certain area. This is because existing datasets for land use analysis have not been conceived in order to provide such information. It is, however, important to emphasize that more-detailed and precise images of land use are increasingly needed for many other territorial policy activities, which calls for better integration and collaboration in data production and processing among different categories of potential final users.

From the cost side, in an economic perspective, the main goal of a decision-making process is to be able to identify the best possible mitigation measures able, individually, to maximize the reduction in damage amounts, and the risk if possible. Better knowledge of the variety of structural and nonstructural risk and damage mitigation measures seems increasingly important considering the variety of potential stakeholders and interventions. Multidisciplinary research groups could ameliorate the knowledge system starting from data collection activities and processing to the use of data in CBA models. It is also important to enhance the capabilities of all involved public and private subjects to distinguish the better options compared to a variety of different starting environments for the design of more effective risk and damage mitigation measures. Moreover, the multidisciplinary perspective in teams working on data collection and the implementation of CBA in risk management has proved many times to be very important. The scientific and technical capabilities needed to design risk and damage scenarios must better interact (and vice versa) with those needed to recognize and deal with the whole of the socioeconomic elements and values involved, their specificities, and measure units. This particularly refers to the need for better integration of the economic thinking when assessing the economic impacts of mitigation, also in light of other, economic-based, solutions to tailor to the economic subjects' needs, such as "better fitting" insurance options or incentives programs to develop mitigation measures at the firm level.

A smart city is a city able to face and cope with disasters in a way assigning a high value to safety as a highly valuable resource for territorial development and, increasingly important, a resource to attract new and smart economic subjects and individuals. Safety, therefore, has to be recognized as an urban service and a characteristic of being smart and, of course, intrinsically more resilient. The production of such a service, mainly (but not only) by the public institutions (because safety is a public good), deserves the right governance, action and management model, investments, and economic efforts according to local specificities and capabilities and resources.

References

Botzen, W.W.J., Monteiro, E., Estrada, F., Pesaro, G., Menoni, S., 2017. Economic assessment of mitigating damage of flood events: cost—benefit analysis of flood-proofing commercial buildings in Umbria, Italy. The Geneva Papers on Risk and Insurance — Issues and Practice forthcoming.

Cochrane, H.C., 2004a. Economic loss: myth and measurement. Disaster Prevention and Management 13 (4), 290—296.

Cochrane, H.C., 2004b. Indirect losses from natural disasters: measurement and myth. In: Okuyama, Y., Chang, S.E. (Eds.), Modeling the Spatial and Economic Effects of Disasters. Springer, New York.

Mechler, R., 2016. Reviewing estimates of the economic efficiency of disaster risk management: opportunities and limitations of using risk-based cost–benefit analysis. Natural Hazards 81 (2016), 2121–2147.

Meyers, V., Becker, N., Markantonis, V., Schwarze, R., van den Bergh, J.C.J.M., Bouwer, L.M., Bubeck, P., Ciavola, P., Genovese, E., Green, C., Hallegatte, S., Kreibich, H., Lequeux, Q., Logar, I., Papyrakis, E., Pfurtscheller, C., Poussin, J., Przyluski, V., Thieken, A.H., Viavattene, C., 2013. Review article: assessing the costs of natural hazards – state of the art and knowledge gaps. Natural Hazards and Earth System Sciences 13, 1351–1373.

Nanto, D.K., Cooper, W.H., Donnelly, J.M., Johnson, R., 2011. Japan's 2011 Earthquake and Tsunami: Economic Effects and Implications for the United States, Congressional Research Service Report for Congress.

Pesaro, G., 2005. La conservazione dei centri storici in zona sismica: un approccio economico. In: Lagomarsino, S., Ugolini, P. (Eds.), Rischio sismico, territorio e centri storici. FrancoAngeli, Milano.

Pesaro, G., 2007. Prevention and mitigation of the territorial impacts of natural hazards: the contribution of economic and public-private cooperation instruments. In: Aven, T., Vinnem, J.E. (Eds.), Risk, Reliability and Societal Safety – Vol. 1 Specialisation Topics, London. Taylor & Francis, London 603-612.

Pesaro, G., Bezzam, V., Botzen, W.W.J., Hudson, P., Mendoza, M., Menoni, S., Minnucci, G., Erika Monteiro, E., Russo, F., 2016. Cost-benefit Analysis of Mitigation Measures to Pilot Firms/Infrastructures in Italy, IDEA Project, Deliverable D.4. Available at: http://www.ideaproject.polimi.it.

Rose, A., Huyck, C.K., 2016. Improving catastrophe modelling for business interruption insurance needs. Risk Analysis. https://doi.org/10.1111/risa.12550.

Shreve, C.M., Kelman, I., 2014. Does mitigation save? Reviewing cost-benefit analyses of disaster risk reduction. International Journal of Disaster Risk Reduction 10 (2014), 213–235.

Van der Veen, A., Vetere Arellano, A.L., Nordvik, J.-P. (Eds.), 2003. In Search of a Common Methodology on Damage Estimation, Workshop Proceedings, European Commission - DG Joint Research Centre, EUR 20997 EN European Communities.

Further Reading

European Environment Agency, 2010. Mapping the Impacts of Natural Hazards and Technological Accidents in Europe. An Overview of the Last Decade. EEA Technical Report. 13/2010, Copenhagen.

European Environment Agency, 2016. Floodplain Management: Reducing Flood Risks and Restoring Healthy Ecosystems (EEA Report, Copenhagen).

Hawley, K., Moench, M., Sabbag, L., 2012. Understanding the Economics of Flood Risk Reduction: A Preliminary Analysis. Institute for Social and Environmental Transition-International, Boulder.

Howe, C.W., Cochrane, H.C., 1993. Guidelines for the Uniform Definition, Identification, and Measurement of Economic Damages from Natural Hazard Events: With Comments on Historical Assets, Human Capital, and Natural Capital, Program on Environment and Behavior. Special Publication No. 28. Institute of Behavioral Science University of Colorado.

National Research Council, Committee on Assessing the Costs of Natural Disasters, 1999. The Impacts of Natural Disasters: A Framework for Loss Estimation. The National Academies Press, Washington DC.

Rose, A., 2017. Defining and Measuring Economic Resilience from a Societal, Environmental and Security Perspective. Springer Science+Business Media, Singapore.

CHAPTER 32

Flood Resilient Districts: Integrating Expert and Community Knowledge in Genoa

Daniele F. Bignami, Emanuele Biagi
Fondazione Politecnico di Milano, Milan, Italy

32.1 URBAN CONTEXT, ENVIRONMENTAL CHARACTERISTICS AND CITIZENS' ENGAGEMENT

The field investigation concerns the districts of Piazza Adriatico and Ponte Carrega, located in Val Bisagno in the upstream of the city of Genoa. The study area is located in Genoa-Staglieno, in the basin of the Bisagno River, the most important stream that crosses the city. The blocks involved in the study, with an altitude between 60 and 30 m AMSL, are bordered to the north by Via Ponte Carrega, to the south by the most southern part of Piazza Adriatico, to the east by the steep hills of the upper part of Piazza Adriatico, and to the West by Via Lungobisagno Dalmazia (Fig. 32.1). These areas are particularly exposed to flood risk, like most of the Ligurian cities; the characteristics of Genoa's streams have significantly changed due to the present urban setting of the city, especially in their final stretches within the built territory. In fact, most of the time such watercourses have been forced to flow in small canals, almost always covered, which in most cases are insufficient to contain the flow project (Rosso, 2014). If compared to other zones within the Genoa context, Piazza Adriatico and the area of Ponte Carrega are historically among the areas that are more at risk from the hydraulic point of view. In fact, the area is crossed not only by the western stream of the Bisagno River, a torrent sadly known because of its frequent floods (Faccini et al., 2015) but also by two secondary tributary canals, the Mermi and Torre creeks, which are the main and direct cause of the flooding that occurred in recent years in the two selected districts.

In the days right after the floods of 2011, local residents, many of them seriously stricken by the event, were active initially trying to address the main risk issues of the area; therefore, they founded the association "Amici di Ponte Carrega" (http://www.amicidipontecarrega.it/). This association operates in the field of prevention and flood risk culture, organizing meetings and conferences, and promoting activities focused on citizens' participation in land management; the association works also in collaboration with universities and associations at local and national levels. Because of the recent floods, the association "Amici di Ponte Carrega" contacted, among others, "Fondazione Politecnico di Milano" for an investigation of the area and to suggest ideas and solutions for the defense of citizens and goods of the area and to transfer them, bottom-up, to the decision makers.

32.2 MAIN METEOROLOGICAL PHENOMENA AND THEIR IMPACTS

The climate of the city of Genoa, as also reported in the section "Climate and microclimate" of P.U.C. (City Urban Planning), is strongly influenced by its geographical location and by a few other factors including the immediate and extended contact with a deep, open sea, the southern exposure, and the close presence of mountains that provide protection from northerly winds and a barrier to the humid flows from the south.

The precipitation regime, which represents an essential element in the definition of each climate area, contributes in a peculiar way to characterize the climate of Genoa. In fact, the annual total average and the monthly distribution of rainfall does not deviate from the norm of the Italian Mediterranean climates; however, the presence of rainy

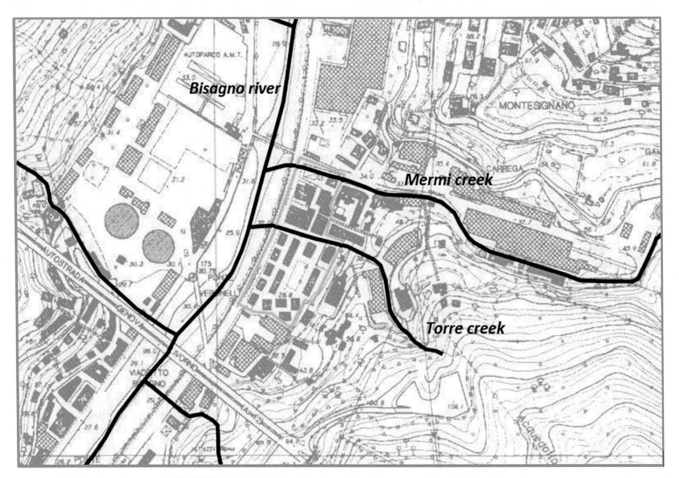

FIGURE 32.1 Hydrographic network of the area. *Authors' elaboration.*

events of great violence, that occur at highly irregular intervals, is an important distinctive characteristic. During such events, precipitation intensities are always very high, so that they can cause serious damage to the city, as they did in October 1970, and, recently, in November 2011 (but also in October 2014 and September 2015). Among the extreme events, the most interesting one is the 948 mm/24 h recorded by Bolzaneto station (rain gauge station located about 10 km away from the city center) during the flood event in October 1970, which is to date the highest value of precipitation in 24 h recorded in a European city.

A matter of great importance, especially in today's context, is to be able to assess and predict the variations of the characteristics of meteorological events that affect the Genoa area. The analysis of rainfall recorded between 1833 and 2008 reveals that even if the cumulative values of annual precipitation show a steady trend, there is a significant negative trend concerning the number of days of annual rain. Both of these trends show how the daily precipitation rate tends to increase along the duration of the registration period, which is symptom of a possible change toward a more extreme climate regime (Fig. 32.2).

32.3 MAIN FRAGILITIES

The area of intervention has been subjected to many flooding events, especially during the most recent years (Fig. 32.3). These phenomena are mainly affected by the neighborhood's urban layout, which forced the secondary creeks (Mermi and Torre) to flow in artificial riverbeds across the study area. Some stretches of these creeks are covered and insufficient to regulate the flow of flood discharge. The Torre creek, in particular, is characterized in its terminal part by two long stretches covered and separated by an open, short stretch, which tends to suffer high pressure for flows having a return period of less than 50 years: this causes floods that mainly affect the area of Piazza Adriatico, which is built under the road network level.

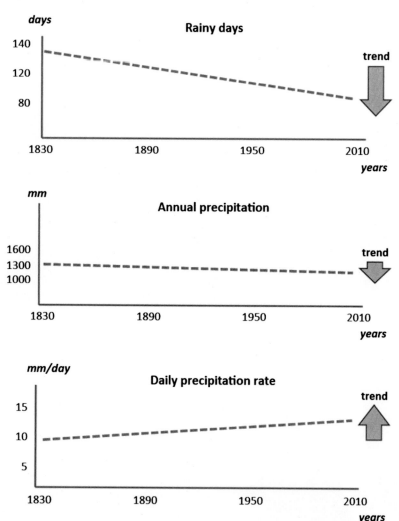

FIGURE 32.2 Trend of number of rainy days (first graph), annual precipitation rate (second graph), and daily precipitation rate (third graph) obtained by data recorded at the University of Genoa Station between 1833 and 2006; red line represent the trend of the data used. *Authors' elaboration based on Brandolini, P., Cevasco, A., Firpo, M., 2012. Geo-hydrological risk management for civil protection purposes in the urban area of Genoa (Liguria, NW Italy). Natural Hazards and Earth System Sciences 12, 943–959. https://doi.org/10.5194/nhess-12-943-2012.*

Concerning the Mermi creek, in recent years it has been the target of some hydraulic adjustment measures, which were completed in 2015, after the construction of a new commercial building in the northwest area. The goal of such actions is to guarantee a proper disposal of the flow with a return period of 200 years; however, they did not take into account the runoff volumes that derive from the upstream slopes, which unfortunately, tend to spread inside the area thanks also to new crossing bridges built on the Mermi creek. This is what actually happened during the flood of September 2015 (Fig. 32.3 right).

Another critical area is Piazza Adriatico, since it is located at a lower level with respect to the street and just above the riverbed of Bisagno. This area therefore constitutes a potential risk zone for flooding, due to runoff from Ponte Carrega.

As far as the hydraulic risk is concerned, these districts belong to the highest risk category (R4: very high risk) as assessed by the Bisagno river basin management plan updated to 2015 (Fig. 32.4). Consequently, in order to reduce this level of risk, taking into account the morphological characteristics of the land, it appears that the possibility of adopting temporary nonstructural flood-proofing techniques is useful, in addition to the classical intervention methods already put in place. This complementary strategy aimed to increase risk culture and awareness on the matter and, simultaneously, actively involve local residents in risk control strategies. All the costs should be included in the urbanization costs of new commercial building construction, either using resources coming from the decrease of the starting price of the call for tenders for works on the Torre creek or from the savings from the works on Bisagno River (this is what emerged during the public meetings between citizens and the local public institutions) (Comune di Genova, 2016).

FIGURE 32.3 2011 (A, *left*) and 2015 (B, *right*) flood event within the area. *Authors.*

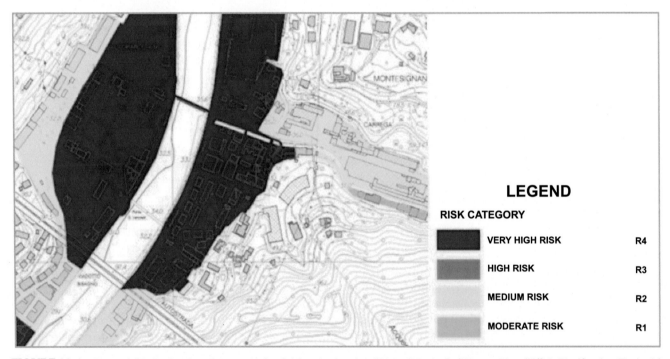

FIGURE 32.4 Extract of the hydraulic risk map. *Authors' elaboration based on "Piano di Bacino" of Bisagno River (Ufficio Pianificazione Territoriale della provincia di Genova, 2015a,b).*

32.4 THE SUGGESTED LOCAL STRATEGY OF DISASTER RISK REDUCTION AND PLANNING OF ACTIONS

Flood proofing consists of a series of interventions aimed to reduce the local impact of floods; in particular, temporary flood-proofing techniques are put in place and used only in case of major events that can cause the flooding of rivers. Such techniques, compared to the simplest and generally applied traditional solutions (such as sandbags, wooden dams, etc.) usually require less improvisation in action, in the preparation and in the study of the area, but give more effective results together with lower costs in the medium term. The use of flood-proofing techniques in highly urbanized and vulnerable contexts like this may be the only possible solution to reduce the flood risk, as the conventional disaster risk reduction (DRR) options cannot be implemented for technical, social, or economic reasons (levees, relocations, etc.). Several analyses have been carried out in order to achieve the goal, from the hydraulic and economic point of view, taking into account the activation times of the barriers. This allowed us to define the most suitable solution for the context, using the amount of 112.000 € as planning fees (Comune di Genova, 2014) of the new commercial building to be used, and to send a synthesis of our results to the local administration, via the association Amici di Ponte Carrega.

32.4.1 Hydraulic Analysis

In order to realize a hydraulic adjustment project in the area, using only interventions of temporary flood proofing and works of microurbanization, it is first necessary to define the access points of water in the neighborhood and assess the water levels reached by the Mermi and Torre creeks, related to the return time of flood events. Such evaluations have been derived from the hydraulic reports that were available for the two creeks, and from the results provided by a hydraulic model, specially designed by the authors, a model that regards the Torre creek but also considers the presence of Piazza Adriatico. The results obtained, checked also through the analysis of the pictures of the last three flood events and on ground investigations, are shown in the following images (Fig. 32.5 and 32.6).

Based on the results obtained from the hydraulic assessments, three different scenarios have been defined. Such scenarios take into account the use of temporary flood-proofing techniques, which, if properly implemented, would lead to a considerable improvement in the hydraulic buoyancy of the neighborhood, helping to avoid much of the flooding to which the area is still periodically subjected (Fig. 32.7).

In order to further reduce the hydraulic risk and to guarantee the best performances of the flood-proofing techniques used, interventions of microurbanization and urban design, involving some elements strategic from the hydraulic point of view, have been defined.

32.4.2 Economic Evaluation

Thanks to the available information and the analysis that has been carried out, it was possible to evaluate technically and economically the interventions that have been proposed, considering three different time windows (10, 15, and 30 years), aiming to quantify from an economic point of view the benefits and effectiveness that these

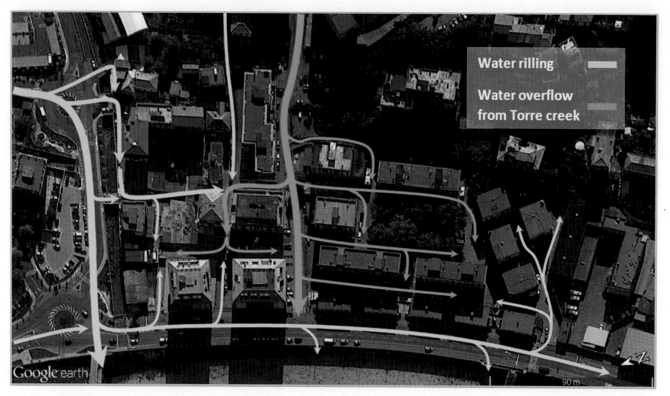

FIGURE 32.5 Route of the water within the area (after adjustment measures completed in 2015). *Authors' elaboration.*

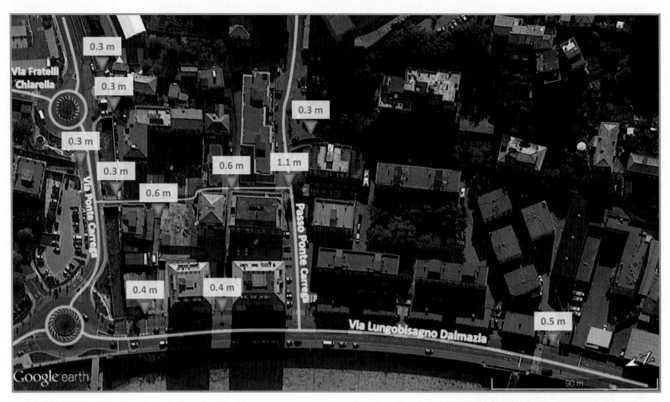

FIGURE 32.6 Estimate of water depth within the area (after adjustment measures completed in 2015) in proximity of the barriers position. *Authors' elaboration.*

interventions could provide. In particular, given the estimates received, a simulation of the total restoration costs has been carried out, considering the case of an event with a return period of 200 years, and taking into account possible insurance policies against natural disasters for all the floors and basements of all the buildings in the area. The cases that have been considered for the economic evaluation are the following:

- Case 1: NO flood proofing, NO insurance;
- Case 2: NO flood proofing, YES insurance;
- Case 3: YES flood proofing, NO insurance;
- Case 4: YES flood proofing, YES insurance.

Examining the simulations' results, it can be noticed that most of the times Case 4 is the most advantageous from the economic point of view. However, when it comes to long periods of time (15 and 30 years) and taking into account a good barrier efficiency, the most favorable is Case 3. These results therefore show how a correct use of such barriers can lead to major benefits from the economic point of view, with respect to the ones deriving from the combined use of insurance contracts.

Another way to evaluate the effectiveness of flood-proofing interventions is to assess the hydraulic risk for the area (in this case, for sake of simplicity, this evaluation was done only for Piazza Adriatico), with and without flood barriers. In this study, the risk (R) was considered according to the definition provided by UNESCO (Fournier D'Albe, 1979), and was assessed through the convolution of three components (dangerousness: H; vulnerability: V; exposed value: W), each of them characterized by a certain probability distribution. After defining an integration period and introducing appropriate assumptions, the risk formula could be written and then rewritten as:

$$R = \int_0^{200} H \cdot V \cdot W = \int_0^{200} 1/T \cdot D(T) \cong \sum_0^{200} 1/T \cdot D(T)$$

where T is the return time and D the expected damage, calculated as the product of vulnerability and exposed values.

These last two terms can generally be evaluated starting from the curves of damage that can be found in the standard method, which is used for the evaluation of flood damage on the microscale, developed by the Dutch

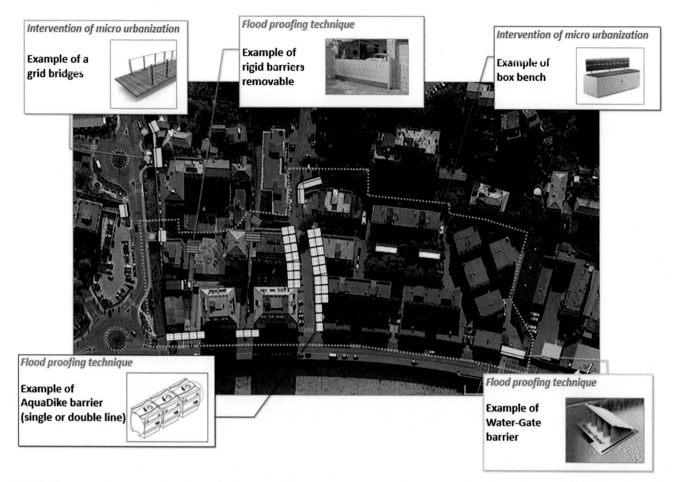

FIGURE 32.7 One of the plans of flood-proofing (and microurbanization) scenarios elaborated to protect the study area. *Authors' elaboration.*

Ministry of Water Resources (Kok et al., 2009) and from data provided by the "Osservatorio del mercato immobiliare dell'agenzia delle entrate" of the Italian government.

Starting from this data, it is therefore possible to calculate the hydraulic risk as a function of the return time. The results show how the use of flood barriers can lead to significant benefits, for this case study; in particular, the hydraulic risk could hypothetically be reduced up to 97% (Fig. 32.8).

32.4.3 Evaluation of the Activation Time

Another analysis that is needed to evaluate the effectiveness of this kind of intervention is the analysis of the implementation times, which allows us to see if the implementation times are compatible or not with the alert times of the Regione Liguria's warning systems. In particular, it is first of all important to estimate the time that usually elapses between the emanation of the alert messages and the actual occurrence of the flooding events, because this will be the maximum time interval useful for the installation of the barriers. The estimated alert time has been identified on the basis of the alert messaging provided by the Civil Protection of the Regione Liguria from 2009 to 2015. This also allowed us to know all the events for which there was a high alert ("Allerta 2") in Genoa, and to verify if for such events real flooding had actually occurred, also taking into account the analyses provided by Event's Meteo-hydrological Reports ARPAL and the time elapsed between the emanation of the alert's time and the actual flooding of the area (Δt_{alert}) (Table 32.1).

The Δt_{alert} considered for the following analyses were the average, the median, and the minimum values recorded from 2009 to 2015.

After evaluating the warning times for the area, it was necessary to identify the installation times of flood-proofing techniques. These could be obtained from the information provided by retailers, which in addition to

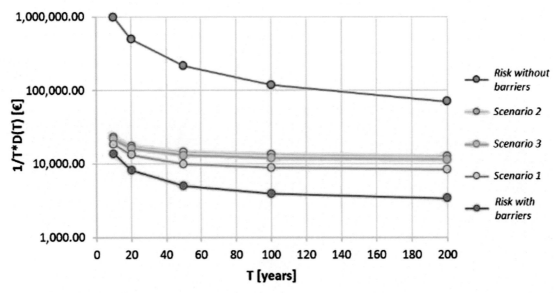

FIGURE 32.8 Evaluation of hydraulic risk in presence or not in presence of barriers; the red line does not consider the cost of barriers. *Authors' elaboration.*

TABLE 32.1 Evaluation of Time Between the Issuance of the "Allerta 2" and the Real Time of Arrival of Flood (Δt_{alert})

Events	Starting "Allerta 2" (dd/mm/yyyy hh:mm)	Time of Flooding for the Area (dd/mm/yyyy hh:mm)	Δt_{allert} (dd:hh:mm)
27–30.11.2009	29/11/2009 15:00	30/11/2009 11:51	00:20:51
03–05.10.2010	04/10/2010 12:00	04/10/2010 20:21	00:08:21
02–09.11.2011	04/11/2011 06:00	04/11/2011 12:52	00:06:52
23–26.12.2013	25/12/2013 10:30	26/12/2013 07:37	00:21:07
08–13.11.2014	10/11/2014 06:00	11/11/2014 06:37	01:00:37
13–16.11.2014	14/11/2014 21:00	15/11/2014 13:52	00:16:52
11–14.09.2015	13/09/2015 07:00	14/09/2015 00:07	00:17:07

the installation times, also define the number of people necessary to install the barriers and all the necessary equipment (the time of transportation of the barriers on site was considered null, since it has been assumed that it was possible to have a complete storage of material equipment within the area in question). Starting from this information, it was then possible to evaluate the installation times of the barriers for different scenarios, depending on the number of available workers, and compare them with the previously defined Δt_{alert}. The results show how only Scenario 1 had installation times lower than the minimum Δt_{alert} (with a number of field operators at least equal to 8), while for the other scenarios the installation times always proved longer.

32.4.4 Evaluation of Effectiveness

A final confirmation of the effectiveness of this type of barriers, as considered in this case study, is provided by the index assessing the resistance susceptibility of buildings to flood risks of Certification of the Predisposition of Resistance of Edifices to Disasters (Bignami, 2014). In fact, the score it reached for the three examined "sample" buildings was substantially different in case of using or not using flood-proofing techniques, thus highlighting the possibility of being able to certify such buildings, an action that in absence of these barriers would not be possible since the buildings would not have been able to reach the minimum threshold of certification.

32.5 CONCLUSIONS AND CRITICAL ANALYSIS

The experience and analyses that were carried out for the case of Piazza Adriatico and Ponte Carrega confirmed how, from the economic and methodologic points of view, taking into account the possibility of using temporary flood-proofing and microurbanization techniques at the neighborhood level can have considerable importance, being useful for the promotion of disaster-resilient communities (Chou & Wu, 2014). Such nonstructural interventions strategy shows its feasibility and constitutes a valid option, additional or alternative to classical intervention methods, also considering the citizens' involvement and the resulting growth of awareness and prevention culture.

Specifically, from a technical point of view, the selection and deployment of these solutions, however, cannot be improvised but must be evaluated and based on specific technical, hydraulic, and economic analyses, considering, in addition, the time of activation of the barriers.

This way, our chapter has shown how the proposed DRR strategy in the study area can be rightly defined resilient, since even if the area will continue to suffer floods, it would be able to better adapt to the present waterways, suffering less damage and finding it easier to recover to normal living conditions after each event.

Finally, our experience represents an interesting perspective for the widespread use of self-protecting actions and flood-proofing techniques at the neighborhood scale, since it shows the possibility of taking advantage of the social capital represented by the inhabitants and their new kind of relationship with decision makers, in this way being able to put in place a real bottom-up process. Participation and volunteering will encourage the local population to work closely with experts and institutions, and will also allow the city administration to start a pilot experiment to test the suggested strategy to evaluate and possibly extend the solution to the rest of the city. In so doing, a new practice could be institutionalized and part of the current obstacles that today make it difficult to protect people from floods could be overcome.

References

Bignami, D.F., 2014. Towards a Territorial Multi-disaster Buildings' Resistance Certification, Springer. In: Springer Briefs in Applied Sciences (Peer Reviewed Series). ISBN:978-88-470-5222-2.

Brandolini, P., Cevasco, A., Firpo, M., 2012. Geo-hydrological risk management for civil protection purposes in the urban area of Genoa (Liguria, NW Italy). Natural Hazards and Earth System Sciences 12, 943–959. https://doi.org/10.5194/nhess-12-943-2012.

Chou, J., Wu, J., 2014. Success factors of enhanced disaster resilience in urban community. Natural Hazards 74 (2), 661–686. https://doi.org/10.1007/s11069-014-1206-4.

Comune di Genova, 2014. Piano Urbanistico Comunale Norme Tecniche di Attuazione.

Comune di Genova, Municipio IV Media Val Bisagno (POR Liguria 2014–2020), 2016. Report sintetico dei tavoli di lavoro, 10 Novembre 2016 Centro Civico V.Bobbio, Risposte formulate dai civici uffici.

Faccini, F., Luino, F., Sacchini, A., Turconi, L., De Graff, J.V., 2015. Geohydrological hazards and urban development in the Mediterranean area: an example from Genoa (Liguria, Italy). Natural Hazards and Earth System Sciences 15 (12), 2631–2652. https://doi.org/10.5194/nhess-15-2631-2015.

Fournier D'Albe, E.M., 1979. Objective of volcanic monitoring and prediction. Journal of the Geological Society London 136, 321–326.

Kok, M., Huizinga, H.J., Vrouwenvelder, A.C.W.M., Barendregt, A., 2009. Standard Method 2004 Damage and Casualties Caused by Flooding, Ministerie van Verkeer en Waterstaat.

Rosso, R., 2014. Bisagno. Il Fiume Nascosto, Marsilio.

Ufficio Pianificazione Territoriale della Provincia di Genova, 2015a. Relazione Generale – Piano di Bacino Stralcio del Torrente Bisagno per la Tutela dal Rischio Idrogeologico.

Ufficio Pianificazione Territoriale della provincia di Genova, 2015b. Sottobacini e Parametri Caratteristici – Piano di Bacino Stralcio del Torrente Bisagno per la Tutela dal Rischio Idrogeologico.

Further Reading

Abebe, A.J., Price, R.K., 2005. Decision support system for urban flood management. Journal of Hydroinformatics 7 (1), 3–15. Available at: http://jh.iwaponline.com/content/ppiwajhydro/7/1/3.full.pdf.

Berkes, F., 2007. Understanding uncertainty and reducing vulnerability: lessons from resilience thinking. Natural Hazards 41 (2), 283–295. https://doi.org/10.1007/s11069-006-9036-7.

De Moel, H., Aerts, J.C.J.H., 2011. Effect of uncertainty in land use, damage models and inundation depth on flood damage estimates. Nat Hazard 58 (1), 407–425. https://doi.org/10.1007/s11069-010-9675-6.

Haigh, R., Amaratunga, D., 2009. Guest editorial. International Journal of Strategic Property Management 13 (2), 83–86.

Silvestro, F., Gabellani, S., Giannoni, F., Parodi, A., Rebora, N., Rudari, R., Siccardi, F., 2012. A hydrological analysis of the 4 November 2011 event in Genoa. Natural Hazards and Earth System Sciences 12 (9), 2743–2752. https://doi.org/10.5194/nhess-12-2743-2012.

Tyler, S., Moench, M., 2012. A framework for urban climate resilience. Climate and Development 4 (4), 311–326.

SECTION VI

TOWARDS A CLIMATE-SENSITIVE URBAN DEVELOPMENT

CHAPTER 33

Drawing Lessons From Experience[1]

Adriana Galderisi[1], Angela Colucci[2]
[1]University of Campania Luigi Vanvitelli, Aversa, Italy; [2]Co.O.Pe.Ra.Te. ldt, Pavia, Italy

33.1 LARGE-SCALE STRATEGIES IN THE FACE OF CLIMATE CHANGE: TRAJECTORIES AND BARRIERS

The large-scale strategies to counterbalance climate change so far undertaken in Europe, United States, China, Africa, and Australia have been individually presented and discussed in Section 2 of this book. A comparison among these strategies is quite difficult, due on the one hand to the significant heterogeneity of the geographical, economic, and technological contexts in which these strategies have been conceived and implemented as well as to the different political systems that characterize these countries; on the other hand, comparison is also difficult due to the different roles played by each country with respect to climate issues. China, the United States, and Europe are currently considered, indeed, the top CO_2 emitters at global scale, with China and the United States accounting for about half of the global greenhouse gas (GHG) emissions (World Resource Institute, 2015). Furthermore, taking into account the close relationships among population, namely urban population, and climate change, the roles of the considered countries will further change in the future, mostly if no adequate emission reduction policies are undertaken. By 2050, in fact, half of the global population growth is expected to occur in Africa and Asia; smaller increases are projected to occur in Northern America, while the European population is expected to shrink. Furthermore, "over the coming decades the level of urbanization is expected to increase in all regions, with Africa and Asia urbanizing faster than the rest" (UN, 2014), while Australia is projected to become 90% urban.

Hence, the role played by the considered countries is already and will be even more crucial both for an effective reduction of global emissions and for ensuring the safety of increasing population in the face of climate-related events.

Despite their significant differences, the paths undertaken by all selected countries to counterbalance climate change allow us to draw some reflections on current trajectories and on the potential obstacles to their effectiveness. Hence, starting from a brief reflection on the heterogeneity of political systems and its implications for the effectiveness of the undertaken climate strategies, we will focus on:

- The relevance of the three considered urban metaphors in framing current large-scale climate strategies;
- The difficult routes toward integrated climate strategies;
- The key role assigned to cities in pursuing both mitigation and adaptation goals.

33.1.1 Heterogeneous Political Systems Dealing With Climate Issues

Despite their numerous differences, both the United States and Australia are characterized by federal political systems, whereas Europe and Africa represent unions of states. These last two share numerous similarities, at least on paper: the African Union (AU) was modeled indeed on the European Union (EU), even though AU is still far from

---

[1] Although the chapter resulted from common reflections, Sections 33.1 and 33.2 were written by Adriana Galderisi; Section 33.3 by Angela Colucci; Section 33.4 by Adriana Galderisi and Angela Colucci.

achieving the same results of EU, mainly in terms of regional integration (Abdulrahman and Abraham, 2016). Finally, the People's Republic of China has a very centralized political system still largely controlled by the Communist Party through the Politburo.

The significant differences in political systems have largely affected the ways these countries have faced climate change and the effectiveness of their large-scale strategies.

For example, in the case of the United States, the federal state has played a very weak role, characterized by a late start and a recent change of route due to the reversal by the political leadership. However, some individuals and groups of states, and above all numerous cities (e.g., New York), have pursued independent and effective initiatives, taking on in some cases a leadership role in the international arena. Australia could have played a stronger role, due to its early commitment (starting from the 1980s) in the fight against climate change but numerous factors have made and still make Australian climate strategies quite weak. Among them, at least the following can be listed: the difficulty of the federal state in pushing and sustaining the different levels of government, local communities, and economic actors in developing integrated and cross-level actions; the short-term perspective of electoral mandates and the reversals of the political will to cope with climate change; and, above all, the contradictions that emerge from pursuing an economic and productive model still largely based on fossil fuels while calling for climate mitigation (see Chapter 9).

Also the EU and AU have acted very differently. The EU has shown, indeed, a significant commitment both in terms of mitigation policies and, more recently, in terms of adaptation policies. The EU has been one of the world leaders in global mitigation policies by promoting, and achieving to a large extent, ambitious energy and climate targets. However, despite its efforts in favoring and sustaining member states and cities in developing their climate strategies, the performances of the individual states have been very heterogeneous, ranging from a very poor commitment of some states to the overachievements of others. As refers to the AU, it is difficult to discuss the effectiveness of its climate strategy since it was released only in 2014 and refers to the wide time span of 2015—35. However, the weakness of the AU does not promise significant short-term results even though African strategy is of great interest, due to its aim "to implement climate change programmes in such a way as to achieve sustainable development, alleviate poverty and attain the Millennium Development Goals, with emphasis on the most vulnerable groups, especially women and children" (AU, 2014).

China, also thanks to its centralized political system, has shown in the last decade a strong political will, as well as an integration of climate issues into the national economic planning. However, even though the undertaken direction is promising for the near future, it is worth noting that China's coal consumption trebled in the period 2000—13, nationally binding targets for energy consumption and carbon emissions have been established only recently, and the national Emissions Trading Scheme was started in 2017.

33.1.2 Large-Scale Climate Strategies and Urban Metaphors

All the examined large-scale climate strategies provide clear references to the Smart City and the Resilient City metaphors.

For example, in Europe, China, and United States, Smart City and Resilient City metaphors are generally used to frame, respectively, mitigation and adaptation strategies. For example, the European initiative aimed at pushing local authorities to start the transition toward low-carbon cities was launched in 2011 under the flag of the Smart Cities and Communities Initiative. Similarly, the cooperation among China and the United States to face climate change, namely the US-China joint Climate Leaders Declarations issued in 2015 and 2016, which engaged the two countries in establishing ambitious emission reduction targets and establishing climate action plans (LBNL, 2016), are explicitly framed in the context of the US-China Climate-Smart/Low-Carbon Cities Summit.

Meanwhile, both the European Strategy on Adaptation to climate change issued in 2013 and the Chinese Sponge City Program issued in 2014 explicitly refer to resilience, aiming at strengthening, in the first case, Europe's resilience to the impacts of climate change and, in the second case, cities' resilience to pluvial flooding in a context of rapid urbanization and climate change.

However, some countries introduce a different perspective: Australia and Africa refer to resilience, for example, as a "convening concept" to integrate mitigation and adaptation (see Chapters 7 and 9). Both the project "Working With Informality to Build Resilience in African Cities" launched in 2013 by the Climate and Development Knowledge Network and the National Climate Resilience and Adaptation Strategy released by the Australian Government in 2015 aim to combine mitigation and adaptation goals. Such an emerging approach seems to be very promising; it confirms the potential of the Resilient City metaphor to establish itself as an umbrella concept, capable to better link mitigation and adaptation strategies and to embed them into the broader goal of sustainable development.

33.1.3 The Difficult Paths Toward Climate Strategies

It is worth noting that most of the considered countries, following international obligations, started their climate policies by focusing on mitigation issues, while all over the world adaptation issues rose to the fore only in the last few years.

For example, Europe, China, the United States, and Australia have initially addressed climate issues by establishing targets and tools for reducing GHG emissions, improving energy efficiency, and increasing the spread of renewable energy sources. However, times and outcomes are significantly different: Europe, starting from 2000, has indeed achieved significant results with respect to the established targets; The United States and Australia have been characterized by drastic route changes along their paths resulting from upheavals in political leadership; China—despite the establishment in 2007 of a Climate Change Department—has introduced binding targets for energy consumption and CO_2 emissions only recently, making evaluation of the outcomes difficult.

With respect to adaptation issues, all these countries have established adaptation strategies in the last few years by providing—mainly in Europe, the United States, and Australia—principles, tools, and platforms for guiding individual states and local authorities in enhancing their capacities to cope with climate impacts, also thanks to the availability of shared guidelines and best practices.

Therefore, mitigation and adaptation have been recognized for a long time as complementary strategies to address climate change (Yohe and Strzepek, 2007), to be carried out separately, taking into account, at most, the cobenefits they produce on each other.

So far, only Europe and Africa have put in place strategies aimed at better integrating mitigation and adaptation goals. In 2013, the Climate and Development Knowledge Network launched the project "Working With Informality to Build Resilience in African Cities," specifically addressed to combine mitigation and adaptation goals along with the broader ones of poverty reduction and sustainable development. For example, the project was intended to ensure basic services and infrastructure to the urban poor in a context of changing climate, driving meanwhile local development toward low-carbon emissions. It is worth noting indeed that, although nowadays Africa accounts for only 2%–3% of the world's CO_2 emissions from energy and industrial sources, being in contrast largely affected by climate-related hazards, the expected growth of urban population and the rapid urbanization processes in some areas might determine in the near future a significant contribution of this country to climate change on a global scale.

In the same vein, in 2015 Europe launched the Covenant of Mayors for Climate & Energy aimed at achieving mitigation goals, by accelerating the decarbonization process, improving energy efficiency, encouraging the use of renewable sources, and pursuing meanwhile adaptation goals, by strengthening local capacities to cope with climate impacts.

The shift from an approach to mitigation and adaptation as separate strategies toward an integrated approach, capable to maximize synergies and trade-offs between them, seems currently to be the most promising one also at global level. In June 2016, in fact, the UN General-Secretariat and the world's most important city networks (C40, ICLEI, etc.) announced the new Global Covenant of Mayors for Climate and Energy, officially started in January 2017 and aimed at tackling three key issues: climate change mitigation, adaptation to its adverse effects, and access to secure, clean, and affordable energy.

33.1.4 The Pivotal Role of Cities

The new Global Covenant of Mayors for Climate and Energy, besides marking a further step forward along the global scale route to counterbalance climate change, emphasizes another key aspect common to most of the considered large-scale strategies: the key role assigned to cities, whose voice is more and more relevant in shaping international climate policies and whose actions are often effective far beyond the local or regional boundaries. As emphasized by Ulrich Beck (2017), in the face of global risks, cities become pioneers in finding answers to the problems of the world at risk, regaining a central position like the one they had long time ago, in the prenational world.

The pivotal role assigned to cities clearly arises from most of the examined large-scale strategies—Europe, China, and Africa assign a key role to cities in implementing both mitigation and adaptation strategies. US cities are playing a pivotal role also through their direct involvement in international networks, often in contrast with federal and sometimes national strategies.

Specifically, despite some large European cities having directly promoted or joined global initiatives aimed at carrying out mitigation and/or adaptation plans, the EU has significantly supported all cities, including very small ones, in addressing climate issues, with effective results. Also, China launched two important initiatives aimed at testing its mitigation and adaptation strategies in cities: the "pilot cities' low carbon plans" and more recently the "carbon trading pilots" (see Chapter 8), and the Sponge City Program launched in 2014 by the Ministry of Housing & Urban-Rural Development to improve cities' resilience to pluvial flooding in a context of rapid urbanization and climate change (see Chapter 19).

Summing up, all the considered countries have devoted increasing attention in the last decade to the development of large-scale strategies to mitigate and adapt to climate change capable to guide and sustain national and, above all, urban authorities in carrying out climate strategies and actions tailored to their specific contexts. Nevertheless, their effectiveness is often difficult to prove, since in some cases (China and Africa) large-scale strategies are still at an early stage, while in other countries (United States and Australia) have been affected by more than one change in political willingness to tackle climate change.

Moreover, the effectiveness of these announced or undertaken strategies largely depends on the capacity of these countries, still to be demonstrated, to radically change current socioeconomic development models, shifting toward low-carbon development models crucial to achieve environmental, economic, and social sustainability goals.

33.2 CLIMATE POLICIES AT LOCAL LEVEL: DRAWING UPON CITIES' EXPERIENCES

Climate change is not only modifying existing geographies of risks (Beck, 2017), landscapes, and economies all over the world but it is also bringing new actors to the global arena; through direct involvement in international initiatives or thanks to national initiatives aimed at promoting and supporting local actions, cities are increasingly establishing themselves as key players in the face of climate change. The more and more tangible effects of climate change in cities encourage, indeed, innovation and cooperation within and among them, driving their actions far beyond national and supranational borders and constraints, and leading them to undertake relevant decisions affecting both the root causes and the impacts of climate change.

In this book (Section 3), numerous European and non-European cities have been examined in order to draw upon their current experiences, providing a better understanding of the driving forces pushing climate policies in the different contexts, the main conceptual and organizational innovations introduced at city scale, and the main barriers hindering the effectiveness of local actions.

33.2.1 Case Studies: General Overview

The selected case studies do not provide an exhaustive picture of the heterogeneous landscape of cities' actions in the face of climate change; the well-known best practices, both in Europe (e.g., Rotterdam, Copenhagen, London, etc.) and in the United States (e.g., New York, Seattle, etc.) have been left out, giving room to less-investigated examples both outside Europe (Africa, China, and Australia) and within Europe (Southern and Eastern Europe). The initiatives launched by the EU starting from 2008 to address climate change had, indeed, different outcomes—they have been less effective, in terms of number of plans submitted and implemented, precisely in the Mediterranean countries and in Eastern Europe.

The focus on "weak" contexts allows us to look not at the spearheads but to shed light on difficulties, delays, and future needs for strengthening cities' capability to counterbalance climate change in disadvantaged areas.

According to this premise, 6 out of the 11 selected case studies are located in Europe, mainly in Southern and Eastern Europe (Latvia, Serbia, Italy, Spain and Greece) and four outside Europe (Turkey, Africa, China, Australia). Despite the large differences among them in terms of population size (ranging from the 200,000 inhabitants of Portsmouth to the 13 millions of Guangzhou), most of them play an important economic role at regional or national scale. Moreover, most of the selected case studies are coastal and/or fluvial cities, which have already experienced climate impacts; not by chance, indeed, floods (including fluvial, coastal, and pluvial floods) are the most widely experienced (and therefore considered) climate impacts, although heat waves and sea level rise are also considered as significant issues.

Furthermore, most of the considered case studies, despite acting as individual entities, are involved in large European or international networks (sometimes both) (e.g., C40, 100 Resilient Cities, Global Covenant, World Energy Cities, ICLEI). According to Bulkeley et al. (2012), in the last decade it is possible to count at least 60 transnational initiatives addressing climate issues in a direct or indirect way. Thus, urban climate initiatives often result from transnational alliances that overcome and sometimes conflict with national climate policies. These networks, although very different from each other, generally play an important role in driving and shaping cities' initiatives providing both conceptual frameworks and operational tools (e.g., 100 RC, Global Covenant, etc.).

However, in some cases (United Kingdom, China) national policies play a key, although very different, role; in the case of the United Kingdom, for example, according to the National Planning Policy Framework 2012), it is a statutory requirement for local planning authorities to proactively address climate change whereas in China cities' actions are strongly shaped by the national state that establishes rules and guidelines to be applied and tested on urban scale.

Finally, it is noteworthy that numerous cities' initiatives have been developed thanks to external funds and supporting actors. For example, besides the support provided by the international initiatives mentioned herein (e.g., 100 RC), most of the European case studies have been undertaken thanks to European programs (e.g., Life+, Interreg, etc.) and Funds (e.g., EU fund for Regional Development, EU Social Fund), whereas outside Europe significant economic support has been provided by development agencies (e.g., the Japan International Cooperation Agency or the French Development Agency, AFD, which is currently playing a key role in promoting climate investments all over the word).

33.2.2 Case Studies: Conceptual Framings

Similarly to what has been emphasized in respect to large-scale climate strategies, also at urban scale the Smart City metaphor is generally used to frame initiatives and actions addressing mitigation issues; most of them, indeed, have joined Smart City initiatives or adopted Smart City strategies to undertake and guide actions aimed at reducing CO_2 emissions and energy consumptions by mostly acting on urban mobility and building sectors.

In contrast, the Resilient City metaphor has been generally used to frame local adaptation policies and actions. However, even though numerous initiatives explicitly mention or refer to resilience, the meanings attributed to this term are very different, ranging from the wider interpretation of resilience to stresses and shocks adopted by the initiatives undertaken within the 100 RC Network, to the narrow focus on the cities' capacities to withstand, absorb, and timely recover from the impacts of climate-related hazards.

Moreover, it has to be emphasized that in many cases, both mitigation and adaptation strategies/initiatives are framed into the wider conceptual framework of sustainability. The case study of Athens (see Chapter 16), for example, clearly highlights how current planning tools, by addressing sustainability goals, indirectly address important climate change goals by promoting sustainable mobility or by improving the creation of green networks. Hence, the reference to the theoretical framework of sustainability allows embedding climate actions, even when no specific strategies and plans to address climate change are locally carried out.

Finally, most of the selected case studies include a clear reference to disaster risk reduction; for example, with specific reference to floods, local adaptation strategies generally address flood risk management, including actions aimed at improving hydrological risk assessment, early warning systems, green or grey infrastructures for disaster prevention, etc. In particular, local climate adaptation strategies and plans often complement traditional flood risk management, since most of the considered initiatives mainly address "pluvial" floods, whereas the DIRECTIVE 2007/60/EC on the assessment and management of flood risk only refers to "floods from rivers, mountain torrents, (...), and floods from the sea in coastal areas."

33.2.3 Climate Urban Policies Between Persistence, Adaptability, Transformability, and Learning Capacity

Despite the widespread references to the resilience concept, the examined practices show a tendency to be based on a narrow conceptualization of this concept.

As highlighted in previous chapters of this book (see Chapters 2 and 4), so far a significant distance between the theoretical interpretation of the resilience concept and its translation into practice can be noticed. On the one hand, indeed, the idea of "evolutionary resilience" introduced by Davoudi et al. (2012) aims to drive cities toward proactive and tailored-to-the-site climate strategies, based on urban systems' capacities to continuously learning from experience (learning capacity), to cope in the short term with climate impacts (persistence), to continuously adapt in the face of changing conditions through incremental adjustments (adaptability), and to innovate in the long term (transformability) by introducing fundamental changes within and across urban systems.

However, most examined practices promote strategies and measures aimed at enhancing, above all, urban systems' persistence and adaptability in the face of climate change—induced challenges. For example, current measures to counterbalance climate change are mainly addressed to increase efficiency of urban systems (e.g., by reducing energy consumption and transport-related emissions); to optimize their capacity to withstand recurrent climate-related threats
(e.g., by improving early warning systems, by maintaining and strengthening existing rivers' embankments, etc.); to increase their flexibility in the face of the increasing climate-related impacts (e.g., by preserving and improving green and blue infrastructures within and outside urban areas); and, although to a lesser extent, to enhance learning capacity (e.g., by providing tools and procedures for monitoring and evaluating achieved outcomes). Hence, based on the examined practices, so far they are often more addressed to optimize existing urban systems than to innovate them, according to a long-term perspective. Furthermore, learning capacity is still scarcely considered, since few case studies have set up decision-making processes, including a clear definition of implementing actors, time frames, expected outcomes, and means to monitor implementation and review goals and outcomes, and very few of them have adequate internal expertise to implement and constantly review such processes.

33.2.4 Climate Urban Policies Between Sectoral and Integrated Approaches

Despite the shift from a sectoral toward an integrated approach to climate issues, aimed at maximizing synergies and trade-offs between mitigation and adaptation measures, which has recently occurred in some countries (e.g., Europe, Africa) and is currently promoted at global scale by the new Global Covenant of Mayors for Climate and Energy, most of the selected case studies have carried out separate mitigation and adaptation strategies and plans. For example, current practices have generally addressed climate issues by focusing first on mitigation, shifting attention only in the last decade to adaptation issues. Obviously, as already noted in respect to large-scale strategies, the paths followed by the different case studies are very heterogeneous; while in Newcastle (Australia) (see Chapter 20), for example, climate policies were started in the 1980s focusing on mitigation goals, namely on energy efficiency, and only in the early 2000s the focus was shifted to adaptation policies, African cities were pushed toward an integrated approach to climate issues, capable of combining mitigation and adaptation goals along with the broader ones of poverty reduction and sustainable development by the project "Working With Informality to Build Resilience in African Cities" launched in 2013 by the Climate and Development Knowledge Network.

Furthermore, it is worth noting that, mirroring the prevailing sectoral approach, most of the examined case studies have carried out different plans addressing specific issues (from water management to infrastructure and coastline protection) at different scales (from large scale to building regulations); unfortunately, since these plans were developed over a long time period by different institutions or even different departments of the same institution, they are not always adequately coordinated with one another.

Another important aspect that deserves to be discussed here is the increasingly recognized need to embed climate mitigation and adaptation strategies into urban planning tools. The key role of spatial planning processes to improve cities' response to climate change, favoring an integrated approach to mitigation and adaptation, balancing and mediating "trade-offs between them and other social and economic goals" (Davoudi et al., 2009) and, above all, framing these issues as parts of the broader goal of sustainable development, has been clearly recognized in planning literature in the last decade (Priemus and Davoudi, 2012; Bulkeley, 2013). Despite this increasing awareness, current practices still show significant difficulties in embedding climate issues in urban planning tools. Apart from some case studies (Portsmouth, UK; Newcastle, Australia), the examined practices show several barriers to pursue such an integration, related for example to the rigidity of current urban planning tools as well as to the slow times of their updating. Moreover, even though climate issues are sometimes included in strategic planning tools, operational tools for putting these issues into practice are not clearly envisaged.

Finally, due to the close linkages between adaptation policies and flood risk management, mentioned earlier, it is noteworthy that the relationships between the mandatory flood risk management plans, introduced by the EU Flood Directive and generally carried out by the river basin authorities, and local adaptation plans, generally carried out on a voluntary base by municipal authorities, are not clearly explored.

33.2.5 Governance Models: Cues of Innovation and Signs of Weaknesses

The governance models adopted in the examined case studies are quite heterogeneous with a prevalence of top-down models, with a limited involvement of local communities and other stakeholders. For example, only four case studies (Portsmouth, Belgrade, Genoa, and Newcastle) developed their climate strategies by involving, although with a strong institutional leadership, local stakeholders (e.g., landowners, nongovernmental organizations, local communities, universities).

However, some of them, despite involving only institutional stakeholders, have promoted significant innovations in current organizational structure of local governments, often arranged into "silos," acting on specific issues (e.g., risk reduction, water management, energy, and transport). Such a "siloed" structure does not encourage the required cross-sectoral approaches and strategies to address the complex challenges posed by climate change (Galderisi, 2017). Hence, some case studies, by overcoming current organizational structure, have promoted strong horizontal cooperation among different sectors of local public administration (e.g., Guangzhou) as well as multilevel cooperation (e.g., Genova) among different institutional actors (national authorities, regions, metropolitan cities, municipalities), bringing out new links between both disciplinary and competence areas (e.g., urban planning, water management, urban ecology, etc.).

Focusing on the main weaknesses of current practices, it is worth noting that where local initiatives have been activated thanks to strong local leadership or have been strongly guided by external actors (e.g., Gaziantep, Athens), they have not been effective in empowering local institutions and, especially, local communities. Hence, when local leadership changed or international support was over, the process of implementation and monitoring of the adopted strategies, which should have been managed locally, has slowed down or even stopped.

This highlights the importance of "capacity building" both at institutional (e.g., upgrading of technical competences, improvement of current procedure, innovation of organizational frameworks, etc.) and societal levels (e.g., raising awareness on climate issues, encouraging active involvement in decision-making, etc.) to ensure effectiveness and temporal continuity of climate policies.

Moreover, whereas all the considered urban metaphors strongly emphasize the importance of comprehensive knowledge (including both experts' and communities' knowledge) as well as of continuous learning based on the effective monitoring of implemented strategies and actions, the examined case studies are generally based only on expert knowledge, as the engagement of local stakeholders is more addressed to inform and communicate than to knowledge coproduction. In respect to monitoring issues, it has to be noted that in some cases climate strategies and plans are very recent and the implementation process is still at an early stage (e.g., Guangzhou); in other cases, implementation and monitoring failed, due to the lack of knowledge and expertise of the municipality staff (e.g., Gaziantep).

33.2.6 Barriers to Effective Climate Urban Policies

The factors currently hindering the paths of cities toward effective climate policies are numerous and heterogeneous. The most relevant factors arising from the examined practices can be grouped according to three main constraints:

- Theoretical Constraints

    In the examined case studies, climate change has not been considered so far a priority in the urban policy agenda; low levels of perception and awareness make climate change still perceived by communities, practitioners, and decision makers as a geographically and temporally distant phenomenon. Also the increasingly frequent shocks due to the impacts of climate-related phenomena are attributed to external pressures to be faced at global scale and not clearly linked to the internal weaknesses of urban systems and above all of their current development patterns.

    Furthermore, the long economic crisis that affected Europe, especially Southern Europe, in the last decade, as well as the widespread poverty and the substantial delay in economic development in other countries still led to prioritize economic growth, despite its numerous environmental externalities, over other objectives. In many cases, the fear of hindering economic growth condemns to ineffectiveness the adopted climate policies: the lack of a radical breakthrough in economic development models, still based on high rates of energy consumption and GHG emissions, leads climate policies to fail in effectively addressing the root causes of climate change.

- Organizational Constraints

    Despite some attempts to overcome the "siloed" organization of decision-making processes within the examined case studies, this represents even today the most widespread constraint at local scale, leading to a scarce level of coordination between climate change policies and other developmental policies. Some case studies clearly show that also the few attempts addressed to promote organizational changes (e.g., the establishment of a "climate change bureau" under the environmental office of the Gaziantep Municipality in Turkey, or the establishment of a team devoted to environmental issues, including sustainability and CC in Newcastle, Australia) were dismantled a few years later, following changes in political leadership.

Other relevant and interconnected constraints arising from the examined practices are that climate change is still a political sensitive issue both at national and local scales and climate policies are often undertaken on a voluntary basis and scarcely rooted in national and local political agendas; these factors allow new political leadership (at local as well as at national levels) to easily assign priority to other conventional or emerging issues. Moreover, the scarce engagement of citizens and community groups in policy making represents a further obstacle to the continuity of climate policies.

- Operational Constraints

Among the main operational constraints, the inadequacy of current urban planning tools has been noted in most of the examined practices; despite the numerous attempts to mainstream climate policies into planning tools, indeed, their prevailing role of regulation and control leaves scarce room for innovation, preventing the adoption of a "transformational" perspective that, allowing long-term innovation, is crucial to reverse current unsustainable urban development patterns. Urban planning tools generally just introduce new "climate-oriented standards" (e.g., the runoff coefficient), too rigid with respect to the uncertainties that characterize the future of climate and urban development.

Moreover, it is worth noting that climate strategies may result as ineffective both when they are carried out through devoted sectoral plans and when they are integrated into ordinary urban planning tools: in the first case, indeed, climate plans are intended as voluntary tools, without any binding effect, whose success largely depends on the willingness and strength of political leadership; in the second case, namely in the Southern and Eastern European countries, the effectiveness of planning tools is strongly limited by poor implementation mechanisms as well as by the poor compliance of private interventions with the rules imposed by these tools.

As well, some of the examined practices highlight that, despite the key role currently assigned to urban planning tools in counterbalancing climate issues, planners as well as planning departments are often unprepared to cope with climate and risk issues, both because climate policies require the development of specific competencies (e.g., vulnerability analysis and assessment methodologies) and because they require strong interdisciplinary and interinstitutional cooperation, based on common languages and agreed-upon goals and objectives.

Finally, a remarkable constraint is represented by the economic incentives aimed at promoting climate policies; in some examined case studies, indeed, climate policies have been undertaken thanks to the availability of supports and financial incentives provided by the national state or international organizations, for a given time span. Hence, financial incentives attract the attention of local authorities to take actions, but once the incentives are over, local actors do not have the required strength to carry out and complete the undertaken programs that remain in the background, surpassed by other priorities.

33.3 TRANSITION INITIATIVES IN THE FACE OF CLIMATE CHANGE: HINTS FROM CURRENT PRACTICES

Policies and strategies facing climate change at international, regional, and local levels require stronger engagement of local communities to their effective implementation and need to be supported by a societal transition able to integrate mitigation and adaptation issues (led by public polices) in individual and community behaviors and values. A societal and cultural innovation is needed to spread the awareness on climate issues and to strengthen the capabilities of communities to deal with change and uncertainties (adaptation capabilities).

The exploration of grassroot innovations and transition initiatives provides promising experiments of integration of climate issues in (and with) local community activation processes toward renovated governance approaches able to enhance synergies and coproduction (cooperation) between experts, institutions, and communities facing climate changes. A focus was dedicated to the Transition Town Network (three are directly connected to the Transition Town Network); the others present different aspects of local communities' activation giving the opportunity to highlight approaches, principles, and tools developed in very different contexts and facing different critical phenomena.

The common thread connecting the examined transition initiatives is the resilience metaphor that emerges as an umbrella concept able to integrate social dimension and environmental issues.

33.3.1 Transition Initiatives: General Overview and (Local) Scale Focus

Transition initiatives could be referred to the local scale or local dimension also if the initiative itself is performed in a metropolitan context. The Monteveglio and Totnes Transition initiatives were developed in small/medium towns. The experiences regarding Obrenovac (in the metropolitan area of Belgrade), Barcelona, Visakhapatnam, and Newcastle are located in metropolitan contexts but the presented initiatives involve a single neighborhood, an urban place, or an urban community within the metropolitan context.

The local dimension, as prevailing scale of intervention, depends on different reasons. For sure, it is related to the community dimension of the initiative itself; communities are acting on local contexts, activating different activities based mainly on volunteers and on citizen engagement aiming to give impact and visible results (improvement of urban environment or of urban functions and services) in the local context. Furthermore, the community-led experiences are characterized by strong (physical) relationships among citizens and places they belong to and on which they directly act. The strong relationship with (local) place is a commonality of the presented transition initiatives and underlines the "human-scale" dimension characterizing the community-led initiatives where interventions and actions are tangible and can be experienced by citizens in their everyday lives. The attention to human dimension, the strong relationship between people and places, and the tangible effects of the implemented actions on citizens' everyday lives is an interesting aspect that could be assumed as a principle able to support the implementation of urban polices and measures facing climate issues.

Investigating the role of public institutions in the partnership leading transition initiatives is interesting in order to understand if and how public institutions and community-led initiatives can find possible synergies in climate actions implementation. In particular, the initiative of Monteveglio (see Chapter 23) is one of the few (the first) Transition Town initiatives where the municipal administration assumed a key-stakeholder role since the initial phase of activation and in the transition initiative official launch. In other Transition Town initiatives presented, such as Newcastle (see Chapter 28) and Totnes (see Chapter 22), the role of municipal institutions is less relevant and not crucial in the early phase of community activation. In Totnes the municipal institution became partner in supporting single actions, thematic strategies promoted by "transitioners," but, for instance, in Newcastle the transition initiative did not establish any agreement and structured alliances with city administration.

Other community-led transition initiatives presented (external to the Transition Town Network), activated dialogues and alliances with public institutions of small towns or with a neighborhood administration within a metropolitan area. The capability to activate alliances with different administrative/government levels, as in the Pedda Jalaripeta slum practice (see Chapter 27), is not very common, and the activation of dialogues with public institutions at different scales still remains sporadic, representing sometimes a critical aspect (or even a barrier) in community-led initiatives.

The networking or the partnership among actors at different scales permits transition initiatives, otherwise embedded at local scale, to be effectively multiscale. The setting up of alliances with other community-led practices that are developing initiatives acting on similar topics, common aims, and perspectives or using similar tools represents an opportunity for potential upscaling of local initiatives. Networks such as Depave and Repair Cafè (see Chapter 21) are associations or networks among initiatives aimed to share experiences and to provide mutual support (technical assistance, sharing values and action principles, sharing common rules and operative toolboxes, etc.). In sharing experiences to identify transferable tools or action principles, local practices are committed to get involved on larger scales, from metropolitan to regional and global scales, and understand how urgencies and solutions at local scale can intercept global phenomena and which action principles could be generalized. Networking and alliances of local transition initiatives developed discourses focused on their contribution to global phenomena, as climate mitigation and adaptation, as well to public policies and measure frameworks. Hence, local initiatives could intercept and dialogue with the existing public policies and could provide contributions in facing global issues such as climate change.

The case of Barcelona (see Chapter 25) demonstrates the potentiality in activating synergies between the rich and complex "panorama" of transition initiatives (social innovation) and institutional (public and also private) actors, representing the public and economical/social interests. Barcelona municipality launched a public policy engaging local initiatives in public spaces' stewardship. The public authority was able to launch this policy thanks to an existing network of grassroots initiatives (supported and engaged in the public decision-making process). The empowerment and engagement of local community-led practices is a precondition for their upscaling, and, at the same time, a rich and stabilized network of community-led initiatives can support the implementation of institutional climate policies.

Networks among local practices can provide a larger scale or a multiscale perspective to local practices, providing advantages (in terms of constancy and complexity) both to punctual context of the individual community-led initiative and to the larger urban context.

The Transition Town Network (see Chapter 21) is a consolidated network acting at global scale with an organizational structure based on regional/national hubs that guarantee a multiscale approach. From one side, Transition Network supports global awareness on climate issues; from the other side, local transition initiatives (and activities) are strictly related to the local dimension in defining both topics for actions and local strategies. The Newcastle, Monteveglio, Totness, and other transition initiatives are characterized by different issues and addressed to activate a process that is locally identified in respect to local urgencies and interests. The awareness in contributing to global issues (as climate change) is a specific assignment of (and developed by) the Transition Network.

33.3.2 Transition Initiatives: Conceptual Framings

As mentioned earlier, the Resilient City metaphor (or, more precisely, the resilience concept) emerges as a common umbrella to which almost all transition initiatives here presented refer when their general principles and actions criteria are outlined. However, individual practices refer to the resilience concept according to different approaches.

Transition Town Network refers to the resilience concept by emphasizing the capacities of local communities to deal with uncertainty and to adapt and evolve toward preferable scenarios in relation to peak oil, natural resources scarcity, and climate change. In these terms, resilience is not assumed as a conceptual model to understand and cope with environmental phenomena but is referred to features and capacities of local communities to activate responses, actions, and reactions locally based, flexible, and adaptable in order to adapt and navigate the transition toward more sustainable communities. Strengthening the resilience of its social components, local complex systems will be able to adapt and evolve, dealing with the potential impacts of climate change.

In the metropolitan area of Barcelona, existing initiatives focus on resilience in relation to adaptation and evolution of complex systems (evolutionary resilience). In particular, the Can Masdeu community garden project and the case of Pla BUITS focus on specific resilience properties related to the enhancement of local and social resources as opportunities for change. The resilience is used as an umbrella concept, able to connect different locally based initiatives (acting on different topics in relation to local urgencies and opportunities) toward a long-term strategy for urban environment and urban public life improvement.

In the case of Obrenovac (see Chapter 24), the project explicitly refers to the Making Cities Resilient Campaign launched by the UN Office for Disaster Risk Reduction (UNISDR) and aimed at strengthening resilience of complex urban systems by improving, in particular, social and organizational components (adaptation of local communities and local institutions in the face of risks).

In the Pedda Jalaripeta case, resilience is used as a key concept to understand if (and how) a poor urban neighborhood can retain cultural and local identity when combating disasters and external pressures. As emerged also from the large debate after the New Orleans disaster (Bankston et al., 2010), the Pedda Jalaripeta case underlines a challenging aspect of resilience, reflecting on the "recognizability" of the system and on the capabilities of local urban communities to guarantee the maintenance of cultural, social, and organizational identity of urban neighborhoods. The resilience of Pedda Jalaripeta community is demonstrated by the safeguard of the local identity, of the social, economic, and cultural characteristics and of the landscape characterizing the neighborhood (that is inseparable from the community itself). The cultural and social activities of Pedda Jalaripeta community are strongly connected with the built (the slum structure) and natural environments (the beach) and with its social places, in a intimate coevolutionary relationship between the community and the environment. The strong cultural identity and social (informal) network emerged as key factors of resilience and allowed Pedda Jalaripeta community to maintain its characteristics and identity and overcome fire events and gentrification pressures.

The Smart City metaphor is not explicitly mentioned in transition initiatives and grassroots innovation initiatives. In the presented case studies only a few social innovation initiatives (at networking organizational level) in the Barcelona metropolitan area explicitly refer to this metaphor. In the large number of local initiatives promoted by local communities to tackle food, mobility, housing, or energy issues, it is possible to highlight initiatives dealing with integrated and transversal approaches to climate change through social transformation and referring to smartness, as the Smart Citizen Barcelona (a participatory educational platform). The initiatives focusing on social innovation and digital fabrication such as FabLab Barcelona, Rosa Sensat, and Hangar (oriented to art research and production) also approached the issues of climate change and environmental awareness in Barcelona context through Information and Communications Technologies (ICTs) (Smart metaphor is used, for instance, in a project promoting a participative process for engaging citizens in the management and implementation of sensors and software for environmental monitoring).

In the presented transition initiatives, a large range of IT/ICT tools have been developed for coproduction and codesign activities, for learning and capacity building of communities, and for improving networking (sharing practices and narrative of experiences). For example, the Transition Town Network developed a specific strategy for improving ICT capacities (training) in order to promote the use of ICTs in local transition initiatives that are asked not only to launch a website but also to be aware and trained in digital innovation as powerful instruments for local issues dissemination, citizen engagement, and online forums organization.

33.3.3 Persistence, Adaptability, and Transformability in Transition Initiatives

Applying the concepts of persistence, adaptability, and transformability to community-led practices requires an adjustment of their common definition: persistence, when transition initiatives aim to cope with local urgencies related to climate impacts in the short term; adaptability, when transition initiatives aim for incremental adjustments; transformability, when transition initiatives experiment with innovative approaches in a long-term perspective able to promote changes at the local level, by influencing and activating meanwhile societal transition of the whole urban system.

In general, transition practices are more oriented to adaptability or transformability. At the same time, it is possible to underscore a disconnection between some general principles characterizing the level of network—in which conceptual reframing and more complex visioning are usually developed—and the level of single intervention, where actions and project implementation represent the core of activities generally characterized by a low level of theoretical and conceptual awareness. For instance, the Transition Town Network implies a long-term strategic vision toward energy descent scenario that includes also strategies for local resilience strengthening and adaptation to climate change. At the Transition Town Network level, transformability characterizes the whole movement, but shifting the focus on the analyzed transition initiatives, only Totnes (that is the first transition town) developed the descent plan and comprehensive visioning and scenarios. In the Monteveglio and Newcastle experiences, activities are more oriented to implement and manage a set of interventions/actions (focused on food in Monteveglio, more heterogeneous in Newcastle). Also, in Barcelona the numerous transition initiatives (and the presented case of Pla BUITS) are promoting visions for societal changes (in social and environmental components of complex urban system and in their organizational structure).

Only the case of Pedda Jalaripeta refers to persistence, being mostly focused on how the local community is able to survive dealing with shocks and pressures.

33.3.4 Governance Models in Transition Initiatives: Hints for Innovation

The selected case studies refer to activated or ongoing processes of societal transition allowing underscoring some aspects related to the adopted governance models. The latter can be referred to as single local initiatives (process of citizen engagement, internal decision-making process, etc.) and to the networks and alliances among single initiatives. In relation to climate issues, the network level is a promising and interesting aspect to be investigated.

Networks and alliances are activated for different reasons and show different characteristics and structures:

- Networks related to common issues/aims (e.g., Depave, food, climate action, social innovation, etc.); these kind of networks provide tools for mutual support (expert contributions, knowledge sharing, solutions sharing, etc.), are characterized by a horizontal structure, and aim to coordinate local initiatives without a defined leadership;
- Networks sharing principles and aims and referring to global issues, as the Transition Town Network; these networks are more structured and provide to local transition initiatives a long-term vision, training, and supporting tools; networking is based on shared values and on the assumption of a common vision, and usually the network has a clear organizational structure and assumes a clear role of leadership and coordination of local initiatives;

- Networks of local/regional practices addressing similar topics and issues, but strictly limited to a metropolitan or regional context; these kinds of networks promote knowledge sharing, mutual support in common initiatives (as Barcelona networking), and can also include local institutions, with a coordination role, in the partnership.

In transition initiatives networking is a very important organizational level of governance, and usually the networks activate dialogues with institutions and private actors. It is also an innovative organizational structure of the governance framework of urban complex system able to develop and provide original tools and conceptual reframing to support local initiatives.

The degree of involvement of urban public institutions (or other public and private actors) in transition initiatives is very heterogeneous. The Transition Town Network suggests and promotes the activation of alliances and synergies with public institutions and existing local associations. Locally, single transition initiatives show very different approaches in activating alliances and partnerships. In the case of Newcastle, for example, the lack of connection with local associations has been underlined as a critical aspect. The Monteveglio initiative developed a strong partnership both with the municipality and with other associations and local initiatives (e.g., local community gardening associations, citizen association for green and public spaces stewardship). Also in Totnes several projects and actions have been promoted in collaboration with other local associations.

In contrast, in Barcelona the municipality played (and still plays) a coordination role in supporting local transition initiatives through the launch of the public call for local public spaces stewardship and management of public green spaces.

The relationships with institutional policies and programs are not explicitly mentioned in most of the transition initiatives. The Obrenovac case—an applied research project promoted by academics and engaging a large range of stakeholders—clearly refers to the Making Cities Resilient Campaign promoted by the UNISDR and develops an institutional policy framework aimed to highlight mutual synergies. Other initiatives are randomly connected to planning or policies.

33.3.5 Transition Tools for Climate Urban Actions

Transition initiatives developed a large range of solutions, innovative ideas, and implementation management models. In particular, the numerous tools for learning/capacity building and for knowledge and design coproduction represent promising and useful instruments that could be transferred in the framework of urban climate actions policies and measures.

The awareness of citizens and communities is a fundamental goal of the Transition Town Network to be built through training activities, films, and community events able to activate debate and stimulate interest of single citizens and community. The innovative aspect is that awareness is generally built not through a mere transfer of information but thanks to knowledge coproduction, capable of increasing citizen awareness and allowing as well local communities to share skills and competencies. Knowledge coproduction is also used as a tool for empowering the local community.

Knowledge coproduction tools are developed in almost all the transition initiatives. In the Obrenovac case study, it was intended as a first step to promote local institutions' and stakeholders' awareness—a fundamental tool for engaging stakeholders from different sectors and backgrounds in a process of codesign solution for resilience.

An innovative and inspiring aspect is also the biunivocal process of knowledge coproduction promoted in transition initiatives; sharing knowledge implies recognition of the value related to different kinds of knowledge (expert, organizational or institutional, social, and cultural knowledge, etc.) toward the identification of opportunities and local resources. These biunivocal processes go largely beyond the traditional "training" from experts to citizens, activating mutual exchanges of competencies and resources from different "sources" that are not usually connected.

Transition initiatives emphasize the relevance of the empowerment of citizens and local communities that are usually involved in city urban climate polices but not empowered; in transition initiatives citizens and communities have the responsibility to manage actions toward more sustainable scenarios.

The role of individual citizens is also strengthened thanks to communication tools based on a storytelling approach in order to emphasize the idea that each single citizen can "make the difference," which gives them the feeling of being part of the change and key players in the transition process. The Transition Town Network provides specific strategies of communication (and training modules for local transition initiative) based on "transition stories" that are addressed to promote the achieved results.

33.3.6 Barriers and Opportunities in Transition Initiatives

Focusing on the contribution of transition initiatives to urban climate policies, barriers and opportunities emerging from the examined transition practices will be discussed, focusing on theoretical, organizational, and operational aspects (see paragraph 33.2.6).

- Theoretical Constraints and Opportunities

    In the examined case studies, climate change is one of the core issues of transition initiatives, demonstrating a widespread awareness of local communities on climate issues. It is quite clear (apart from the Pedda Jalaripeta case study) that sustainability is the common reference for defining local principles and long-term envisioning (if explicitly produced).

    Although there is significant awareness of climate-related issues in transition movements, it is possible meanwhile to see the lack of clear definitions (e.g., mitigation and adaptation are always used as overlapping concepts) and of robust theoretical frameworks in transition initiatives.

    The topics that characterize transition initiatives, mostly when they do not belong to wider networks, often arise from local urgencies and depend on the local community's prioritization.

    The transition initiatives represent, however, an opportunity in terms of trained and aware citizens as well as of energies devoted to improve urban sustainability and to counterbalance climate change.

- Organizational Constraints and Opportunities

    Numerous organizational barriers emerge from the selected case studies. A critical barrier, common to numerous practices, is related to the difficulties in citizens' engagement as well as in enlarging the number of citizens participating in transition practices. The Transition Town Network (and other networks) developed tools and training modules supporting transition initiatives in improving local capabilities to engage citizens, ensuring in so doing the growth and stabilization of the transition process. At the same time, it is possible to see a certain degree of fragility of local transition initiatives (also when they are part of a wider network); in several cases, indeed, the actors promoting the initiative feel isolated. This "island effect" of transition initiatives could refer to the difficulty for a single local initiative to be effectively connected with institutions and organizations acting at a larger (urban) scale. Hence, the networking of local initiatives could play a key role in favoring alliances with public institutions and relevant stakeholders and in developing tools and training for citizens' involvement. The presented networks are often related to a very specific topic or solution (for example, Depave or Repair Cafè) and mostly addressed to promote replication and transfer of adopted organizational models and solutions. This is, on the one hand, an opportunity (transferable principles, action criteria, and tools); on the other hand, it is a barrier, since it duplicates efforts and networks on similar issues that are all addressed to counterbalance climate change but that still remain isolated from each other. The result is the flourishing of numerous initiatives, the redundancy of networks, and, meanwhile, the lack of dialogue, synergies, and alliances among them.

    Climate issues require integrated and intersectoral strategies and actions; the "island effect" implies a low level of integration and dialogue not only among the participants to the transition networks but also among them and public institutions.

    The last aspect (that could become a further research path) is related to the degree of transition community democracy. It is a common assumption that if a practice is led by the community, it is "per se" inclusive. Unfortunately, the internal decision-making process characterizing this initiative is often not very inclusive and participative, being led by the active actors that manage the transition initiative. Solutions and actions (in particular if transition initiatives are acting on public spaces or public services) are often defined without a participative and inclusive process involving all the interested citizens and stakeholders. The lack of participation may also induce difficulties and conflicts in projects' implementation or even a process of gentrification of some "undesired" sectors of users and citizens.

    Climate change also requires stronger integration and coordination between public policies (climate measures and actions plans) and community-led initiatives. The examined case studies, in particular those characterized by a strong community-led approach, show difficulties in coordination with institutional policies and existing planning tools.

    As discussed in Chapter 3, in order to guarantee a societal transition toward sustainable scenarios able to face climate issues, the "innovations niches" have to be part of a larger policy framework in which there are multiscale and multiactor processes. Community-led initiatives at local scale need strong coordination to outline shared visions and long-term strategies. To reach societal transition and changes at macroregime or

landscape levels, alliances and dialogues between local transition initiatives and the mesoregime, trends and actors are required.
- Operational Constraints and Opportunities

In transition initiatives the lack of well-defined monitoring and assessment tools and above all of common monitoring frameworks represents a barrier for their integration in urban climate policies and a difficulty in looking at single transition initiatives as "actions" to be framed into wider urban climate strategies. Without any assessment/monitoring tool it is difficult indeed to understand and assess the benefits arising from the transition initiatives to local urban context both in terms of measurable positive impacts toward specific environmental targets (e.g., climate mitigation and adaptation) and in terms of enhancing urban resilience (social and organizational aspects). Few transition networks have so far developed monitoring and evaluation tools and provided indicators and updated data (as the case of Depave network) to demonstrate the positive impacts to local community given by the implemented practices carried out by different actors of the network. The few networks that have developed monitoring and assessment tools are generally coordinated or promoted by scholars, researchers or by experts and practitioners.

The enhancement of local communities' resilience for adapting to climate change is surely one of the main results achieved by transition initiatives where awareness, training, and capacity building tools as well as knowledge and solutions coproduction tools are usually developed and activated. The community empowerment in contributing to urban climate issues is often a key issue of transition initiatives.

33.4 FINAL REMARKS

Summing up, the examined case studies allow us to observe that in the last decade climate change has progressively entered the urban agenda all over the world, albeit pushed by heterogeneous factors and addressing different priorities and goals.

Climate policies in the examined case studies have been promoted by international or European initiatives, by international cooperation agencies, or even pursued by community-led initiatives acting locally but globally distributed. Universities are also playing an important driving role; in some cases, indeed, the engagement of academic and institutional actors in research projects has allowed both to increase local awareness on climate issues and to promote and support local authorities in carrying out and implementing climate strategies. Universities and experts also play a key role in promoting and supporting transition initiatives, mostly in urban and metropolitan contexts, by sharing knowledge and expertise and transferring technological innovation to local associations and communities.

All the examined practices reveal traces of innovation on both conceptual and organizational levels that deserve to be further developed.

On a conceptual level, climate issues seem to be increasingly framed into the broader path of sustainable development, although a radical shift toward a green and circular economic development model is still far from being adopted. However, transition initiatives are often acting as "niches of innovation" by spreading local (and sometimes fragile) seeds of innovative economic models (experimenting green and circular economy, social economy, and social innovation), providing opportunities and resources for a societal transition.

Moreover, the increasing awareness of the close linkages between climate-related hazards and urbanization dynamics (land take and soil sealing, urban sprawl in flood-prone areas, inadequacy of sewage networks, etc.) and the growing (although still limited) coordination between climate change and disaster risk reduction is leading to bring out risk issues from the sectoral perspective that has for long guided disaster risk reduction and to assign a key role to spatial planning in counterbalancing climate change and, in a broader perspective, risks. The importance attributed to planning tools is also pushing toward a shift from the traditional engineering-based measures to prevent and mitigate hazards to risk-aware planning strategies, leading to the development of large- and small-scale green infrastructures, to the widespread use of permeable materials and surfaces, as well as to the development of de-sealing measures in compact urban areas.

Transition initiatives are significantly contributing to activate projects and to implement locally based actions mostly oriented to climate adaptation (green roofs, urban agriculture and local food chains, water ecosystem services regulation, increasing permeable and green surface, etc.) but also to increase citizens' and local communities' awareness on climate issues.

Regarding the organizational level, it is worth emphasizing that climate issues are bringing out new links among different disciplinary fields (e.g., urban planning, hydrology, urban ecology) and testing new paths toward a multilevel and multisectoral cooperation and collaboration among different institutional actors (national, regional, and local authorities, river basin authorities, etc.) or among different sectors within the same institution (planning department, civil protection, environment, etc.). Resilience offices, climate teams, and intersectoral agencies are only some examples so far developed to cope with the complex challenge of climate change. Furthermore, transition initiatives are bringing to the fore new elements for the innovation of organizational models. By promoting the active engagement of local communities, they favor the coproduction of innovative and locally shared solutions, increasing their social acceptability and facilitating their implementation. These initiatives are often characterized by a targeted scope, although many of them, as the Transition Town Network, are based on a multi issue approach, combining low-carbon transition, environmental sustainability, and resilience enhancement. The integration between environmental issues and social resilience is also a promising aspect that characterizes numerous transition initiatives both at network level and at local level (including isolated transition practices); most of them address, indeed, a specific result (e.g., local food chain, shared renewable energy, etc.) through the empowerment of citizens and local communities.

Finally, it is worth emphasizing that, although cities currently play a pivotal role in counterbalancing climate change, their actions may lead sometimes to provide fragmented and ineffective responses to large-scale climate-related phenomena, such as coastal erosion or rising sea levels. This clearly emphasizes that, although cities have to play a key role, the effectiveness of their actions requires adequate frames and coordinating visions at upper levels (regional, national) that so far are not always available. Cities are also the common (natural) place to activate transition initiatives; they provide, indeed, the most promising "environmental conditions" for creating synergies among different initiatives aimed at counterbalancing climate change. However, the fragmentation of individual initiatives and the diffuse lack of integration among transition practices and institutional policies are often relevant barriers hindering both the development of long-term visions and the effective implementation of the foreseen actions and measures. The Barcelona case study clearly demonstrates the potential of a coordinated and integrated metropolitan strategy that is capable of including and empowering local transition initiatives, establishing synergies and alliances toward the common goal of counterbalancing climate change.

References

Abdulrahman, A., Abraham, M.P., 2016. Comparative analysis of African Union (AU) and European Union (EU): challenges and prospects. International Journal of Peace and Conflict Studies (IJPCS) 3 (1), 46–57. Available at: http://www.rcmss.com. ISSN:2354-1598(Online) ISSN: 2346-7258 (Print).

African Union, 2014. Draft African Union strategy on climate change. AMCEN-15-REF-11. Available at: http://www.un.org/en/africa/osaa/pdf/au/cap_draft_auclimatestrategy_2015.pdf.

Bankston III, C.L., Barnshaw, J., Bevc, C., Capowich, G.E., Clarke, L., Das, S.K., Esmail, A., 2010. The Sociology of Katrina: Perspectives on a Modern Catastrophe. Rowman & Littlefield Publishers.

Beck, U., 2017. Le metamorfosi del mondo. Editori Laterza. ISBN:9788858129777.

Bulkeley, H., Andonova, L., Bäckstrand, K., Betsill, M., et al., 2012. Governing climate change transnationally: assessing the evidence from a database of sixty initiatives. Environment and Planning C 30 (4), 591–612.

Bulkeley, H., 2013. Cities and Climate Change. Routledge, New York.

Davoudi, S., Crawford, J., Memhood, A., 2009. Climate Change and Spatial Planning Responses. In: Davoudi, S., Crawford, J., Memhood, A. (Eds.), Planning for Climate Change. Strategies for Mitigation and Adaptation for Planners. USA, Earthscan, UK.

Davoudi, S., Shaw, K., Haider, J.L., Quinlan, A.E., Peterson, G.D., Wilkinson, C., Fünfgeld, H., McEvoy, D., Porter, L., Davoudi, S., 2012. Resilience: a bridging concept or a dead End? "Reframing" resilience: challenges for planning theory and practice interacting traps: resilience assessment of a pasture management system in Northern Afghanistan urban resilience: what does it mean in planning practice? Resilience as a useful concept for climate change adaptation? The politics of resilience for planning: a cautionary note. Planning Theory & Practice 13 (2), 299–333. https://doi.org/10.1080/14649357.2012.677124.

Galderisi, 2017. The Nexus Approach to Disaster Risk Reduction, Climate Adaptation and Ecosystem Management: New Paths for a Sustainable and Resilient Urban Development. In: Colucci, A., Magoni, F., Menoni, S. (Eds.), Peri-urban Areas and Food-Energy-Water Nexus. Sustainability and Resilience Strategies in the Age of Climate Change. Springer. ISBN:978-3-319-41022-7.

LBNL (Lawrence Berkeley National Laboratory), 2016. US-China climate Smart cities initiative. Declaration. Available at: https://ccwgsmartcities.lbl.gov/declaration.

Priemus, H., Davoudi, S., 2012. Introduction to the special issue. European Planning Studies 20 (1), 1–6.

UN (United Nations) Department of Economic and Social Affairs, Population Division, 2014. World urbanization prospect. The 2014 Revision. (ST/ESA/SER.A/366). Available at: https://esa.un.org/unpd/wup/publications/files/wup2014-highlights.pdf.

World Resources Institute, 2015. Infographic: What Do Your Country's Emissions Look like? Available at: http://www.wri.org/blog/2015/06/infographic-what-do-your-countrys-emissions-look.

Yohe, G., Strzepek, K., 2007. Adaptation and mitigation as complementary tools for reducing the risk of climate impacts, Mitigation and Adaptation Strategies for Global Change 12, pp. 727–739. https://doi.org/10.1007/s11027-007-9096-3.

CHAPTER

34

Future Perspectives: Key Principles for a Climate Sensitive Urban Development[1]

Adriana Galderisi[1], Angela Colucci[2]

[1]University of Campania Luigi Vanvitelli, Aversa, Italy; [2]Co.O.Pe.Ra.Te. ldt, Pavia, Italy

34.1 INTRODUCTION

While climate change is increasingly recognized as a priority issue both in the political agenda and in the scientific debate, doubts about the effectiveness of the undertaken strategies to counterbalance it and a growing alarm about the likely failure of these strategies are emerging, due to the limited advances so far recorded.

The 2016 Global Risk Report identifies indeed such failure as one of the top global risks; looking at the ranking provided by the Report in respect to the time span 2007–16, it is worth noting that "climate change", ranked as one of the Top 5 Global Risks, in terms both of likelihood (fifth position) and impact (second position) in 2011, has turned into "failure of climate change mitigation and adaptation" in 2016, placing itself in the third position in terms of likelihood and in the first one in terms of impact (WEF, 2016).

The emergence of such an alarm should not be surprising in light of the numerous case studies presented in this book that have clearly emphasized delays, afterthoughts, and barriers that significantly undermine the effectiveness of the strategies so far undertaken at different scales to counterbalance climate change (see Chapter 33).

The wide range of case studies presented in this book has been extended, indeed, far beyond Europe, including strategies and practices carried out in economically advanced countries (United States, Australia, and Europe), emerging countries (China, India), and countries that are still lagging behind (Africa). This choice depends on the awareness, firstly, that climate change and consequently the strategies to counterbalance it have to be faced at a global scale since, as clearly remarked by Pope Francis in the Encyclical Laudato Sii "climate change is a global problem with serious environmental, social, economic and political implications" all over the world; secondly, that strengths and weaknesses arising from the different case studies, despite having some common features, present also a significant heterogeneity and provide different insights to the development of future trajectories.

Hence, based on the lessons learned from the wide range of analyzed case studies and being aware that there are no single paths and rules suitable to all contexts for pursuing the common goal to counterbalance climate change, in the following paragraph some key principles that could better guide planners and decision makers in overcoming current criticalities and building up climate sensitive urban development processes will be discussed.

34.2 THE KEY PRINCIPLES FOR CLIMATE-SENSITIVE URBAN DEVELOPMENT

34.2.1 Integration

The concept of integration has been often referred to in this book with respect to both theoretical and operational perspectives. Hereinafter, the importance of this concept for shaping a climate-sensitive urban development will be

---

[1] Although the chapter is the result of a common reflection, paragraphs 34.1 and sub-pagraphs 34.2.1 and 34.2.2 and 34.2.3 have been written by Adriana Galderisi; sub-paragraph 34.2.4, 34.2.5 by Angela Colucci.

discussed with reference to three main aspects: integration of approaches, integration of knowledge, and integration of policies and strategies.

Integrated Approaches—In the first section of this book, challenges and opportunities related to the integration of the three widespread urban metaphors—Smart City, Resilient City, and Transition Towns—currently largely used when referring to the need for empowering cities in the face of climate change, have been intensely explored. As already remarked in Chapter 4 and confirmed by current practices (Sections 3–5), these three urban metaphors—despite underlying different approaches and strengthening different characteristics of urban systems in the face of the multiple challenges they have to cope with and, above all, in the face of climate issues—present numerous commonalities and largely contribute to recognizing cities as complex adaptive systems. Moreover, the Resilient City metaphor has been largely confirmed by current practices as the most commonly used, being the ultimate goal of numerous initiatives aimed at increasing urban smartness as well of many transition practices.

Hence, by embracing an integrated perspective, the Resilient City metaphor can be identified as a "convening concept" in the face of climate issues (see Chapter 4). However, this requires a shift in current interpretations of the Resilience concept itself, promoting a transition from the "bouncing back" perspective, which has so far largely guided current initiatives and practices, toward an "evolutionary" perspective, allowing to base the building up of climate sensitive cities on adaptive decision-making processes, allowing constant learning, crucial to cope with complexity and uncertainty, and looking meanwhile at different time horizons: short term, by improving the efficiency and optimization of existing urban systems in order to reduce greenhouse gas emissions and better withstand current impacts of climate-related hazards; medium term, by increasing flexibility and redundancy in order to better adapt to uncertain circumstances; and long term, by promoting innovation and creativity to drive urban transition toward novel urban development patterns.

Finally, it is worth reminding that in order to promote an integrated approach to climate issues, the latter have to be better framed into the wider conceptual framework of sustainability and the close relationships between climate change and disaster risk reduction have to be recognized and deepened. As clearly underlined by the Outcome Document of the United Nations Conference on Sustainable Development, the Future We Want, indeed, climate change is a crosscutting issue, capable of undermining the ability of all countries to achieve sustainable development (UN, 2012). Also, the Global Sustainable Development Report (2015), focusing on the idea of sustainable development goals "as an interconnected system," requiring a holistic, multisectoral, and multidimensional approach to their implementation, has further remarked on the linkages among sustainable cities, climate change, and disaster risk reduction (DRR). Hence, as already said by Kelman et al. (2015), climate change should be analyzed into the wider framework of DRR, placing the latter as "a subset of wider development and sustainability processes."

Integrated Knowledge—The importance of developing integrated knowledge, capable to bring together different disciplinary perspectives (see Chapter 31) as well as experts' and communities' knowledge (see Chapter 32), for building up climate sensitive cities also has been discussed in detail in this book. As emphasized by Loevinsohn et al. (2014), in fact, the "disconnect between the different scientific communities and related knowledge and practice hampers comprehensive diagnosis of the problems at stake and the mounting of more effective actions to address it." Thus, as seen in numerous case studies, current segmentation of knowledge, due to the persisting difficulties in overcoming both disciplinary boundaries and the "silo" organization of competencies at different governmental levels, still represents a barrier toward both the effective understanding of drivers and consequences of climate phenomena and the development of adequate solutions.

Integration in knowledge is also crucial to better understand the numerous interdependencies among different phenomena and processes as well as the feedback derived from the strategies/actions that are put in place: hence, it represents a key principle to tackle uncertainty (Scott et al., 2015) and to improve learning capacity that is typical of complex adaptive systems.

Moreover, as mentioned earlier, the need for integrating knowledge arising from different areas of concern related to sustainability, comprising risks and climate change, has been clearly documented both by institutional documents and scientific literature. However, integrated knowledge does not automatically result from the availability of larger and larger amounts of data and information arising from different areas and produced by different actors; on the opposite, lacking shared languages and conceptual frameworks, they could result only in "a chaotic mix of information, analysis and interpretation" (Menoni et al., 2014).

Finally, it has to be noticed that, despite the emphasis currently attributed to the engagement of citizens and communities in knowledge cocreation (Concilio, 2016), being "local knowledge very context specific and sometimes more relevant than the highly formalized but detached from context" produced by experts (Menoni et al., 2014), most current practices are still mostly based on expert knowledge; the engagement of local stakeholders is very limited and generally addressed to inform them, neglecting the crucial role of knowledge cocreation as a basis for an effective coproduction of innovative solutions.

Information and Communication Technologies (ICTs) can significantly contribute both to improve knowledge coproduction—by providing manageable (open and interoperable) data and information allowing different stakeholders to interact—and to codevelop innovative and shared solutions to the problems at stake.

In transition initiatives, for example, knowledge coproduction based on ICT platforms is increasingly emerging as a common practice, contributing to the development of adaptive governance models based on a continuous knowledge exchange among different domains as well as between expert knowledge and experiential/local knowledge.

Integrated Policies—In this book the emerging trend at global scale as well as in some of the examined countries toward a better integration between mitigation and adaptation policies has been clearly pointed out (see Chapters 5 and 7). Despite mitigation and adaptation strategies that have been for long—and in most of the case studies examined in this book are still—considered separately, due to their different aims, time spans of reference and involved actors, nowadays both in scientific literature and in numerous institutional documents, the need for developing integrated climate strategies, designed to maximize synergies and trade-offs between mitigation and adaption policies, is largely emphasized. At global scale, the Global Covenant of Mayors for Climate and Energy, announced in January 2017, is specifically addressed to tackle three interrelated issues: climate change mitigation; adaptation to its adverse effects; and access to secure, clean, and affordable energy. This is an important step forward toward the building up of climate-sensitive urban development, since it emphasizes the positive and mutual capacity of adaptation and mitigation strategies and actions to reinforce each other; this is particularly evident, for example, when nature-based solutions to enhance adaptation are at stake, due to their potential for improving meanwhile carbon sequestration and storage (Pramova et al., 2015).

Furthermore, as climate change represents a crosscutting issue—depending on and affecting different sectors, from transportation to water and land-use management—the need for developing cross-sectoral strategies to cope with this issue, by strengthening both vertical and horizontal cooperation among different government levels, different involved sectors and departments, in contrast with the still prevailing sectoral approach to policy making, has to be emphasized.

Finally, policy integration is crucial to promote a better integration among institutional climate policies and the heterogeneous landscape of community-led initiatives, favoring synergies and alliances toward the common goal of counterbalancing climate change.

34.2.2 Mainstreaming

The term *mainstreaming* has gained growing importance in the last two decades both at the European and international levels. The concept is generally used to indicate a process aimed at embedding a particular issue or value into broader decision-making processes or, as remarked by Gupta (2010), aimed at bringing "marginal, sectorial issues into the center of discussions, thereby attracting more political attention, economic resources and intellectual capacities." The term found prominence, for example, in the Sustainable Development Document of the United Nations Conference held in 2012, with reference to different topics, from gender issues to climate change (UN, 2012). In respect to climate change, it is worth noting that starting from the 1990s the need for mainstreaming mitigation issues into energy and transport policies has been emphasized, while only since early 2000 has the concept been used with reference to climate adaptation (Klein et al., 2005). Nowadays, there is widespread agreement on the idea that, as climate change is a crosscutting issue, closely linked to all development sectors, the mainstreaming of climate policy (including both mitigation and adaptation) "into all stages of policy-making in 'other' policy sectors" (Rayner and Berkhout, 2012) is required. However, it is important to stress that the concept of "mainstreaming" largely overcomes the mere idea of integrating climate issues in existing policies and programs. As remarked by Gupta (2010), indeed, it implies the adoption of a transformative approach, aimed at rethinking current development policies under the climate change lens. Moreover, the need for systematically evaluating the expected impacts of all development policies on climate change mitigation and adaptation has to be noted too.

Based on the case studies analyzed in this book, it is worth emphasizing that, despite the large emphasis put on the need for mainstreaming climate change in development policies both in institutional documents and in scientific literature, current practices do not show significant steps forward along this direction.

For example, according to Davoudi et al. (2009), planning tools could play a key role in mainstreaming climate policies into urban development strategies, enabling mitigation and adaptation strategies to be framed into the broader perspective of sustainable development. Moreover, the importance of mainstreaming DRR and climate risks in urban planning has been clearly emphasized by the Outcome Document of the United Nations Conference, the Future We Want (UN, 2012). Unfortunately, the attempts to mainstream climate issues into urban planning tools so far undertaken show limited results, mostly because of their prevailing regulative role, which leaves scarce room for the adoption of a "transformational" perspective that is crucial to reverse current urban development patterns.

Hence, the need for better translating into practice the concept of mainstreaming and for effectively mainstreaming climate change into urban planning tools is of paramount importance for achieving climate-sensitive urban development; however, a radical change of current urban planning tools aimed at redesigning them according to the new lens of climate change mitigation and adaptation as well as of disaster risk reduction would be required.

34.2.3 Transformative

A transformative perspective is a key principle for climate-sensitive urban development; such a perspective should better guide planners and decision makers to go beyond the goal of enhancing cities' persistence and adaptability in the face of climate change—which generally moves current practices, even when explicitly referred to the Resilient Cities metaphor (see Chapter 33)—addressing more firmly the third characteristic of a resilient city—transformability. Resilience has been interpreted, indeed, as a "dynamic interplay of persistence, adaptability and transformability across multiple scales" (Folke et al., 2010), where transformability is intended as "the capacity to create a fundamentally new system when ecological, economic, or social structures make the existing system untenable" (Walker et al., 2004; Folke et al., 2010).

Hence, this principle becomes crucial to promote innovation and foreseen radical changes, according to a long-term perspective, of current urban development models in the face of a growing uncertainty; even though the latter is inherent to urban systems, and consequently to planning practices, climate change adds further uncertainty to the already difficult anticipation of future scenarios, making the knowledge of past behaviors of urban systems a tool no longer reliable to understand future behaviors, even in apparently similar circumstances. Thus, the growing uncertainty puts a strain on the adequacy of traditional planners' toolkits (Davoudi et al. 2013) and requires "new practices in front of novel problems" (Grøtan 2014).

The inadequacy of current urban planning tools in coping with uncertainties that characterize the future of climate scenarios and urban development has been largely stressed in most of the examined practices (see Chapter 33). Planners and decision makers, indeed, often tend to cope with growing uncertainties by ignoring them, pursuing to continue to impose "certainty". Moreover, planning tools are generally designed to define "what is expected, permitted and prohibited. (…) This very characteristic, however, might actually prohibit more flexible or adaptive forms of practice able to work with contingency and uncertainty" (Hillier, 2017).

34.2.4 Capacity Building

Capacity building is here intended as a process aimed at strengthening the skills of individuals as well as of local institutions to cope with climate change. It represents "a multifaceted approach that addresses the ability of multiple stakeholders and institutions at all levels of governance, and combines the individual, societal and institutional capacity to formulate, implement, enhance, manage, monitor and evaluate public policies for sustainable urban development" (UN, 2016). The capacity-building tools, largely adopted in community-led initiatives (namely in the Transition Town Network), have largely demonstrated their efficacy both in improving communities' and citizens' awareness on climate change and in empowering citizens, allowing them to become key actors in outlining and implementing urban climate policies.

Moreover, it is worth reminding that capacity building represents a powerful tool to the stabilization and continuity of climate policies, whose lack, as remarked in Chapter 33, represents a critical operational barrier for their effectiveness. As pointed out by Pope Francis in the Encyclical Laudato Sii, in fact, "continuity is essential, because policies related to climate change and environmental protection cannot be altered with every change of government.

Results take time and demand immediate outlays which may not produce tangible effects within any one government's term."

In particular, capacity-building processes (characterized by different methodologies) should be addressed both to institutional actors and to the local communities (citizen and community organizations).

With respect to institutional actors, capacity-building processes are crucial to guarantee effectiveness and continuity of institutional strategies and policies they might increase awareness about climate issues, improve knowledge and skills, and facilitate effective exchange among the numerous sectorial departments involved in outlining and implementing climate strategies, enabling public administrations to better manage transitional processes toward low-carbon and adaptive scenarios.

Furthermore, technical staff in public institutions need to be adequately trained in managing ITCs' tools. The latter are, indeed, powerful instruments to support innovative governance processes and integrated approaches to climate change. For example, the widespread implementation of open-data policies might be very important to support and enhance a collaborative exchange among different stakeholders, favoring democratization and inclusion of current decision-making processes. Nevertheless, making data and information 'open source' without adequately training both data managers and users could be ineffective, making the large amount of available data illegible and unusable.

Hence, ICTs require to be better managed by local institutions in order to support "innovative ways of decision-making, innovative administration or even innovative forms of collaboration" (Meijer and Bolivar, 2015); for example, to provide new collaborative environments for adaptive governance processes, based on knowledge coproduction, innovative solutions codesign, continuous monitoring, and learning. Universities could play a key role in developing both innovative ITC tools and capacity building modules for local institutions.

With respect to local communities, capacity-building processes might play a crucial role both in raising civil society awareness on climate issues and, above all, in stimulating citizens' actions, assigning them an active role in increasing climate sensitive urban development.

In brief, the establishment of capacity-building processes involving all the numerous stakeholders that at different levels and with different tasks contribute to define and implement climate strategies (e.g., institutions, civil society, private sector, scholars, practitioners) represents a precondition for activating virtuous cycles, favoring the alignment of sectorial languages and promoting synergies within a "common environment."

34.2.5 Governance

The term *governance* refers to all differentiated mechanisms (self-regulation or informal rules of civil society; coregulation of public and private actors and regulation through institutional government) through which individuals and institutions, public and private actors collaborate in managing common interests and social affairs (Dingwerth and Pattberg, 2006; Borraz and Le Galès, 2010). With respect to urban climate action, governance is here referred to as the whole process starting from the decision-making (including problem setting and the development of strategic visions) to the implementation and management of policies, measures, and actions.

Based on the numerous case studies presented in this book, governance is one of the main obstacles in the implementation of both institutional-led and community-led initiatives, emphasizing the urgency of renovating current governance models, enabling them to better embrace cross-scale interactions and multiscale phenomena and dynamics (temporal/spatial) of complex systems (Ostrom, 1990, 2005), in order to better deal with the complexity arising from environmental, social, and economic challenges and from their interactions. Thus, the development of new governance models currently represents a crosscutting goal of international and regional programs; the concept of good urban governance (that refers to the idea of "good governance" introduced by the United Nations Development Program in 1997) is promoted, for instance, by several UN-HABITAT programs (UN, 2016) and meanwhile by interregional European Union policies focused on the renovation of governance approaches (Knieling and Leal Filho, 2013). The shift from current governance models toward more adaptive and collaborative models is a common challenge for cities all over the world, being them increasingly called upon to integrate both institutional climate strategies and emerging community-led initiatives into the wider process of sustainable development (Colucci 2017).

A renovated adaptive governance framework should be based on principles arising from the theoretical debate on urban metaphors (see Section 1) as well as on promising trajectories emerging from both institutional-led (Section 3) and community-led (Section 4) practices.

As clearly noted in other chapters of this book (see Chapter 33), in the face of climate change cities play a pivotal role, by connecting global goals and large-scale strategies with locally based strategies and actions. At the same time, cities are the preferred (and natural) place where the greatest amount of innovative community-led initiatives arise.

Thus, a new adaptive urban governance framework should allow creating collaborative environments in managing flexible and adaptive processes of decision-making and implementation, in which institutional actors play a coordinating role, facilitating meanwhile dialogue and collaboration among a large range of stakeholders (comprising citizens) in outlining shared long-term visions.

This implies a radical shift from current governance models, generally based on linear processess starting from goals definition, policies and measures prioritization, up to the implementation phase - towards adaptive models (see Chapter 4), capable to engage different stakeholders in outlining multiple visions consistent with likely alternative scenarios and addressed to reach a shared and constantly updated set of goals (Geels 2002, Rotmans et al. 2001, Loorbach, 2010).

An adaptive urban governance framework should hence allow a multistakeholders process capable to firstly outline the long-term scenario (strategic vision). This concept - outlined by the transition management theories — shows some similarities with the concept of "threshold/boundaries" of socio-ecological systems (Berkes et al., 2000; Walker et al., 2004), since the strategic scenario outlining coincides with the definition of the "preferred regime" threshold for complex systems. The strategic scenario—as well the definition of the "preferred regime" thresholds of socio-ecological systems—will leave "space" and opportunities for multiple evolution trajectories of urban systems that are necessary in order to cope with uncertainty and constantly changing environmental, social, economic conditions but also to embrace new and often unforeseen trajectories arising from grassroots initiatives and practices.

Climate issues are often primarily referred to as "urban policies"; however, to effectively deal with climate-induced challenges, a societal transition is required. Such a transition implies a cultural and societal change able to modify societal (civil society and private/economic sectors) behaviors placing the principles of climate action as consolidated cultural and societal values. Innovative and unexpected trajectories as well as community-led initiatives are needed in reaching a societal transition.

The networking of transition initiatives is emerging as a promising organizational structure (see Chapter 33).

Hence, by strengthening the links among the heterogeneous 'niches of innovation' arising from community-led initiatives and the more consolidated organizational structures and involved stakeholders (e.g. institutional, civil society, private actors, universities) that characterize institutional-led initiatives, a innovative adaptive urban governance framework could arise, paving the way to innovative spaces of dialogue among different stakeholders, and promoting meanwhile the stabilization, up-scaling, and improvement of community-led initiatives, and favoring 'creative diversity' and redundancy within urban systems. (Fig. 34.1).

However, the community-led network could significantly contribute to the establishment of an innovative urban governance framework addressed to guide a climate-sensitive urban development, but only if certain conditions occur (Fig. 34.1). In detail:

- Transition networks can include initiatives and practices promoted and acted on by local communities, the private sector, academic actors, and also by local institutions becoming "exchange environments" only if the rigid separation of consolidated stakeholders' categories are avoided and opportunities for alliances and local synergies are ensured;
- Transition networks have to connect and include the heterogeneous landscape of transition initiatives, not only those explicitly oriented to climate issues. In current participatory processes, indeed, very often only associations and civil society actors already coherent or oriented to the goals of institutional policy are involved. The involvement of different transition or community-led initiatives is important to engage a large range of innovation experiments, since an initiative could contribute to urban climate action also when focused on an apparently different issue;
- Transition networks are natural (fluid) places for dialogue and contamination among researchers, communities, and economic/private sectors, giving opportunities for innovation transfer and promoting citizens' awareness and individual behavioral changes, crucial to address climate-induced challenges;

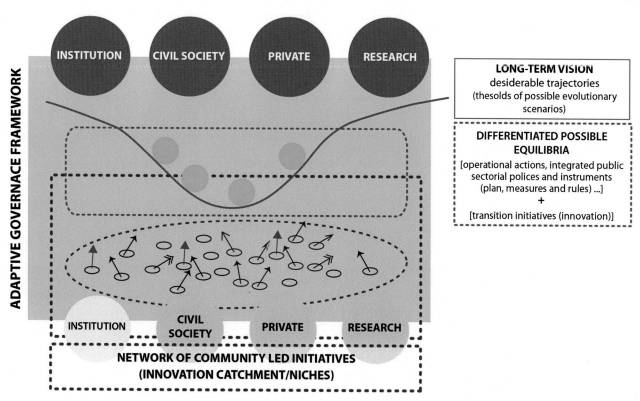

FIGURE 34.1 A The Adaptive Urban Governance Framework: Graphical diagram. *Author's elaboration (Angela Colucci).*

- Transition initiatives, based on ICTs, could promote social activation and social innovation platforms, providing opportunities and resources for increasing citizens' awareness, for engaging multiple stakeholders, and for facilitating collaborative interactions and processes integrating different sources and types of knowledge, from the scientific/expert knowledge, developed by different actors, on different geographical scales and in different domains (Galderisi, 2017), to the experiential knowledge (Albrechts, 2016) of local communities.

Actors from academia should play a significant role in the adaptive urban governance framework, contributing to the process of strategic visioning and leading the continuous monitoring that is a vital part of an adaptive approach (Taanman et al., 2008), based on a continuous learning process. However, monitoring activities should be addressed to evaluate not only individual actions/projects but also their role in the wider transition process, the features of the alliance/network, the social and institutional learning levels, the rate of progress, as well as the barriers and opportunities of the transition process.

References

Albrechts, L., 2016. Strategic planning as governance of long-lasting transformative practices. In: Concilio, G., Rizzo, F. (Eds.), Human Smart Cities, Rethinking the Interplay between Design and Planning. Springer International Publishing, Switzerland, pp. 3–20.
Berkes, F., Folke, C., Colding, J. (Eds.), 2000. Linking Social and Ecological Systems: Management Practices and Social Mechanisms for Building Resilience. Cambridge University Press.
Borraz, O., Le Galès, P., 2010. Urban governance in Europe: the government of what? Pôle Sud 32 (1), 137–151. https://www.cairn.info/revue-pole-sud-2010-1-page-137.htm.
Colucci, A., 2017. Peri-urban/peri-rural areas: identities, values and strategies. In: Colucci, A., Magoni, M., Menoni, S. (Eds.), Peri-urban Areas and Food-Energy-Water Nexus. Sustainability and Resilience Strategies in the Age of Climate Change. Springer-Verlag GmbH, Heidelberg.
Concilio, G., 2016. Urban living labs: opportunities in and for planning. In: Concilio, G., Rizzo, F. (Eds.), Human Smart Cities, Rethinking the Interplay between Design and Planning. Springer International Publishing, Switzerland, pp. V–X.
Davoudi, S., Brooks, E., Mehmood, A., 2013. Evolutionary resilience and strategies for climate adaptation. Planning Practice & Research 28 (3), 307–322. https://doi.org/10.1080/02697459.2013.787695;.

Davoudi, S., Crawford, J., Memhood, A., 2009. Climate Change and Spatial Planning Responses. In: Davoudi, S., Crawford, J., Memhood, A. (Eds.), Planning for Climate Change. Strategies for Mitigation and Adaptation for Planners. Earthscan, UK, USA.

Dingwerth, K., Pattberg, P., 2006. Global governance as a perspective on world politics. Global Governance: A Review of Multilateralism and International Organizations 12 (2), 185–203.

Folke, C., Carpenter, S.R., Walker, B., Scheffer, M., Chapin, T., Rockstrom, J., 2010. Resilience thinking: integrating resilience, adaptability and transformability. Ecology and Society 15 (4), 20. Available at: http://www.ecologyandsociety.org/vol15/iss4/art20/.

Galderisi, A., 2017. The Nexus Approach to Disaster Risk Reduction, Climate Adaptation and Ecosystem Management: New Paths for a Sustainable and Resilient Urban Development. In: Colucci, A., Magoni, F., Menoni, S. (Eds.), Peri-urban Areas and Food-Energy-Water Nexus. Sustainability and Resilience Strategies in the Age of Climate Change. Springer. ISBN:978-3-319-41022-7.

Geels, F.W., 2002. Technological transitions as evolutionary reconfiguration processes: a multi-level perspective and a case-study. Research Policy 31 (8), 1257–1274. https://doi.org/10.1016/S0048-7333(02)00062-8.

Grøtan, T.O., 2014. Hunting High and Low for Resilience: Sensitization from the Contextual Shadows of Compliance. In: Steenbergen, R.D., van Gelder, P.H., Miraglia, S., Vrouwenvelder, A.C. (Eds.), Safety, Reliability and Risk Analysis: Beyond the Horizon. Taylor & Francis Group, London, pp. 327–335.

Gupta, J., 2010. Mainstreaming Climate Change: A Theoretical Exploration. In: Gupta, J., van der Grijp, N. (Eds.), Mainstreaming Climate Change in Development Cooperation: Theory, Practice and Implications for the European Union. Cambridge University Press.

Hillier, J., 2017. Lines of Becoming. In: Gunder, M. (Ed.), The Routledge Handbook of Planning Theory. Routledge.

Kelman, I., Gaillard, J.C., Mercer, J., 2015. Climate change's role in disaster risk reduction's future: beyond vulnerability and resilience. International Journal of Disaster Risk Science 6, 21–27. https://doi.org/10.1007/s13753-015-0038-5.

Klein, R.J.T., Schipper, E.L.F., Dessai, S., 2005. Integrating mitigation and adaptation into climate and development policy: three research questions. Environmental Science & Policy 8, 579–588. https://doi.org/10.1016/j.envsci.2005.06.010.

Knieling, J., Leal Filho, W., 2013. Climate Change Governance: The Challenge for Politics and Public Administration, Enterprises and Civil Society. In: Knieling, J., Leal Filho, W. (Eds.), Climate Change Governance. Springer-Verlag GmbH, Heidelberg.

Loevinsohn, M., Mehta, L., Cuming, K., Cumming, N.A., Ensink, O.H.J., 2014. The cost of a knowledge silo: a systematic re-review of water, sanitation and hygiene interventions. Health Policy and Planning 2014, 1–15. https://doi.org/10.1093/heapol/czu039.

Loorbach, D., 2010. Transition management for sustainable development: a prescriptive, complexity based governance framework. Governance 23 (1), 161–183. https://doi.org/10.1111/j.1468-0491.2009.01471.x.

Meijer, A., Bolivar, M.P.R., 2015. Governing the smart city: a review of the literature on smart urban governance. International Review of Administrative Sciences 0 (0), 1–17. https://doi.org/10.1177/0020852314564308.

Menoni, S., Weichselgartner, J., Dandoulaki, M., et al., 2014. Enabling knowledge for disaster risk reduction and its integration into climate change adaptation. In: Input Paper
Prepared for the Global Assessment Report on Disaster Risk Reduction 2015. Available at: http://www.preventionweb.net/english/hyogo/gar/2015/en/bgdocs/inputs/Menoni%20et%20al.,%202014b.%20Enabling%20knowledge%20for%20disaster%20risk%20reduction%20and%20its%20integration%20into%20climate%20change%20adaptation.pdf.

Ostrom, E., 1990. Governing the Commons: The Evolution of Institutions for Collective Action. Cambridge University Press, Cambridge.

Ostrom, E., 2005. Understanding Institutional Diversity. Princeton University Press, Princeton.

Pramova, E., Di Gregorio, M., Locatelli, B., 2015. Integración de la adaptación y la mitigación en las políticas sobre cambio climático y uso de la tierra en el Perú. Working Paper 189, Bogor, Indonesia, CIFOR. Available at: http://www.cifor.org/publications/pdf_files/WPapers/WP189Pramova.pdf.

Rayner, T., Berkhout, F., 2012. EU Climate Policy Mainstreaming Background Paper for RESPONSES/IEEP Symposium. Available at: http://www.responsesproject.eu/pdf/Symposium_background%20paper_24_06_12.pdf.

Rotmans, J., Kemp, R., Van Asselt, M., 2001. More evolution than revolution: transition management in public policy. Foresight 3 (1), 15–31. https://doi.org/10.1108/14636680110803003.

Scott, C.A., Kurian, M., Wescoat Jr., J.L., 2015. The Water-energy-food Nexus: Enhancing Adaptive Capacity to Complex Global Challenges. In: Kurian, M., Ardakanian, R. (Eds.), Governing the Nexus. Springer International Publishing, Switzerland. https://doi.org/10.1007/978-3-319-05747-7_2.

Taanman, M., de Groot, A., Kemp, R., Verspagen, B., 2008. Diffusion paths for micro cogeneration using hydrogen in The Netherlands. Journal of Cleaner Production 16 (1), S124–S132. https://doi.org/10.1016/j.jclepro.2007.10.010.

UN, 2012. The Future We Want. Outcome Document of the United Nations Conference on Sustainable Development. Available at: https://sustainabledevelopment.un.org/content/documents/733FutureWeWant.pdf.

UN, 2016. Draft Outcome Document of the United Nations Conference on Housing and Sustainable Urban Development (Habitat III/New Urban Agenda) United Nations, A/RES/71/256 General Assembly Distr.: General 25 January 2017 Seventy-first Session Agenda Item 20-16-23021 [Resolution Adopted by the General Assembly on 23 December 2016/without Reference to a Main Committee (A/71/L.23)].

UNDP (United Nations Development Programme), 1997. Governance for Sustainable Human Development. A UNDP Policy Document. United Nations Development Programme, New York.

Walker, B., Holling, C.S., Carpenter, S.R., Kinzig, A., 2004. Resilience, adaptability and transformability in social-ecological systems. Ecology and Society 9 (2), 5. Available at: http://www.ecologyandsociety.org/vol9/iss2/art5/.

World Economic Forum (WEF), 2016. The Global Risk Report 2016, eleventh ed. Available at: http://www3.weforum.org/docs/GRR/WEF_GRR16.pdf.

Index

Note: Page numbers followed by "f" indicate figures, "t" indicate tables, and "b" indicate boxes.

A

Academic Initiative, 197–199
Action Aid, 217–219
Activation time evaluation, 263–264
Adaptability, 13, 16
Adaptation, 23, 40, 44, 62–63, 98–99, 111, 139, 146, 271
   Australian approach, 69–70
   Portsmouth's climate change, 88, 89f
Adaptation action plans (AAPs), 111
Adaptive capacity or resilience, 241
Adaptive cities, 7–8
Adaptive complex system, 24
Adaptive governance, 31, 50–51
Adaptive learning tools, 26
Adaptive urban governance framework, 290, 291f
Advanced Urban Drainage Management, 118
Aesthetics, 219
Africa, mitigation and adaptation initiatives in
   CDKN, 55–57
African Centre for Cities (ACC), 56
African Union (AU), 269–270
Aggregation process, 26
Agriculture
   policy, 44
   production, 44
Alimentazione Sostenibile project, 192
Alliances, 279–280
American approach to climate change
   focusing on northern and arctic regions, 48–51
"Amici di Ponte Carrega" association, 257
Annual forum, 183–184
Anticipated hazards/impacts of CC, 239
Arctic Marine Shipping Assessment, 50
Arctic permafrost, 49
Assets specialization, 227
Athens
   climate change trends in Athens Metropolitan area, 128–129
   facing climate change, 125–134
Athens city center, 126
Attentive-to-climate change recovery after devastating earthquake, 232–234
Attica Region
   in Greece, 126–127
   Strategy for Smart Specialization (RIS3), 131
Australian approach
   climate change issues in, 67–68
   mitigation and adaptation, 69
   for mitigation and adaptation in, 70
Australian government system, 67–69
Australian Renewable Energy Agency (ARENA), 223

B

Banca della Memoria project, 192–193
Barcelona
   community resilience enhancement in, 203–208
   flooding resilience and, 115–116, 116f
   municipality, 277
   storm water tanks and antidischarge unitary system tanks, 119t
Barcelona Ciclo de l'Aigua SA, 120–121
Barcelona Green Plan, 207
Barcelona Olympic model framing contemporary challenges, 118–119
Beautification of urban governance, 219
Belgrade
   Belgrade City Assembly, 105–106
   Belgrade city GUP (2016), 105
   Belgrade Urban Planning Institute, 102
   Belgrade-based NGO Palgo Centar, 103
   City of Belgrade Development Strategy (2007–11), 103
   environmental atlas (1998–2002), 102
   green regulation (2002–07), 102
   local context, 101
   as part of 100 resilient cities network (2015), 104
   regional spatial plan (2003, 2011), 102–103
   suburban municipalities, 106
   urban development, 101
Biophysical-social systems, 243
Bisagno river basin management plan, 259
Bottom-up approaches/initiatives, 5, 115, 221–224
"Bottom-up" governance models, 31
"Bounce forward" perspective, 11, 13
"Bouncing-back" perspective, 12, 286
Building resilience, 89
Building stock in Municipality of Athens, 127–128

C

C40 network, 75
Can Masdeu community garden project, 205
Canton. *See* Guangzhou

Capacity building, 275, 288–289
Carbon dioxide emissions (CO_2 emissions), 47, 130–131
Carbon market, 62
"Carbon Tax", 69
CARIPLO foundation, 177
"Caviar" project, 50
"Central element of controlling flood risk in UK spatial planning" approach, 90
Central Italy Event (2016), 232–234, 233f–234f
Cfb climate zone, 111
Cfsa area, 111
China, climate change in, 61
Chinese Sponge City Program, 270
CIEL, 49
Cities, 55, 63. *See also* Smart city
Citizens' engagement, 257
City Assembly, 103
City of Belgrade Development Strategy (2007–11), 103
City of Belgrade General Urban Plan, 102
City of Belgrade Urban Planning Institute, 103
City Public Health Agency, 102
City Resilience Framework (CRF), 15
City Urban Planning (P.U.C.), 257
City-wide Floodplain Risk Management Study and Plan, 164
Clavegueram de Barcelona, S. A. (CLABSA), 118, 120
Clean Energy Plan reform, 69
Clean Power Plan, 47
Climate Action, 223
Climate adaptation, 43–44, 160
Climate and Development Knowledge Network (CDKN), 55
Climate change (CC), 127, 203–204, 209–211, 222, 229, 237–240, 275
   adapting and mitigating, 203–205
   American approach to, 47–51
   Australian approach to, 67–71
   in China, 61–66
   as contested political issue, 69
   DRR and adaptation to, 240–243
   European cities addressing, 75–84
   European strategies and initiatives to tackle, 39–46

Climate change (CC) (*Continued*)
  Genoa and, 109–114
  in Guangzhou, 154–155
  impacts on coastal urban regions, 95–100
  mitigation and adaptation initiatives in Africa, 53–59
Climate Change Action Plan of Gaziantep, 137
Climate change adaptation (CCA), 163, 178, 228, 237, 242–243, 242f, 244t–245t
  and Australian local governments, 163–167
  and spatial planning responses in UK, 85–93
Climate Change Adaptation Plan (2015), 102, 106–107
Climate Change Adaptation Plan and Vulnerability Assessment (CCAPVA), 103–104
Climate Change Department, 271
Climate change interventions (CCIs), 141–143, 148–149
Climate Change Plan, 207
Climate governance, 61–62, 136–137
Climate initiatives in Genoa, 111–113
Climate monitoring system development, 102–103
Climate policies, 272–276
  in Turkey, 135–136
Climate Sensitive Urban Development, 285–291
Climate urban actions, transition tools for, 280
Climate urban policies, barriers to, 275–276
Climatic and geographical features, 109–111
Climatic diversity, 67
Coastal Defense Strategy, 88
Coastal flooding, 90
Coastal hazard planning, 98
Coastal Urban regions, 95–100
Cocreation, integrated approaches from, 205
Codesign tools, 26
Coevolution processes, 25
Cognitive dimension enhancement, 89–90
Colector de las Rondas, 116–117, 117f
Collaboration, 219
Collective reflexive opportunity, 24
Community, 203–204
Community Strategic Plan, 164–165
Community-based approach, 183
Community-based models of governance, 6–7
Community-led processes, 25, 204
Community-led projects capacity, 175
Complex living system, 85
Complex systems approach, 24
Complexity theory, 31
"Concept of resilience practice", 177
"Conceptual framework" by Transition Town Movement, 31
Conference of Parties (COP), 40b, 41, 135

Construction Land Development and Allocation Program, 105–106
Contemporary cities, 227–229
Continual learning, 13, 15
Continuous learning processes, 7–8
Continuous monitoring, 32
Convening concept, resilience as, 53–55, 270–271, 286
Cooperation, 196–197, 201
COP. *See* Conference of Parties (COP)
Coproduction tools, 26
Corporations, 223
Cost-benefit analysis (CBA), 252–254
Council of Australian Governments (COAG), 67–68
Counterbalance climate change
  European strategies to, 40–44
  international steps to, 40b
Covenant of Mayors for Climate & Energy, 44–45, 75–79, 78f, 111–112
Covenant's signatories, 77, 77f
"Creative class", 4
Cross-cutting land-use planning, 98
Csa area, 111
Cultural identity of PJ, 215–217
CURSA funding, 192
Customization, 16

D

Danube rivers, 101
Daugava River, 97
Decarbonization process, 41
Decentralization enhancement for future of urban flooding resilience, 120–121
Decision-making processes, 6, 21, 32, 41, 249, 251–254
Department of Architecture and Urban Studies (DASTU), 176
Department of Science, Planning and Territorial Policies (DIST), 176
Department of Urbanism (DU), 196
Depave project, 176
Development Control Plan, 164
"Deviation of *Riera Malla*", 116–117, 117f
Digital technologies, 205
Directorate of Environmental Protection and Control, 136
Disaster risk reduction (DRR), 11, 237, 244t–245t, 249, 260–261, 286
  and adaptation to CC, 240–243
  CBA as economic-based tool, 252–254
  coping with disaster risk, 251–252
  integration and adaptation to CC, 243–247
  local strategy, 260–264
  urban systems and disasters from economic perspective, 249–250
Disasters, 215, 229
  from economic perspective, 249–250
  risk, 237–240
  surviving disasters through collaboration, 218–219
Disciplinary boundaries, tracing resilience evolution across, 12–13, 12f

Division of climate change, energy efficiency, and renewable energy (2016), 104
Drainage systems design and urban plans, links between, 116–117
Drought, 238. *See also* Flood(ing)
Dykes, 159–160
Dynamic equilibrium, 12–13
"Dynamic nature of resilience", 15

E

Earth Justice, 49
Eastern Mediterranean region, 129
Eco-house, 138, 138f
Economic
  evaluation, 261–263
  identity of PJ, 215–217
  understanding might enhance DRR action, 251–252
  urban systems and disasters from economic perspective, 249–250
Education, 219
Emergency Management Unit (2012–16), 103, 106
Emissions Trading Scheme (ETS), 63
Empowerment, 219
Energy
  in China, 61
  production, 5
  transition, 21
Energy Descent Action Plan (EDAP), 172, 191
"Energy Independence" document, 47
"Energy Roadmap 2050" (2011), 41
Energy Savings in Urban Quarters Through Rehabilitation and New Ways of Energy Supply (EnSURE), 112
English planning system, 88
Ensanche, 115–117
Environmental and Natural Resource Management, 143
Environmental impact assessments (EIAs), 148
Essential elements, 173–174
EU-GUGLE-Sustainable Renovation Models for Smarter Cities Project, 139
Europe launched Covenant of Mayors (2008), 42
European adaptation strategies, 42–44
European cities
  adaptation issues, 79–80
  embracing integrated climate strategy, 80–81
  mitigation issues, 77–79
  pivotal role of cities in counterbalancing climate change, 75–77
European Climate Change Programme (ECCP), 41
European Commission, 77, 111–112
European Environment Agency, 239
European funds (LIFE+), 97, 120
European initiatives, 76, 80
European mitigation goals and strategies, 41–42

European strategies
  to counterbalance climate change, 40–44
  and initiatives to tackle climate change, 39–46
European Union (EU), 5, 75–76, 111, 130, 135, 269–270
Evolutionary resilience, 13, 273
"Exclusion and inequality", 6
Experimental navigation activities, 50
Exposure
  to climate change, 125–129
  issues, 127–128

F
Fair Trade Group, 193
Federal government of Australia, 67
Fire, 218
Firma Energetica project, 192
Five Year Plan (FYP), 61–62
Flats-for-land system, 126
Flexibility attributes, 25
Flood resilient districts
  local strategy of disaster risk reduction and action planning, 260–264
  meteorological phenomena and impacts, 257–258
  urban context, environmental characteristics and citizens' engagement, 257
Flood Response Plans, 88, 91
Flood(ing), 115, 195–196, 228, 238, 238f
  resilience, 115–116, 153, 156, 159, 161
  risk, 156–157
Food security, 49–50
Food sovereignty, 50
"Formative" approach, 15
Forum of Resilience Practices (RPF), 177–178
Fourth National Report on CC, 129
Fragilities, 258–259
Fragmentation, 228–229
"Freedom of the seas", principle of, 50
French Development Agency, 273

G
Gal Genovese agency, 112
Gaziantep City, 135–136
Gaziantep Metropolitan Municipality, 136
General Urban Plan for Belgrade (GUP), 105–107
General Urban Spatial Plans (GUSP), 130–131
"Generic vulnerability" analysis, 246
Genoa and climate change, 109–114
Genoese streams, 109
"Genova Smart City" Association, 112–113
Gentrification/eviction, 218–219
German Corporation for International Cooperation, 103–104
German planning, 96, 98
Ghana, 141
Ghana National Climate Change Policy (NCCP), 144

Ghana Poverty Reduction Strategy (GPRS I), 143
Ghana Shared Growth and Development Agenda (GSGDA I), 143
Giffinger's model, 3, 5
Global Covenant, 77
Global Covenant of Mayors for Climate and Energy, 271, 287
Global Risk Report (2016), 285
Global warming, 239. *See also* Climate change (CC)
Governance, 6, 19–21, 132–133, 289–291
  models, 274–275, 279–280
  of transition management, 19–21
GRaBS project, 111
Grama-Sabhas for slum, 215
Grassroots
  initiatives, 175
  innovations, 22, 222
"Great Reskilling" events, 183
Greater Kumasi Comprehensive Plan (GKCP), 143
Greater Visakhapatnam Municipal Corporation (GVMC), 218
"Green and blue" infrastructure, 111
Green infrastructure, 104, 127–128
Green regulation of Belgrade (2002–07), 102
"Greener Portsmouth", 88, 90
Greenhouse gas emissions (GHG emissions), 40–41, 42f, 61, 68, 75, 85, 135, 269
Gross domestic product (GDP), 126–127
Growth and Poverty Reduction Strategy (GPRS II), 143
Gruppo d'Acquisto Fotovoltaico e Solare Termico project, 193
Guangdong Provincial 5-Year Plan, 155–156
Guangdong Provincial Sponge City Program, 159–160
Guangzhou, 153
  critical analysis of municipal interpretation, 159–161
  rapid urbanization and exposure to climate change, 154–155
  sponge city, 155–159
Guangzhou Master Plan, 156
Guangzhou SCP, 156
Guangzhou Sponge City Plan (SCP), 154, 157, 159
Guangzhou Water Affairs Bureau, 159
Guide Group, 191

H
"Hardware" infrastructure, 4
Hazards
  climate-related, 67
  and risk in contemporary cities, 228–229
"Head, Heart, and Hands principle", 173
Health and Medicine group, 184
Heat waves, 164
Heterogeneous political systems with climate issues, 269–270

Heterogeneous pressure factors, 11, 29
Holling's contribution, 12–13
Hopkins model, 125, 191
Housing policies, 104
Human centered approach, 4
Human Cost of Weather-Related Disasters (1995–2015), 238
"Human smart city", 4
"Human-scale" dimension, 277
100 Resilient Cities initiative (100RC initiative), 14–15, 76, 131–132
Hydraulic analysis, 261
Hydraulic risk, 263

I
Immaterial dimensions, 21
Individual water footprints, 223
Informality, 56
Information and communications in technology (ICT), 4–7, 29, 278–279, 287
Information flows, 7
Infrastructure Delivery Plan (IDP), 88
Innovation
  for facing urban climate change issues, 24–26
  niches, 281–282
  seeds, 25
Innovative transition initiatives, 19–21, 224
Institutional-driven governance models, 31
Institutional-led processes, 25
Integrated approaches, 112, 205, 286
Integrated climate strategy
  Europe toward, 44
  European cities embracing, 80–81
Integrated knowledge, 286
  in CCA and risk mitigation, 227–235
Integrated perspectives, from sectoral toward, 5
Integrated policies, 287
Integrated strategies, 40, 44
Integrated urbanism master's program
  Academic Initiative, 197–199
  achievements, 201
  in building professional capacity, 196–197
Integration
  of climate change adaptation, 107
  Climate Sensitive Urban Development, 285–287
Intended National Determined Contribution (INDC), 62–63
Intentionality of human action and intervention, 13
Inter American Commission on Human Rights (IACHR), 48–49
Interconnected systems, 16
Intergovernmental Panel on Climate Change (IPCC), 39, 40b, 53–54, 109, 237
Interim Planning Target, 68–69

International Maritime Organization (IMO), 50
INTERREG IVC (EU program), 111
Intervention profitability, 252
Inuit Circumpolar Conference (ICC), 48–49
Inuit Circumpolar Council Alaska, 50
"Island effect" of transition initiatives, 281–282
Istanbul
  economic policies in, 210
  increasing exposure with numbers in, 210t
  Know-4-DRR project in, 211
Italian Observatory of Resilience Practices (ORP), 171
Italy, Transition Towns Network in
  critical analysis of initiative, 193
  Monteveglio transition town, 191–193
IUCN, 50

J
Japan International Cooperation Agency, 273

K
Know-4-DRR project, 211
Knowledge, 7–8, 196
  coproduction, 280, 287
  integrated, 227–235, 286
Kumasi metropolis, 141, 143
  climate change and urban planning in, 147–149
Kumasi Metropolitan Assembly (KMA), 141
Kumasi Urban Forestry Project, 148
Kyoto Protocol, 135

L
La Marina de la Zona Franca project, 120
LaCol, 204
Landscape planning, 98
Landslides, 228
  inventory, 103
Landslips, 103
Langstone Harbor, 85, 91f
Large-scale strategies, 269–272
  heterogeneous political systems, 269–270
  pivotal role of cities, 271–272
Learning capacity, 13, 29, 273–274
Learning process, 32
"Learning-by-doing" approach, 31
"Lighter, Quicker, Cheaper-style Approach", 175
Linking capital, 218
Liquefied Natural Gas (LNG), 130–131
Llobregat River Delta plan actualization, 119
Local Climate Impacts Profile document, 89
Local communities, 222
Local economy, 204
Local Government Act (1993), 141
Local governments areas (LGAs), 163
Local land-use plans development, 96
Local Planning Strategy, 164
Local strategy of disaster risk reduction, 260–264
Local Transition Initiatives, 172–173
Local water company, 91
Local-global dimensions, 25–26
Low public awareness, 133

M
Mainstreaming, 287–288
Making Cities Resilient campaign, 14
Maladaptations, 243
Map of Limitations to Urban Growth, 105
"Mapping Smart City in EU", 5
"Mapping" concept, 177
Maritime traffic issue, 49
Marmara Earthquake (1999), 211
Master Plan of Athens/Attica 2021 (MPA), 130–131
Mayors Adapt initiative, 43, 76, 79–80, 80f
Mecklenburg-Western Pomerania, 96
Medieval ages, 115
Mediterranean climate, 111
Medium-term development plans (MTDPs), 141–143, 145–146
Mercatino del Riuso project, 192
Merewether Beach Reserves Public Domain Plan, 164
Mermi creek, 259
Metaphors, 32
Metropolitan, municipal, and district assemblies (MMDAs), 141–143
Metropolitan Athens, profile of, 125–129
Microlevel bottom-up initiatives, 19–21
Microscale strategy, 262–263
Millennium Development Goals (MDGs), 6, 145
Ministry of Emergency Situations, 103
Ministry of Environment, Science, and Technology (MEST), 144
Ministry of Housing and Urban-Rural Development (MoHURD), 153
Mitigation, 40, 44, 271
  and adaptation policies, 109–114
  Australian approach, 69–70
  initiatives in Africa, 53–59
Mobility, 5, 227
Model for Integrated Urban Disaster Risk Management (MIUDRM), 195–202
Modernday environmental problems, 101
Modi operandi, 175
Monetization models, 253
Monitoring, 26
Monteveglio City in Transition, 192
Monteveglio transition town, 191–193
Mulling, 174
Multi-agency Response Plan, 88, 91
Multiagency document, 88
Multidisciplinary research program implementation, 102–103
Multisector collaboration. *See also* Climate change adaptation (CCA)
  critical analysis of current initiatives/practices, 106–107
  current initiatives/practices, 102–106
Municipal Deliberation 54/2009, 191
Municipality, 127–128, 191, 193
Muvita, 112–113

N
National Adaptation Strategy (November 2013), 62–63
National and urban planning divergence, 149–150
National carbon market, 63
National Climate Change Adaptation Programme, 68–69
National Climate Change Adaptation Research Facility (NCCARF), 68–69
National Climate Change Adaptation Strategy (NCCAS), 144
National Climate Change Committee (NCCE), 144
National Climate Change Policy Action Programme for Implementation (NCCPAPI), 144
National Climate Change Policy Framework (NCCPF), 144
National Climate Resilience and Adaptation Strategy, 68–69, 164–165
National Development and Reform Commission (NDRC), 61–62
National Development Planning Commission (NDPC), 141–143
National Development Policy Frameworks (NDPFs), 141–147
National Garden, 127
National Greenhouse Response Strategy, 68–69
National Greenhouse Strategy, 68–69
National hubs, 174
National Land-Use-Planning Act, 96
National Observatory of Athens (NOA), 128
National People's Congress, 62
National Plan for the Allocation of gas emission allowances (NAP), 129
National Planning Policy Framework (NPPF), 88
National Strategy for Adaptation to CC (NSACC), 130
National Strategy for Disaster Resilience, 68–69
NATO bombing campaign, 101
Natural ecosystems, 39
Natural hazards, 227–228, 250
New Covenant of Mayors for Climate and Energy, 80–81, 81f–82f
New Orleans disaster, 278
New South Wales (NSW), 163
Newcastle, CCA in, 163–167
  interviewees' codes and main topics, 164t
  Upcycle, 222

Newcastle City Council (NCC), 163–165
Newcastle Employment Lands Strategy, 165–166
Newcastle Environmental Management Strategy implementation, 164–165
Newcastle Local Environmental Plan, 164
Nonclimatic factors, 243
Nongovernmental organization (NGO), 22, 48–49, 103, 210–211, 217–218
Nonlinear transition, 19, 25
Nonstructural flood-proofing techniques, 259
Nourishing Newcastle Urban Tucker Stall (NNUTS), 223
Numerous community-led initiatives, 176
"Nut Capital of England", 184

O

Obrenovac, 280
  background characteristics of municipality, 195
  resilience state in municipality, 195–196
Official Transition Initiative, 174
"Official Unleashing of TTT", 183
Oil industrial activities and reduction, 50
Olympic Games, 118, 227
Operational constraints, 276
Operationalization of climate change adaptation measures (2016), 105–106
"Operationalizing" of resilience building, 16
Ordinary functioning conditions, 13
Ordinary territorial events, 253–254
Organisation for Economic Co-operation and Development (OECD), 69
Organizational constraints, 275–276
ORP. *See* Italian Observatory of Resilience Practices (ORP)

P

"Panarchy", 13
Papua New Guinea (PNG), 246–247
Paris Agreement, 62, 135–136
Paris Climate Agreement, 47
Participatory Energy Plan (PEP), 204
Participatory process, 16
Passive house, 138
Pati La Pau, 204
Pearl River Committee, 160–161
Pearl River Delta (PRD), 154
Pedda Jalaripeta slum (PJ slum), 215–218, 278
People-centric and pro-poor approach, 56
People's Republic of China, 269–270
PEP Dalera, 204
PEP's local energy observatory (OLEPEP), 204
Persistence, 16
  climate urban policies of, 273–274
  in transition initiatives, 279
Piazza Adriatico area, 258
Piedibus project, 192
"Pilot cities' low-carbon plans", 63, 272

Pipe drain. *See* Ramblar Colector
Pla BUITS experience, 205–207, 206f
Plan de Saneamiento y alcantarillado, 117
Plan Especial de Alcantarillado de Barcelona (PECB), 115
Plan Integrado of Alcantarillado de Barcelona (PICBA), 120
Planning, 95
  CCA and planning responses, 88, 89f
  German, 96, 98
Planning Coordinating Units (PCUs), 141–143
Plataforma, 204
Pluvial flooding events, 153
Policy PCS13, 90
Political primacy, 106
Port Environment Energy Plan, 113
Portsea Island Coastal Defense Strategy, 88–90
Portsmouth, 85
Portsmouth Climate Change Strategy, 88
Portsmouth Harbor, 85, 91f
Portsmouth Sustainability Action Group (PSAG), 88
"Portsmouth's Core Strategy", 88
Postdisaster recovery, 230
Precipitation regime, 257–258
"Preferred regime" threshold, 290
Proactive leadership, 135
Productive environment, 175
Promotion of Public Transport (PTP), 204
Protection and Rescue Plan and Risk and Vulnerability Assessment, 105
"Provincia Energia", 112
Public consultation procedures, 132–133
Public Investment Evaluation Unit (UVAL), 232
Public life, 174–176
Public space, 175
Punctuated equilibrium. *See* Nonlinear transition

Q

Quick risk estimation tool, 15
Quintuple Helix model, 30

R

Ramblar Colector, 116, 117f
Rapid urbanization
  climate change and flood vulnerability, 154–155
  demographic, social, economic, and political features, 154
  and exposure to climate change, 154–155
REconomy Centre, 183–184
Reconstruction
  cities as complex entities, 227–228
  dealing with hazards and risk in contemporary cities, 228–229
  knowledge feeding, 230–231, 231f
  resilience, 229–230
  2016 Central Italy Event, 232–234
Reflexive activities, 19–21
Regalami un Albero project, 192

Regeneration, 210
Regeneration in the Disaster Risk Zones, 210–211
Regional climate models (RCMs), 129
Regional GDP per capita, 126–127
Regional Greenhouse Gas Initiative, 48
Regional hubs, 174
Regional Operational Program of Attica (ROPA), 130
Regional spatial plan (RSP), 102–103
Regione Liguria's warning systems, 263
Remote control procedures, 118
Renewable energy sources (RESs), 130
Renovated adaptive governance framework, 289
Research and Innovation Programme Horizon 2020 in (2011), 43–44
Resilience, 12–14, 23, 53–54, 203–204, 229–230, 278, 288
  assets, 125–129
  CC mitigation and adaptation, 53–55
  metaphor, 219
  Obrenovac, 195–196
  from Slum Dwellers' perspective, 215–220
Resilience practices observatory (RPO), 176–178, 177f
Resilient city, 30, 229. *See also* Smart city
  initiatives, 13–16
  metaphor, 11, 29, 270, 273, 286
  tracing resilience evolution, 12–13, 12f
  transition and, 106
Resilient land-use planning
  barriers and opportunities, 98–99
  in urban regions, climate change impacts and, 95–97
Resilient Portsmouth, 89–91
Response-Recovery Prevention-Preparedness chain (RRPP chain), 241
"Rethink Athens" plan, 131
Rio+20 meeting, 139
"Ripple effect", 175
Risk mitigation, integrated knowledge in, 227–235
Rockefeller Foundation, 14, 76

S

"Saving at Home" project, 130–131
Second European Climate Change Programme (ECCP II), 41
Second National Program for CC, 129
Sectoral and integrated approaches, climate urban policies between, 274
Sendai framework implementation, 15
Sensors' networks, 3
Serbia
  climate change legislation, 107
  flooding (2014), 195–196
Sewer system Cerdá's plan organization, 116
"Siloed" city model, 5
Single transition initiatives, 24, 26

Slum dwellers' perspective, resilience from, 215–220
Small-scale businesses, 219
Smart, resilient, and transition cities
  commonalities and peculiarities of metaphors, 30–32
  urban metaphors, 29–30, 33f
Smart Cities and Communities European Innovation Partnership, 5
Smart Cities and Communities Initiative (2011), 5
Smart city, 3–4, 6–7, 30–31. *See also* Cities; Resilient city
  between conflicting approaches and goals, 4–5
  governance, 6–7
  initiatives, 5
  metaphor, 3–4, 29, 270, 278–279
Smart governance, 6, 31
"Smart Specialization Strategy" for Region of Attica, 130
Social identity of PJ, 215–217
Social networks, 217
Social resilience, 13
Social segregation in Athens, 125–126
Social Strategy (2016–19), 164–165
Societal transition models, 19–21, 20f, 24, 29, 32
Socioeconomic vulnerability, 240–241
Sociotechnological transition models, 31
South Stockton Reserves Public Domain Plan, 164
Spatial planning, 85, 97, 246
  in new flood governance, 159
Sponge City Program in Guangzhou, 153–162
Spontaneous interventions, 22, 175
Squatting, 205
Stakeholder
  coordination and collaboration, 210–211
  engagement, 6
Standardization, 25–26
Stern Review on Economics of Climate Changes, 113
Storms, 238
Strade in Transizione project, 192
Strategic environmental assessment (SEA), 148
Strategic Flood Risk Assessment (SFRA), 88–90
Strategic Implementation Plan, 5
"Strategic long-term scenario" concept, 290
Streccapogn, 192
Street Tree Masterplan, 164
Subnational planning law, 96
"Summative" approach, 15
Surface Water Management Plan (SWMP), 88–90
Sustainable Design and Construction, 88
Sustainable drainage, 120–121
Sustainable Energy Action Plan (SEAP), 75, 111–112

Sustainable Energy and Climate Action Plan (SECAP), 80
Sustainable mobility, 131
Sustainable river basin management, 178
Sustainable Urban Drainage System, 120
SWOT analysis, 199
Systemic vulnerability of CC, 210

T

Technical Guideline for the Construction of Sponge Cities, 159–160
"Techno-practice", 6
"Ten Essentials", 15
Territorial resilience, 252–254
Theoretical constraints, 275
"Threshold/boundaries" of socioecosystems, 290
Top-down approach, 5, 161
Torre Baro project, 120
Totnes economy, 183–184
Totnes Energy Descent Plan, 184, 185t–190t
Totnes transition town (TTT), 183–184
Tram system in Gaziantep, 138–139, 138f
Transformability, 13, 273–274, 279
Transformative capacity, 219
Transformative resilience lens, 95
Transition, 19, 25, 221–222
  arena, 26
  management, 19–21
  metaphor, 24, 171
  network, 22–24, 26, 171–174, 183, 221–222, 290
  and resilient city, 106
  tools, 26, 280
Transition 2.0, 23
Transition Initiative of Totnes, 172
Transition initiatives, 21–22, 24–26, 171, 192–193. *See also* Academic Initiative
  acting on urban commons and public life, 174–176
  complex systems approach, 24
  critical analysis of, 193
  in face of climate change, 276–282
  governance of transition processes, 25
  institutional-led and community-led processes, 25
  local-global dimensions, 25–26
  process as focus of transition metaphor, 24
Transition Newcastle (TN), 221–223
Transition Street program, 222–223
Transition Town Initiatives, 174
Transition Town Network, 32, 171, 276, 278–280
  evolution, 172–174
  in Italy, 191–194
  in United Kingdom, 183–190
Transition towns, 30, 32, 221–222
Trévol, 204
Turkey
  CC adaptation and urban resilience in, 209–213
  climate policy in, 135–136

U

UN-Habitat's Governing Council, 175
United Kingdom, 41
  adaptation and spatial planning responses, 85–93
  transition towns network in, 183–190
United Nations Development Programme (UNDP), 6
United Nations Environment Programme (UNEP), 40b
United Nations Framework Convention on Climate Change (UNFCCC), 40b, 41, 62, 69, 104, 135, 237
United Nations International Strategy for Disaster Reduction (UNISDR), 14, 76, 197, 201, 278
United Nations Secretariat, 14
University of Cape Town (UCT), 56
Unmanageable information, 7
"Upcycling", 222
Upper-floor apartments, 132
"Upscaling", 26
Urban
  decentralization enhancement for urban flooding resilience, 120–121
  fabric, 153
  form in Municipality of Athens, 127–128
  land-use planning in urban regions, 95–97
  metaphors, 29–30, 33f, 270–271
  policies, 290
  population, 96
  systems and disasters from economic perspective, 249–250
  transition initiatives, 174–176
Urban climate
  change interventions, 146–147, 146t
  governance in Turkey, 136
Urban development plans, CC interventions in, 148–149
Urban disaster risk management (UDRM), 195–196
  initiative by integrated urbanism master's program, 196–201
  at local level, 197
Urban Drainage Systems, 90
Urban governance system, 201
Urban heat island (UHI), 127
Urban innovations, transition approach in, 19–28
Urban Management at Technical University of Berlin, 197
Urban planning, 104
  in Kumasi metropolis, 147–149
  links between drainage systems design and, 116–117
  mainstreaming of mitigation and adaptation practices, 106–107
Urban Planning Bureau, 156
Urban resilience, 14
Urbanization, 109, 227–228

Printed in the United States
By Bookmasters

U

US Climate Alliance (2017), 48
US Coast Guard, 49
US Environmental Protection Agency, 47

V

Valsamoggia, 193
Vertical social segregation, 126
Visioning and positive envisioning, 30
Vulnerability, 68–69, 125–129, 210, 240, 250, 252

W

Wastewater, 117
Water, 120
water-efficiency measures, 88, planning, 98
Weather-related hazards and disasters, 238
"Western Regional Climate Action Initiative", 48
Working Group II (WG II), 53–55
"Working with Informality to Build Resilience in African Cities" project, 270–271
"Working with Informality" project, 57
World Meteorological Organization, 209